Backscattering Sources, Volume 1

Theoretical framework and Thomson backscattering sources

Online at: https://doi.org/10.1088/978-0-7503-5974-0

Backscattering Sources, Volume 1

Theoretical framework and Thomson backscattering sources

Alessandro Curcio
INFN, Frascati, Italy

Giuseppe Dattoli
ENEA, Frascati, Italy

Emanuele Di Palma
ENEA, Frascati, Italy

IOP Publishing, Bristol, UK

© IOP Publishing Ltd 2024

All rights reserved. No part of this publication may be reproduced, stored in a retrieval system or transmitted in any form or by any means, electronic, mechanical, photocopying, recording or otherwise, without the prior permission of the publisher, or as expressly permitted by law or under terms agreed with the appropriate rights organization. Multiple copying is permitted in accordance with the terms of licences issued by the Copyright Licensing Agency, the Copyright Clearance Centre and other reproduction rights organizations.

Permission to make use of IOP Publishing content other than as set out above may be sought at permissions@ioppublishing.org.

Alessandro Curcio, Giuseppe Dattoli and Emanuele Di Palma have asserted their right to be identified as the authors of this work in accordance with sections 77 and 78 of the Copyright, Designs and Patents Act 1988.

ISBN 978-0-7503-5974-0 (ebook)
ISBN 978-0-7503-5972-6 (print)
ISBN 978-0-7503-5975-7 (myPrint)
ISBN 978-0-7503-5973-3 (mobi)

DOI 10.1088/978-0-7503-5974-0

Version: 20240801

IOP ebooks

British Library Cataloguing-in-Publication Data: A catalogue record for this book is available from the British Library.

Published by IOP Publishing, wholly owned by The Institute of Physics, London

IOP Publishing, No.2 The Distillery, Glassfields, Avon Street, Bristol, BS2 0GR, UK

US Office: IOP Publishing, Inc., 190 North Independence Mall West, Suite 601, Philadelphia, PA 19106, USA

Contents

Preface	viii
Author biographies	xi
List of symbols	xiii

1 Introduction — 1-1

1.1	The Larmor formula and a classical description of Thomson scattering	1-4
1.2	Elements of special relativity: inertial frames and Lorentz transformations	1-10
	1.2.1 Four-vectors and matrix formalism	1-11
1.3	Relativistic kinematics and dynamics	1-14
1.4	The relativistic kinematics of Compton scattering	1-18
1.5	The kinematics of inverse Compton scattering	1-20
1.6	The relativistic Doppler shift	1-22
1.7	Maxwell's equations and special relativity	1-24
1.8	Comments and exercises	1-29
	1.8.1 Electron–photon interactions: a qualitative phenomenology	1-29
	1.8.2 Generalities related to electromagnetic fields	1-32
	1.8.3 Special relativity, notation, Maxwell's equations, and gauge invariance	1-49
	1.8.4 Special relativity: four-vectors and their associated matrix formalism	1-55
	1.8.5 Comments on the gauge invariance of classical electromagnetism	1-63
	References and further reading	1-70

2 Thomson backscattering radiation — 2-1

2.1	Compton scattering and Thomson scattering	2-1
2.2	Electron dynamics under intense wave excitation	2-4
	2.2.1 Electron motion in an intense plane wave	2-4
	2.2.2 The limit of very small field amplitude	2-6
	2.2.3 The limit of very large field amplitude: direct laser acceleration	2-7
2.3	Retarded potentials	2-8
	2.3.1 Liénard–Wiechert potentials	2-9

2.4	Thomson backscattering radiation	2-10
	2.4.1 Linear Thomson backscattering	2-18
	2.4.2 Nonlinear Thomson backscattering: harmonic emission	2-25
	2.4.3 The extremely nonlinear regime: emission of a continuum	2-30
2.5	Analogy with the emission in magnetic undulators and the Fermi–Weizsäcker–Williams approximation	2-34
	2.5.1 The Fermi–Weizsäcker–Williams approximation	2-37
2.6	Comments and exercises	2-38
	2.6.1 Exercises	2-38
	2.6.2 Undulator radiation and Compton backscattering	2-43
	2.6.3 The properties of generalized Bessel functions	2-45
	2.6.4 The Hamilton–Jacobi approach to the dynamics of electrons interacting with a plane wave	2-47
	2.6.5 Beyond the plane wave approximation	2-48
	Further reading	2-51

3 Charged beam transport   3-1

3.1	Introduction	3-1
3.2	Bending and quadrupole magnets	3-7
3.3	Beam envelope evolution	3-13
3.4	Beam matching	3-17
3.5	Comments and exercises	3-19
	3.5.1 Hamiltonian mechanics and beam transport	3-19
	References and further reading	3-30

4 Optical beam transport   4-1

4.1	Introduction	4-1
4.2	The ray matrix method	4-5
4.3	Matrix optics and lens images	4-9
4.4	A phase-space formalism for optical wave transport	4-10
4.5	Gaussian beams and the formal quantum theory of light rays	4-13
4.6	Comments and exercises	4-15
	4.6.1 Exercise	4-15
	4.6.2 Quadratic forms	4-32
	4.6.3 Gravitational lenses	4-35
	Further reading	4-36

5 Beam–beam interactions

5.1 Introduction — 5-1

5.2 Spectral broadening in CBS devices: on-axis contributions from energy spread and emittance — 5-7

5.3 Spectral broadening in CBS devices: off-axis contributions from the divergence of scattered photons — 5-16

5.4 A toy model of a real CBS radiation facility — 5-19

5.5 Comments and exercises — 5-22

    5.5.1 Exercises — 5-22

    5.5.2 Comments on the definition and practical aspects of the brightness of light sources — 5-28

    5.5.3 A phenomenological perspective on CBS brightness — 5-31

    5.5.4 The Gaussian approximants and the FEL — 5-35

    References and further reading — 5-38

6 CBS sources

6.1 Introduction — 6-1

6.2 Further comments on the bandwidth of CBS sources — 6-4

6.3 Laser systems in operational CBS sources — 6-7

6.4 Electron accelerators in operational CBS sources — 6-8

6.5 A comparison between theory and experiment — 6-9

6.6 Applications and the costs of CBS sources — 6-10

6.7 Comments and exercises — 6-12

    6.7.1 X-ray tubes — 6-12

    6.7.2 Exercises — 6-14

    6.7.3 CBS–FEL coupled devices — 6-16

    References and further reading — 6-17

Preface

In this book and its companion volume, we treat the problems associated with the theoretical and experimental foundations of photon–electron interactions and the achievement of producing radiation sources based on the Compton backscattering (CBS) mechanism. Our decision to separate the exposition into two parts is due to the fact that a sharp split in the relevant description is imposed by the nature of the process itself, which can be dominated by classical or quantum effects (title of the second volume: *Backscattering Sources: Compton backscattering, non-linear QED processes and calculation tools*). Even though such Solomonic decisions are largely arbitrary in physics, we followed this direction to describe: (a) CBS devices dedicated to the production of intense x-ray sources and (b) 'hard' photon–electron interactions, such as the production of gamma photon beams, where quantum electrodynamics (QED) plays a central role. The first volume is devoted to the classical aspects of the photon–radiation interaction, known as Thomson scattering and backscattering, and to the relevant technological consequences, which have enabled the worldwide boom in high-intensity x-ray sources, nowadays dedicated to different applications in pure and applied science. Röntgen radiation (also known as x-rays) was discovered at the beginning of the last century. The eponymous discoverer was the first scientist to be awarded a Nobel Prize in Physics. The radiograph of Röntgen's wife's hand became the symbol of the social impact that sources of x-ray radiation have had in science and society since their discovery. They are indeed pivotal in applied research and medical diagnostics and provide a unique tool for applied research and technological development in different fields. Their use in the health domain (for high-resolution imaging and cancer therapy) was one of their first and best-known applications. X-rays can penetrate dense matter, thus offering the possibility of exploring characteristics that are otherwise unobservable. High-quality x-ray beams have indeed opened important research perspectives in macromolecular crystallography, in the biosciences, and in the structural study of advanced materials such as nanomaterials, magnetic materials, intelligent polymers, etc. In environmental science, x-rays find applications in the determination of atmospheric conditions or even in deepening our understanding of the energy-exchange processes underlying cleaner industrial production technologies. The best-performing x-ray sources are synchrotrons and free-electron lasers (FELs). These devices require significant investment in terms of budget and infrastructure. Furthermore, given the intense and growing interest shown both by academic researchers and private users, they have become increasingly difficult to access. The demand for other platforms that are more accessible in terms of costs and dimensions is therefore becoming compelling. A change of paradigm has accordingly been proposed to overcome this bottleneck. The emergence of high-power lasers operating in the near-infrared region and the availability of high-quality electron beams (not exceeding hundreds of MeV) has paved the way for the generation of quasi coherent x-ray beams obtained via CBS. CBS is the mechanism through which high-energy electrons interacting with low-energy photons transfer some of their energy to the photons.

Accordingly, in a CBS process, an infrared photon is 'transformed' into an x-ray photon. X-ray CBS devices are therefore realized by colliding a laser beam (made of photons, i.e. particles of light) with a beam of electrons. As a result of the collision, a burst of x-rays (monochromatic and short-lived) is generated. Sources of monochromatic and ultrashort x-rays are challenging devices in terms of technological complexity, cost, and human resources. Although CBS-based sources are significant technological and scientific enterprises, they require less demanding infrastructure and fewer financial resources. The CBS process, known since the second half of the last century, has become more interesting not only for its scientific uses but also for the possibilities it offers in applied science. Today, many CBS x-ray sources employing low- and medium-energy accelerators and powerful lasers are in operation, under construction, or have been proposed and designed. Their uses range from medicine to cultural heritage. They represent a cheaper and valid alternative to ordinary synchrotron/FEL radiation sources. Since they depend on several parameters of both the photons and the electron beams, x-ray sources based on CBS can be used in different configurations; thus, they inherently exhibit a unique degree of tunability/versatility. CBS, sometimes also referred to as inverse Compton scattering (ICS), theoretically proposed about 60 years ago, is a flavor of Compton scattering (CS), which has been known since the early 1920s. The difference between the two effects can be summarized as follows: initially, a photon acquires energy at the expense of the energy of a relativistic electron in a head-on collision. Second, an electron acquires recoil momentum after being scattered by a sufficiently high-energy photon. The development of high-quality, low- or medium-energy electron accelerators and high-powered lasers has made it possible to propose and construct CBS-based sources of intense monochromatic x-rays. The interest in CBS photon 'factories' has manifold motivations ranging from practical applications (medicine, cultural heritage, etc.) to advanced QED studies aimed at exploring new phenomena, such as pair production in strong electromagnetic fields, vacuum instability, and photon–photon scattering. A book dedicated to CBS devices and their underlying physics should cover different aspects of the physics and technology of accelerators and lasers. CBS sources and their associated technologies have developed over the last twenty years. The original ideas of the last century have found a concrete realization. The associated literature is now extremely rich and covers a fairly large number of topics. The theoretical analysis of the mechanisms underlying these sources has been enriched by in-depth academic works, developed simultaneously with the construction and commissioning of radiation facilities. We have reached an appropriate moment at which to gather all this information into a book covering the relevant facets from theory to application, accompanied by a rich and complete bibliography. Like the study of FELs, the study of CBS is highly interdisciplinary and demands the combination of different 'cultures' from theoretical physics (classical physics, quantum optics, QED, hard QED processes, etc.) to accelerator and laser physics and their associated technologies. Unlike the FEL field, the CBS field still lacks a book that treats all the different topics, offering a 'gentle' introduction to scientists entering the field and/or functioning as a textbook for an advanced course on the subject. Any such book requires the elaboration of a pedagogical effort, which becomes more evident during the preparation of the exercises. We have tried to merge theoretical and practical notions to move in the direction of providing an effective and useful book. Some parts may appear redundant. We have, for example, devoted the first chapter to a careful discussion of the electromagnetic processes underlying Thomson scattering. This has necessarily required us to review some aspects of special relativity and classical electromagnetism. An understanding of an actual CBS device also requires a description of laser and charged beam transport. To avoid the appearance of redundancy, we have chosen to treat these topics using a unified point of view based on the phase-space formalism. This book has been designed to meet the previously mentioned needs. It benefits from our experiences as researchers and authors of books on both FELs and accelerators. Therefore, our effort has not just been that of presenting an overview of the theoretical and experimental aspects but also of providing the elements required for the design and handling of a CBS-based x-ray source. This volume is, however, not the end of the story. In the following chapters, we 'just' treat classical CBS. A companion volume will be devoted to more advanced topics related to the design of CBS sources conceived to probe hard QED processes. The matter treated here and the relevant presentation are the result of the different experiences of the authors, who have contributed as active researchers to most of the topics discussed in this monograph. The conception, assembly, and writing of the different chapters has been the result of joint work and a painstaking revision of the different points of view to achieve a balanced exposition. We should finally like to mention that chapters 3 and 4 are the outcome of a previous collaboration with the late Drs Franco Ciocci and Amalia Torre, whose competence in light and charged beam transport has always been a guide for our work. We owe our gratitude to many colleagues who, over the course of the years, have shared work experiences and discussions with us. We should like to mention colleagues from the LADON project (in particular, professors Angelo Marino, Lorenzo Federici, Luigi Casano, and Gianfranco Giordano) who, more than half a century ago, started their work which led to the production of the first gamma beam, produced by laser photons backscattering with the ADONE storage ring at Frascati. Finally, we would like to express our gratitude to Prof. Danilo Giulietti for his interest in our work and to Prof. Sultan Dabagov for the discussion we had during various editions of the Conference 'Charged & Neutral Particles Channeling Phenomena', whose warm and stimulating environment created the conditions for the gestation of the ideas which led to the compilation of this book.

Author biographies

Alessandro Curcio

Dr Alessandro Curcio started his career at the University of Pisa with a master's degree thesis on ultracompact and ultrabright laser-based x-ray sources. He then obtained a position in the PhD school of accelerator physics at the University of Rome 'La Sapienza,' concluding with a thesis on innovative concepts for ultrabright radiation sources and particle acceleration based on the laser–matter interaction at relativistic intensities. In November 2017 he won a research fellowship at the CERN Linear Accelerator for Research (CLEAR), in which he was responsible for scientific work on the generation of strongly accelerating fields via the emission of coherent radiation from ultrashort electron beams and on the use of terahertz radiation for ultrashort particle diagnostics. At the beginning of 2020, he joined the National Polish Synchrotron, SOLARIS, in Kraków as section leader in beam diagnostics and instrumentation. He subsequently served as a senior scientist at the Pulse Laser Center (CLPU) of Salamanca (Spain), where his main role was the design and development of beamlines and diagnostics for particles and radiation produced in laser-plasma accelerators. Currently he is a senior scientist at the National Laboratory of Frascati (Rome) of the Italian National Institute for Nuclear Physics (INFN-LNF). His main activity is related to the application of compact plasma accelerators for secondary radiation sources.

Giuseppe Dattoli

Giuseppe Dattoli was born in Lagonegro, Italy, in 1953. He received a PhD degree in physics from the University of Rome 'La Sapienza,' Italy, in 1976. He is a researcher for the Italian National Agency for New Technologies, Energy and Sustainable Economic Development (ENEA) and has been involved in different research projects, including high-energy accelerators, FELs, and applied mathematics networks since 1979. Dr Dattoli has taught in Italian and foreign universities and has received the FEL Prize Award for his outstanding achievements in the field.

Emanuele Di Palma

Emanuele Di Palma received a *laurea* degree in mathematics from the University of Rome 'La Sapienza,' Italy, in 1996. He started his research work on the development of advanced finite-difference time-domain (FDTD) numerical modeling for electromagnetic simulation at the Italian Institute of Mathematics (INdAM). He joined ENEA Laboratories in 2003, where he started his research into FEL applications and microwave tubes. He received a master's degree in 'Fusion Energy: Science and Engineering' from Tor Vergata University of Rome, Italy, in 2013 and a PhD degree in 'Fusion Science and Engineering' from the University of Padua, Italy, in *cotutelle* (joint enrollment) with the Instituto Superior Técnico (IST), University of Lisbon, Portugal in 2018. His research interests are in the fields of physics (namely the applications of intense electron beams, computer-aided design, and the development of CARM (Cyclotron Autoresonance Maser) devices for various novel applications such as solar energy harvesting in space), fusion energy for high-field tokamaks, and the development of compact devices for nuclear diagnostics in biomedical applications.

List of symbols

$q_e = -|e|$, $|e| \simeq 1.6 \times 10^{-19}$ C $\equiv$ elementary charge
$h \simeq 6.262$ J $\cdot$ s $\equiv$ Planck's constant
$c \simeq 299\,792\,458$ m s^{-1} $\equiv$ the velocity of light
1 MeV $= 10^6$ eV $\equiv$ 1 mega electron volt
1 eV $= 1.6 \times 10^{-19}$ J $\equiv$ 1 electron volt

$m_e = \frac{0.511 \text{ MeV}}{c^2} = 9.109 \times 10^{-31}$ kg $\equiv$ electron mass
$\varepsilon_0 = 8.854 \times 10^{-12}$ F $\times$ m^{-1} $\equiv$ vacuum permittivity

$r_0 = \frac{e^2}{4\pi\varepsilon_0} \frac{1}{m_e c^2} \simeq 2.818 \times 10^{-15}$ m $\equiv$ classical electron radius

IOP Publishing

Backscattering Sources, Volume 1
Theoretical framework and Thomson backscattering sources
Alessandro Curcio, Giuseppe Dattoli and Emanuele Di Palma

Chapter 1

Introduction

There are different flavors of the interaction between radiation and electrons, commonly known as:
1. Thomson scattering,
2. Compton scattering, and
3. Inverse Thomson/Compton scattering.

A pictorial view of these is shown in figures 1.1(a)–(c).

In the first case, an electromagnetic field interacts with an electron at rest, which oscillates in response to the electric field transported by the wave and diffuses the incident radiation.

In case (b), a photon 'hits' and transfers its energy to an electron at rest.

In case (c), a high-energy electron 'hits' a photon and transfers its energy to the photon.

There is a clear difference between the description of the first process and the descriptions of the other two. The first is explained in classical terms, while the other interactions are viewed as the scattering of two particles that have their own momenta. Even though rather naïve, we use this loose description for our introductory discussion and consider a more appropriate view in the forthcoming sections.

During recent decades these topics have attracted growing interest within the context of a worldwide effort aimed at realizing sources of high-brightness x-ray beams based on either free-electron lasers (FELs), inverse Compton scattering (ICS), or—more frequently—Compton backscattering (CBS).

Along with the possibility of producing laser-like beams with previously unachieved characteristics, the progress in laser and electron beam technologies has opened the possibility of testing fundamental questions in nonlinear electrodynamics, which represents the heart of physics itself.

This book is devoted to electron–photon diffusion and to the development of radiation sources based on the associated physical mechanisms. The amount of

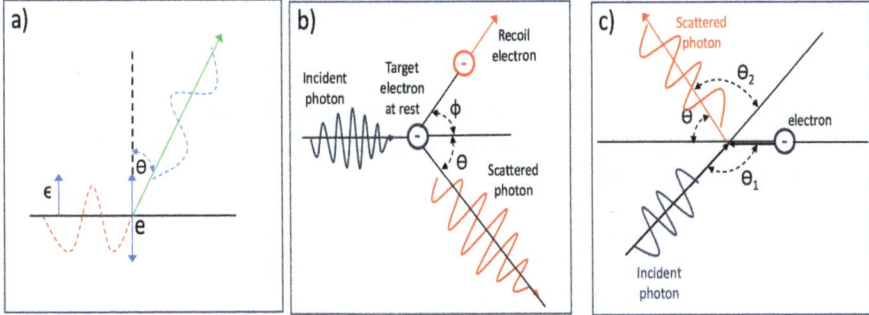

Figure 1.1. (a) The Thomson scattering process, showing an incident electromagnetic field (red), ε polarization of the associated electric field, and diffused radiation (blue). (b) Compton scattering: an incident photon is scattered by a photon at rest, which acquires a recoil momentum. (c) Compton backscattering: a high-energy electron is scattered by a photon, which acquires energy after the collision.

relevant literature has undergone a notable increment in the last two or three decades. Most of the theoretical contributions have their roots in the physics of the nineteenth and twentieth centuries. The CBS mechanism was suggested in the second half of the last century as a tool for generating high-energy photons through the scattering of laser photons and high-energy (ultrarelativistic) electrons. Intense beams of high-energy gamma rays were then generated and used as probes in high-energy physics experiments.

The processes shown in figure 1.1, although defined in different terms, can be traced back to paradigmatic mechanisms involving archetypal entities, namely electromagnetic radiation/photons and electrons (or any point-like particle).

The peculiar difference between processes (a), (b), and (c) is that in the first case, the outgoing radiation has the same wavelength as that of the incident field. In (b), the distinctive feature is a wavelength shift determined by the kinematic conditions. In (c), the process is characterized by a double Doppler upshift, which is duly discussed in the forthcoming sections and chapters.

Before we get into the specifics, we return to the abovementioned item (a).

By Thomson scattering, we mean the mechanism of the *classical* diffusion of electromagnetic radiation by an electron (see figure 1.1). The electron is a light charged particle with mass and charge given by:

$$q_e = -|e|, \quad |e| \simeq 1.6 \times 10^{-19} \text{ C}$$
$$m_e \simeq 0.511 \text{ MeV}. \tag{1.1}$$

The photon is a neutral massless particle with energy:

$$E_f = h\nu = \hbar\omega$$
$$\hbar = \frac{h}{2\pi}, \quad \nu = \frac{\omega}{2\pi} = \frac{c}{\lambda} \tag{1.2}$$
$$h \simeq 6.262 \text{ J} \times \text{s} \equiv \text{Planck's constant}$$
$$c \simeq 299\,792\,458 \text{ m s}^{-1} \equiv \text{the velocity of light}$$

where $\hbar$ is the reduced Planck constant and ν and λ are the photon frequency and wavelength, respectively.

There are three fundamental constants (h, e, c), involved in this process, and we can play with numbers to infer the characteristic quantities of the interaction itself. In this book, we will mostly use the SI unit system. The electron mass in kilograms can be inferred as indicated below:

$$\begin{aligned} 1 \text{ MeV} &= 10^6 \text{ eV} \\ 1 \text{ eV} &= 1.6 \times 10^{-19} \text{ J} \\ m_e &= \frac{0.511 \text{ MeV}}{c^2} = 9.109 \times 10^{-31} \text{ kg} \end{aligned} \quad (1.3)$$

Furthermore, if $[a]$ denotes the physical dimensions of a (a given quantity) and by taking into account that:

$$\begin{aligned} \left[\frac{e^2}{4\pi\varepsilon_0}\right] &= [EL] \\ \frac{1}{4\pi\varepsilon_0} &= 9 \times 10^9 \frac{\text{N m}^2}{\text{C}^2} \rightarrow \left[\frac{1}{4\pi\varepsilon_0}\right] = \left[\frac{EL}{Q^2}\right] \\ \varepsilon_0 &= 8.854 \times 10^{-12} \text{ F} \cdot \text{m}^{-1} \equiv \text{vacuum permittivity} \end{aligned} \quad (1.4)$$

where E denotes energy and L length, we infer that:

$$\left[\frac{e^2}{4\pi\varepsilon_0} \frac{1}{m_e c^2}\right] = [L]. \quad (1.5)$$

The quantity in the square bracket is indeed recognized as the classical radius of the electron, namely:

$$r_0 = \frac{e^2}{4\pi\varepsilon_0} \frac{1}{m_e c^2} \simeq 2.818 \times 10^{-15} \text{ m} \quad (1.6)$$

which plays a central role in our description of Thomson scattering, since it specifies the relevant cross section (see the following section for a thorough discussion).

The energies involved in Thomson scattering are those carried by the photon and the electron; their ratio yields:

$$\begin{aligned} \frac{E_f}{E_e} &= \frac{h\nu}{m_e c^2} = \frac{\lambda_e}{\lambda_f} \\ \lambda_e &= \frac{h}{m_e c} \equiv \text{the electron Compton wavelength.} \end{aligned} \quad (1.7)$$

We will see in the following that the 'transition' between Thomson and Compton scattering is ruled by this ratio. A process is referred to as Thomson scattering if:

$$\frac{E_\mathrm{f}}{E_\mathrm{e}} \ll 1. \qquad (1.8)$$

If not, it belongs to the Compton phenomenology. Even though this is a qualitative statement, to be substantiated during this chapter, we note that condition (1.8) rules out quantum corrections in Thomson scattering.

In this section we have fixed the terms of the discussion we would like to develop in this introductory chapter. In the following section we use classical electromagnetism to discuss a few details of Thomson scattering, namely the scattering of radiation of 'classical' electromagnetic radiation from a charged particle *at rest*. The case we discuss in the following section is doubly classical in the sense that it involves neither relativistic nor quantum mechanics.

1.1 The Larmor formula and a classical description of Thomson scattering

The preliminary discussion of the previous section allows an explanation of the Thomson scattering process, which, in slightly more accurate terms, can be worded as follows:

1. Suppose that an electromagnetic wave is incident on a **'free,' 'nonrelativistic,'** and **'charged'** particle.
2. The charged nature of the particle determines its acceleration by the wave's electric and magnetic fields.
3. The acceleration induces a motion of the charge in the direction of the oscillation of the electric field component of the wave.
4. The consequence is an oscillating dipole that emits radiation via the Larmor mechanism.

According to the points outlined above, the charged particle 'absorbs' electromagnetic power from the incoming field, and the associated acceleration determines the emission of electromagnetic radiation. This is the bare essence of a scattering process. Above, we used three adjectives:

- *free*, which denotes that the electrons are not bound in atomic or molecular environments;
- *nonrelativistic* underscores that the particle moves at velocities much smaller than that of light (see section 3);
- *charged*, which ensures that the particle is sensitive to the effects of electric and magnetic fields (for a more appropriate discussion, see section 1.4).

Under point (4) of the earlier list, we mentioned the Larmor formula, which accounts for the fact that an accelerated charge loses energy by emitting electromagnetic radiation. The treatment we report here is based on classical arguments due to J J Thomson and, more recently, to M Longair (for further reading, see the bibliography at the end of the chapter).

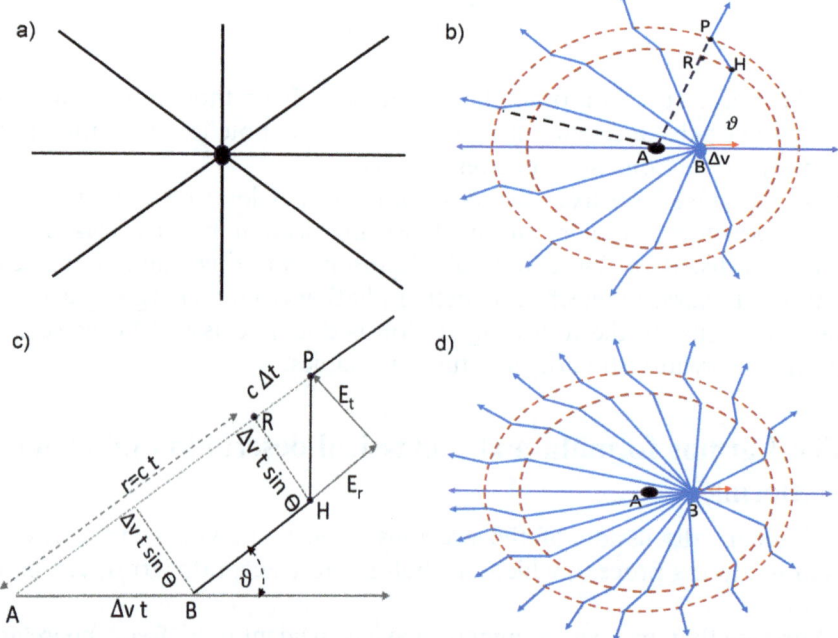

Figure 1.2. (a) Electric field lines for a charge at rest. (b) Electric field lines for an accelerated charge. (c) and (d) The radiation travels outward in the form of a kink pulse.

With reference to figure 1.2, we consider a charge to be at rest *until* a certain time. The associated electric field lines point away from the charge in all radial directions. If, at a certain time, say $t = 0$, the charge is acted upon by a kink force that produces a velocity variation Δv in a time interval Δt, then at a later time ($t > 0$), the charge is shifted by an amount $\Delta v t$ (this treatment is limited to nonrelativistic dynamics).

The action of the force not only produces a mechanical effect but the charge shift also perturbs the field itself. The tangential motion of the charge determines a kind of rearrangement of the outgoing lines, an effect which is responsible for a local time dependence of the field itself. The perturbation due to such a field line reshuffling is interpreted as an electromagnetic field moving forward for a distance $c \, \Delta t$ in the radial directions. The disturbance, moving at finite velocity, modifies the field lines it reaches, while the field lines at larger distances remain the same as established by the charge before the action of the force at time $t = 0$ (we invite the reader to be careful with the distinction between t and Δt).

We underscore that the perturbation propagates in the radial direction as a kink electromagnetic pulse characterized by an electric field component orthogonal to the propagation direction. The tangential component E_θ in the direction orthogonal to the radial line is due to the velocity component $v_\theta = \Delta v \sin(\theta)$, and the distance traveled in time t is:

$$Y_\theta = v_\theta t. \tag{1.9}$$

The discontinuity in the field that appears between the line outside the causality region and the line inside it is just due to the emergence of the tangential field component in the time Δt over which the kink acceleration acts. Upon inspecting figure 1.2(b), it is easily understood that:

$$E_\theta = \frac{Y_\theta}{X_\theta} E_r$$
$$X_\theta = c\Delta t \quad (1.10)$$
$$E_r = \frac{1}{4\pi\varepsilon_0} \frac{q}{r^2}, \quad r = ct.$$

Combining the previous relations, we find:

$$E_\theta = \frac{\Delta v\, t\, \sin(\theta)}{c\Delta t} \frac{1}{4\pi\varepsilon_0} \frac{q}{ct\, r} \quad (1.11)$$

where, for future convenience, we have set $r^2 = ct\, r$. After simplifying and taking the limit

$$\dot{v} = \lim_{\Delta t \to 0} \frac{\Delta v}{\Delta t} \quad (1.12)$$

we eventually find:

$$E_\theta = \frac{\dot{v}\, \sin(\theta)}{c^2} V$$
$$V = \frac{1}{4\pi\varepsilon_0} \frac{q}{r} \quad (1.13)$$

where V is the charge-associated potential. The power density (namely the energy per unit area and per unit time) carried by the electromagnetic pulse propagates in the longitudinal direction and is given by the modulus of the Poynting vector, namely (for more specific comments, see the exercises at the end of the chapter):

$$|\vec{S}| = c\varepsilon_0\, |E_\theta|^2. \quad (1.14)$$

Therefore, the use of equations (1.6) and (1.13) allows us to express equation (1.14) as:

$$|\vec{S}| = c\varepsilon_0 \frac{q^2 \dot{v}^2 \sin(\theta)^2}{16\pi^2 c^4 \varepsilon_0^2 r^2} = c\varepsilon_0 \left(\frac{q}{e}\right)^2 (m_e c^2)^2 \frac{r_0^2}{r^2} \dot{v}^2 \sin(\theta)^2. \quad (1.15)$$

The equation we have just obtained is the Larmor formula, which states that the power radiated by an accelerated charge is proportional to the square of the acceleration experienced by the charge itself. In the case of Thomson scattering, we note that the acceleration is induced according to the mechanism shown in figure 1.1(a). The electric field of the oscillating wave is specified by:

$$\vec{E} = Im(i\vec{\varepsilon}\, E_o e^{i(\vec{k}\cdot\vec{r} - \omega\, t)})$$
$$\vec{k} \equiv \text{wave vector} \tag{1.16}$$
$$\omega = |\vec{k}|c$$

where $\vec{\varepsilon}$ and E_o are the polarization and the amplitude of the electric wave, respectively. These are transverse to the propagation direction and therefore:

$$\vec{\varepsilon} \cdot \vec{k} = 0. \tag{1.17}$$

The motion of the charge under the action of the field (1.9) is determined by the equation:

$$m\dot{\vec{v}} = qIm(i\vec{\varepsilon}\, E_o e^{i(\vec{k}\cdot\vec{r} - \omega\, t)}). \tag{1.18}$$

Keeping the time-dependent part only, we are left with:

$$\dot{\vec{v}} = \frac{q}{m}\vec{\varepsilon}\, E_0 \sin(\omega t) \tag{1.19}$$

which yields the following average value over one oscillation period:

$$\langle \dot{\vec{v}}^2 \rangle = \frac{1}{2}\frac{q^2 E_0^2}{m^2}. \tag{1.20}$$

The modulus of the scattered Poynting vector can therefore be written as:

$$\langle S \rangle_s = \left(\frac{c\varepsilon_0 E_0^2}{2\, r^2}\right) \sin\theta^2 r_0^2. \tag{1.21}$$

After replacing q, m with the electric charge and mass, we can write the scattered power:

$$|\langle S \rangle_s| = |\langle S_0 \rangle_i|\frac{d\sigma}{r^2 d\Omega},$$
$$|\langle S_0 \rangle_i| = \frac{c\varepsilon_0 \left|E_0^2\right|}{2} \tag{1.22}$$
$$\frac{d\sigma}{d\Omega} = r_0^2 \sin(\theta)^2$$

where i and s are the incident power and the scattered power, respectively.

The previous equation comprises all the physics of the problem we are studying. The relevant meaning is better understood by inspecting figure 1.3. The incident wave carries the time-averaged Poynting vector $\langle S_0 \rangle_i$, a fraction of which, $\langle S \rangle_s$, is absorbed and re-emitted (almost **instantaneously**) by the electron. The characteristic quantity that governs this process is the differential cross section $\frac{d\sigma}{d\Omega}$. The main result of this section, reported in equation (1.22) (albeit derived in very simplified terms—we provide more rigorous details in the Comments and exercises section at the end of

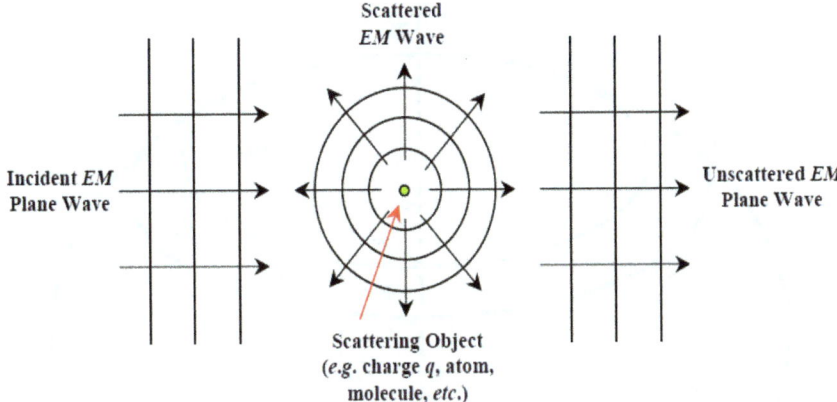

Figure 1.3. An incident wave and the diffused wave created by a scattering point.

the chapter), captures the correct mathematical definition of the differential cross section, which reads:

$$\frac{d\sigma}{d\Omega} = \lim_{r \to \infty} \frac{r^2 \langle S \rangle_s}{\langle S \rangle_i}. \tag{1.23}$$

It is evident, even though not yet explicitly stated, that both the differential cross section and the cross section σ have the dimensions of a surface. The element of solid angle $d\Omega$, shown in figure 1.4, is defined as:

$$d\Omega = d\cos(\theta) d\phi. \tag{1.24}$$

We find, therefore, that:

$$\sigma = \int_\Omega \frac{d\sigma}{d\Omega} d\Omega \tag{1.25}$$
$$d\Omega = 2\pi \sin(\theta) d\theta.$$

In conclusion, we obtain:

$$\sigma = 2\pi \, r_0^2 \int_0^\pi [\sin \theta]^3 \, d\theta = \frac{8}{3}\pi \, r_0^2. \tag{1.26}$$

A naïve consequence of the previous discussion is that Thomson diffusion can be viewed as a dipole emission. The typical angular distribution is shown in figure 1.5 along with the transverse section, which reveals an antenna-like emission pattern.

Therefore, it is worth underscoring how the radiated power is explicitly written in terms of the dipole oscillating in the z-direction. If the dipole moment is denoted by

$$\vec{d} = |e|z\hat{k} \tag{1.27}$$

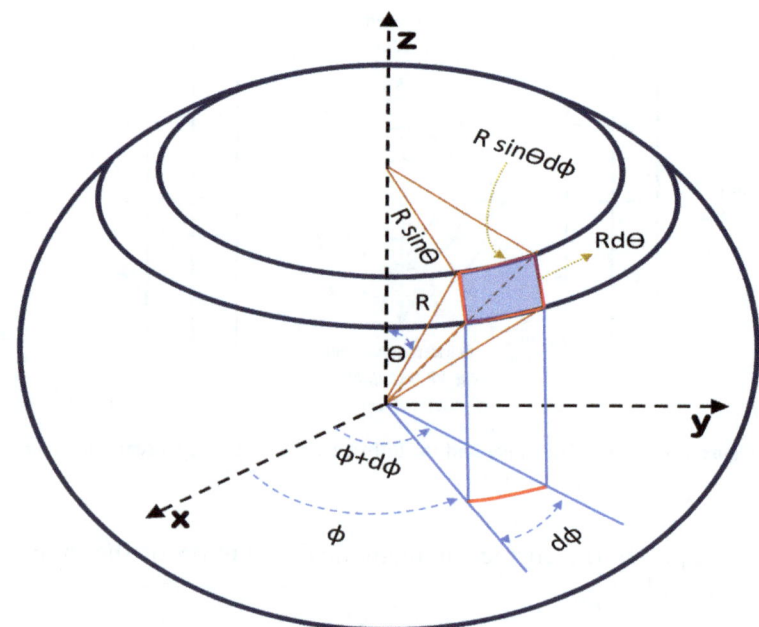

Figure 1.4. A solid angle in spherical coordinates.

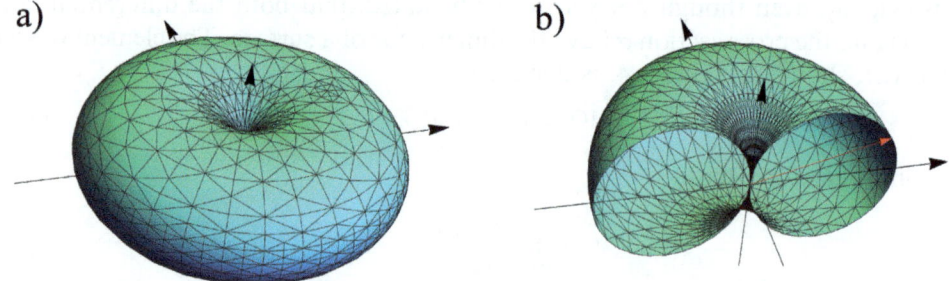

Figure 1.5. (a) The angular distribution of dipole radiation. (b) Details of a dipole: transverse section and angular direction.

we get:

$$\vec{S}_s = c u_s \hat{r}$$
$$u_s = \frac{\mu_0 |\ddot{\vec{d}}|}{16\pi^2 c^2} \left(\frac{\sin\theta}{r} \right)^2 \quad (1.28)$$
$$c = \frac{1}{\sqrt{\varepsilon_0 \mu_0}},$$
$$\mu_0 = \text{vacuum permeability}.$$

The discussion we have developed so far has allowed a point of view which, albeit simplified, yields the tools with which to start a more thorough discussion of the scattering of radiation by a free electron at rest. This discussion and the formulae we have derived are relevant to the scattering of polarized radiation; the unpolarized counterpart can be treated using similar means. The relevant details are presented at the end of the chapter in section 1.8, which is dedicated to comments and exercises. In the next section, we summarize a few elements of relativistic mechanics which are necessary to understand some of the distinctive features of the physics of processes (b) and (c).

1.2 Elements of special relativity: inertial frames and Lorentz transformations

In this section we summarize a few elements of special relativity that are necessary to understand Compton scattering and CBS.

We assume that the reader is familiar with the basic concepts of special relativity, and we therefore avoid any reference to the derivation of specific features such as Lorentz transformation, velocity addition, real and supposed paradoxes, and so on. In order to avoid any misunderstanding, we report the essential tools we will exploit in the following. In figure 1.1 we show two inertial frames, namely two reference frames with relative constant motion in the x-direction.

The Lorentz transformations are a generalization of the Galilean transformations (GTs) of nonrelativistic mechanics. The two transformations (the primed system versus the unprimed and vice versa) are completely equivalent, provided that the sign of v is reversed.

$$
\begin{aligned}
&x' = \gamma(x - vt) & \gamma &= \frac{1}{\sqrt{1 - \beta^2}} & x' &= \gamma(x + vt) \\
&y' = y & & & y' &= y \\
&z = z' & \beta &= \frac{v}{c} & z &= z' \\
&t' = \gamma\left(t + \frac{vx}{c^2}\right) & c &\equiv \text{velocity of light} & t' &= \gamma\left(t - \frac{vx}{c^2}\right)
\end{aligned}
\quad (1.29)
$$

where β and γ are the reduced velocity and the relativistic factor, respectively (see figure 1.6).

The distinctive features of the Lorentz transformations are essentially two:
1. They include either space or time 'coordinates'.
2. The light velocity remains unaffected by the transition from one frame to the other.

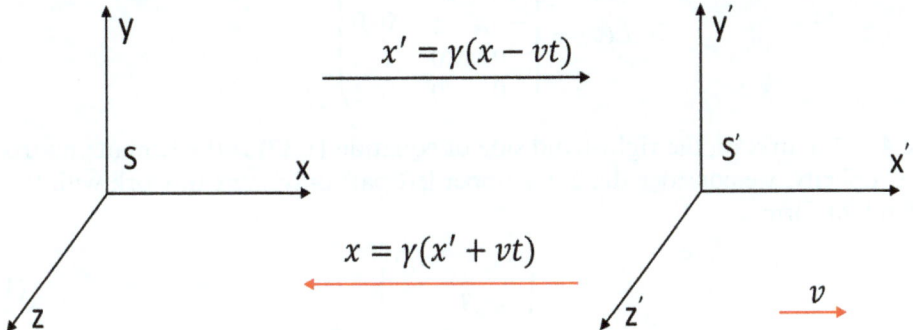

Figure 1.6. Lorentz transformations between inertial frames.

The above points implicitly contain or (better) are an immediate consequence of the postulates of special relativity:
1. The velocity of light is the same in all inertial frames.
2. All inertial frames are equivalent.

The above two statements are sufficient to justify the logical environment of special relativity. The consequences of the Lorentz transformations are length contraction and time dilatation, as sketched in figure 1.7 (for further details, see the exercises at the end of the chapter).

1.2.1 Four-vectors and matrix formalism

In special relativity, space and time coordinates are represented using an extension of the ordinary vectors, known as four-vectors. They are denoted by the following column vector, which has four components:

$$x_\mu = \begin{pmatrix} ct \\ x \\ y \\ z \end{pmatrix}, \mu = 0, 1, 2, 3. \tag{1.30}$$

The above form is said to be covariant, while its contra-variant version reads:

$$x^\mu = \begin{pmatrix} ct \\ -x \\ -y \\ -z \end{pmatrix}, \mu = 0, 1, 2, 3. \tag{1.31}$$

Within this formalism, the Lorentz transformation can naturally be expressed in the form reported below:

$$x'_\mu = \hat{L}(v) \begin{pmatrix} ct \\ x \\ y \\ z \end{pmatrix}$$

$$\hat{L}(v) = \begin{pmatrix} \gamma & -\gamma\beta & 0 & 0 \\ -\gamma\beta & \gamma & 0 & 0 \\ 0 & 0 & 1 & 0 \\ 0 & 0 & 0 & 1 \end{pmatrix}. \tag{1.32}$$

The 4 × 4 matrix on the right-hand side of equation (1.32) is the Lorentz matrix. If, for simplicity, we consider the 2 × 2 upper left part only, we can work with the less redundant form

$$\hat{L} = \begin{pmatrix} \gamma & -\gamma\beta \\ -\gamma\beta & \gamma \end{pmatrix}. \tag{1.33}$$

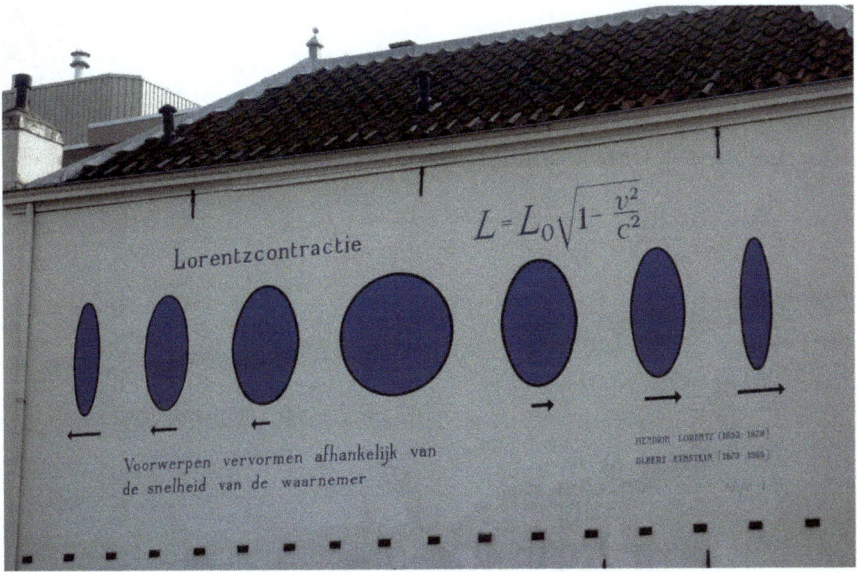

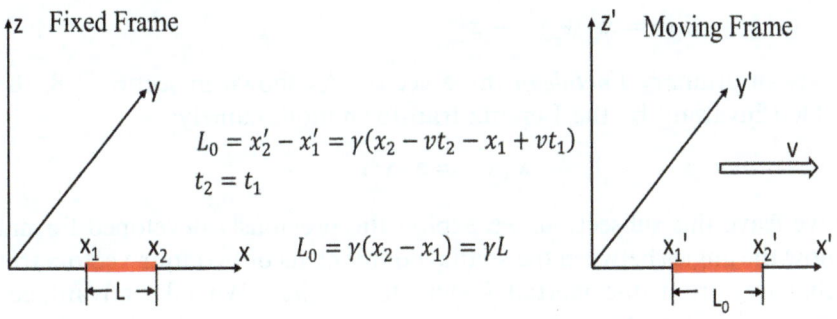

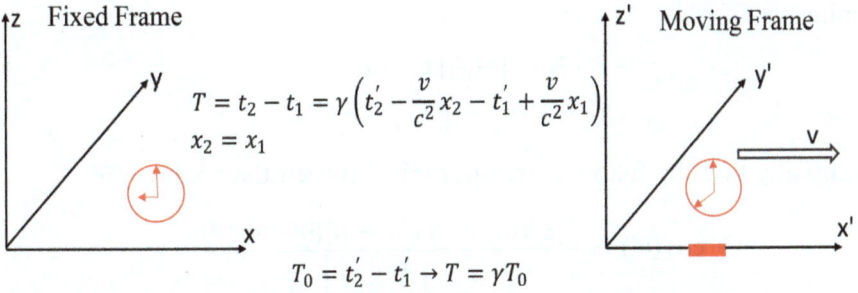

Figure 1.7. (top) Length contraction, a pictorial view at Leiden railway station. This Formula of Lorentz (& George Francis FitzGerald) painted on a wall of a house at Haagweg, Leiden by Stichting Tegenbeeld image has been obtained by the authors from the Wikimedia website where it was made available by Vysotsky (2017) under a CC BY 4.0 licence. It is included within this chapter on that basis. It is attributed to Vysotsky. (middle) The length of a ruler (or of any object) appears contracted in the direction of motion in a 'rest' frame. (bottom) The time in a moving frame is seen to run slower in a 'rest' frame as a consequence of the Lorentz transformations.

Its determinant is easily seen to be unitary:

$$|\hat{L}| = \gamma^2 - \gamma\beta^2 = 1. \tag{1.34}$$

Therefore, the inverse transformation can be written as:

$$x_\mu = \hat{L}^{-1} x'_\mu \tag{1.35}$$

where

$$\hat{L}(v)^{-1} = \begin{pmatrix} \gamma & \gamma\beta \\ \gamma\beta & \gamma \end{pmatrix} = \hat{L}(-v). \tag{1.36}$$

The norm of the four-vector (1.30) is defined by the product $x^\mu x_\mu$, which can be explicitly written as:

$$x_\mu x^\mu = \begin{pmatrix} ct \\ x \\ y \\ z \end{pmatrix}(ct\ -x\ -y\ -z) = (ct)^2 - |\vec{r}|^2$$
$$\vec{r} = (x, y, z) \tag{1.37}$$
$$|\vec{r}|^2 = x^2 + y^2 + z^2$$

where $\vec{r}$ is an ordinary **Euclidean** three vector. As shown in section 1.8, the norm (1.37) is left invariant by the Lorentz transformation, namely:

$$x'_\mu x'^\mu = x_\mu x^\mu. \tag{1.38}$$

Before we leave this subsection, we exploit the previously developed formalism to derive how the angles between the spatial components of two four-vectors transform when changing from one inertial system to another. We take advantage of the invariance under Lorentz transformations of the 'scalar' product:

$$(x_1)'_\mu x_2'^\mu = x_{1\mu} x_2^\mu. \tag{1.39}$$

Assuming that

$$\vec{r}_1 \cdot \vec{r}_2 = |\vec{r}_1||\vec{r}_2|\cos(\theta)$$
$$\vec{r}'_1 \cdot \vec{r}'_2 = |\vec{r}'_1||\vec{r}'_2|\cos(\theta') \tag{1.40}$$

we eventually find the following relationship between the angles θ, θ':

$$\cos(\theta') = -\frac{c^2(t_1 t_2 - t'_1 t'_2) - |\vec{r}_1||\vec{r}_2|\cos(\theta)}{|\vec{r}'_1||\vec{r}'_2|} \tag{1.41}$$

This is a general identity that is valid for transformations with different velocities and different orientations between them.

A simpler derivation valid for less general conditions is discussed below. For this purpose, we consider the composition of the velocity between the two inertial frames, which can be directly inferred from the Lorentz transformations. The time derivative in space in the primed reference frame is written:

$$\frac{dx'}{dt'} = \frac{(dx - vdt)}{dt - \frac{vdx}{c^2}} = \frac{\frac{dx}{dt} - v}{1 - \frac{v\frac{dx}{dt}}{c^2}} \tag{1.42}$$

assuming that, in the unprimed system, the velocity is $dx/dt = u$, we end up with:

$$\begin{aligned} u' &= \frac{u - v}{1 - \frac{vu}{c^2}} \\ u &= \frac{u' + v}{1 + \frac{vu'}{c^2}}. \end{aligned} \tag{1.43}$$

As a by-product of the previous identities, we confirm that the velocity of light remains invariant in any inertial frame. It is evident that if $u = c$, then $u' = c$; in other words, the velocity of light remains the same under the Lorentz transformation.[1]

It should be underscored that the velocity composition in equation (1.42) holds if the velocities u and v are parallel. For a more general discussion, see section 1.8.

Now consider a photon traveling along the x component with a velocity $u' = c \cos(\theta')$. From the second equation of (1.43), we find (see section 1.8 for further comments):

$$c \cos(\theta) = \frac{c \cos(\theta') + v}{1 + \frac{v \cos(\theta')}{c}} \tag{1.44}$$

and, eventually:

$$\cos(\theta') = \frac{\cos(\theta) - \beta}{1 - \beta \cos(\theta)}. \tag{1.45}$$

1.3 Relativistic kinematics and dynamics

The second law of Newtonian dynamics expresses the variation of the velocity of a massive body in response to an external force, namely:

$$m_0 \vec{a} = \vec{F} \tag{1.46}$$

where m_0 is said the 'rest' mass of the body acquiring the acceleration $\vec{a}$ under the action of the force $\vec{F}$.

---

[1] It should be noted that the invariance of the velocity of light is not a consequence of the Lorentz transformations but a consequence of assuming postulates 1 and 2.

A more convenient formulation for our purposes is one that involves the time derivative of the body's linear momentum, namely:

$$\frac{d\vec{p}}{dt} = \vec{F}. \tag{1.47}$$

It is evident that if we use the 'classical' definition of acceleration as the derivative of the velocity equation (1.46), is clearly inadequate to describe the relativistic motion. We find indeed, as $d\vec{v}/dt = \vec{F}/m_0$, that it is in clear contrast with the postulate of c as the maximum achievable velocity.[2]

The formulation in terms of the time derivative of momentum is written:

$$\vec{p} = m\vec{\beta} c \tag{1.48}$$

where m is assumed to be velocity dependent. The mass therefore acquires a dynamical dependence on the velocity and is supposed to increase with the velocity. If we take

$$m = m_0 \gamma \tag{1.49}$$

then the relativistic generalization of the force law reads

$$\frac{d(m\vec{\beta}c)}{dt} = \vec{F} \tag{1.50}$$

Integrating the previous differential equation for constant force (see section 1.8), we find that

$$\vec{v} = \frac{\vec{a}t}{\sqrt{1 + \left(\frac{|\vec{a}t|}{c}\right)^2}}, \quad \vec{a} = \frac{\vec{F}}{m_0} \tag{1.51}$$

which for finite times is always less than c.

Assumption (1.49) implies that the relativistic mass is the product of the rest mass times the relativistic factor.

If we expand γ for low velocities ($\beta \ll 1$):

$$\gamma \simeq 1 + \frac{1}{2}\beta^2 + o(\beta^4) \tag{1.52}$$

we obtain:

$$mc^2 \simeq m_0 c^2 \left(1 + \frac{1}{2}\beta^2\right) = m_0 c^2 + \frac{1}{2} m_0 v^2. \tag{1.53}$$

This result is of extreme conceptual importance. It states indeed that, in the low-velocity limit, the total energy is composed of two parts: a term associated with the

---

[2] For constant acceleration, we indeed find that $\vec{v} = \vec{a}t$; therefore, in the absence of any counteracting contribution, the acquired velocity can become equal to or even larger than c.

body's rest mass and a second contribution, recognized as the classic kinetic energy. It is worth underlining that, within this 'low-velocity' assumption, the meaning of m_0 is understood in the Newtonian sense.

The total energy of a body moving at relativistic velocities is:

$$E = m_0 \gamma c^2. \tag{1.54}$$

Subtracting the $m_0 c^2$ leaves us with:

$$T = E - m_0 c^2 = m_0 (\gamma - 1) c^2 \tag{1.55}$$

which yields the relativistic kinetic energy.

We can now take a step further and reconcile the relativistic kinematics with the four-vector formalism we discussed above. If we put together the definition of relativistic total energy and momentum, it can easily be verified that:

$$E^2 - (|\vec{p}|c)^2 = (m_0 c^2)^2 \tag{1.56}$$

which, compared with equation (1.52), suggests the introduction of a four-vector characterized by energy as component zero and by momentum as a three-part Euclidean component. This four-vector is called four-momentum and is defined below along with its norm:

$$\begin{aligned} p_\mu &\equiv (E, \vec{p} c) \\ p_\mu p^\mu &= m_0^2 c^4. \end{aligned} \tag{1.57}$$

It is interesting to note that an alternative definition of the four-momentum exists:

$$\begin{aligned} p_\mu &= m_0 c^2 u_\mu \\ u_\mu &= \begin{pmatrix} \gamma \\ \gamma \vec{\beta} \end{pmatrix} \end{aligned} \tag{1.58}$$

which is further discussed in section 1.7.

The transformation of p_μ from one inertial frame to another is governed by the Lorentz matrix. Regarding the one-dimensional motion, we find:

$$\begin{pmatrix} E' \\ p'c \end{pmatrix} = \begin{pmatrix} \gamma & -\gamma\beta \\ -\gamma\beta & \gamma \end{pmatrix} \begin{pmatrix} m_0 c^2 \\ 0 \end{pmatrix} = \begin{pmatrix} m_0 \gamma c^2 \\ -m_0 \gamma \beta c^2 \end{pmatrix}. \tag{1.59}$$

We invite the reader to check the identity

$$\begin{pmatrix} m_0 c^2 \\ 0 \end{pmatrix} = \begin{pmatrix} \gamma & \gamma\beta \\ \gamma\beta & \gamma \end{pmatrix} \begin{pmatrix} m_0 \gamma c^2 \\ -m_0 \gamma \beta c^2 \end{pmatrix}. \tag{1.60}$$

The last two identities, although straightforward, add a further physical flavor to the meaning of Lorentz transformations.

In the case of massless particles, the identity (1.56) yields:

$$E = pc \tag{1.61}$$

which, in the case of the photon, yields the following interpretation:

$$E = h\nu$$
$$p = \frac{h\nu}{c}. \tag{1.62}$$

We are now in a position to specify the use of this formalism in problems involving the scattering of two particles moving at different (relativistic) velocities.

The first example we consider is the electron e^-/positron e^+ collision, the formation of a positronium ($e^+ - e^-$) 'atom,' and its consequent decay into two photons. We sketch the process in figure 1.8. Consider an electron at rest and apply the conservation of momentum and the conservation of energy:

$$m_0 \gamma_1 v = M_0 \gamma_2 V$$
$$m_0(\gamma_1 + 1) = M_0 \gamma_2 \tag{1.63}$$

where m_0 and M_0 are the masses of the $e^\pm$ particles. The velocity of the positronium is therefore

$$V = \frac{\gamma_1 v}{\gamma_1 + 1}. \tag{1.64}$$

Regarding the second part of the problem, we apply the same rules and write:

$$M_0 \gamma_2 c^2 = E_1 + E_2$$
$$M_0 \gamma_2 V = \frac{E_1}{c} - \frac{E_2}{c}. \tag{1.65}$$

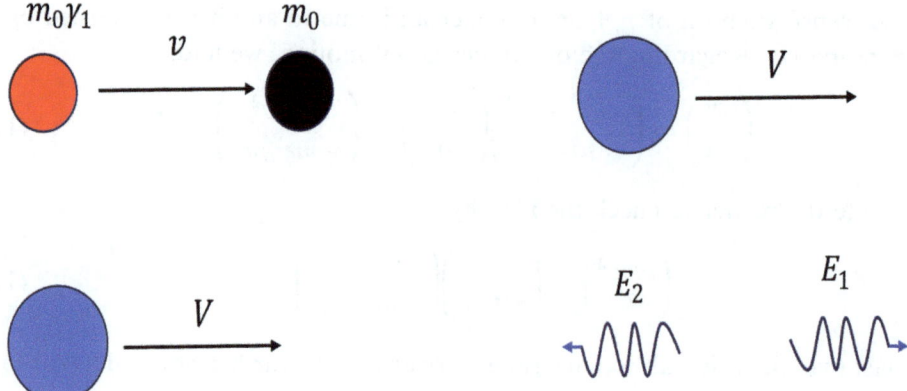

Figure 1.8. The formation and decay of positronium: a positron hits an electron at rest, forming an e^+–e^- bound state which decays into two photons.

Therefore, we find for the energies $E_{1,2}$:

$$E_1 = \frac{M_0 \gamma_2 c^2}{2}\left(1 + \frac{V}{c}\right) = \frac{m_0 c^2}{2}\left[\gamma_1\left(1 + \frac{v}{c}\right) + 1\right]$$
$$E_2 = \frac{M_0 \gamma_2 c^2}{2}\left(1 - \frac{V}{c}\right) = \frac{m_0 c^2}{2}\left[\gamma_1\left(1 - \frac{v}{c}\right) + 1\right]. \quad (1.66)$$

Further examples involving the use of relativistic mechanics are discussed in section 1.8.

In the next section, we apply the previously outlined notions to our understanding of the kinematic details of Compton scattering.

1.4 The relativistic kinematics of Compton scattering

We have described the classical aspects of Thomson scattering, namely the interaction between an electromagnetic wave and an electron initially at rest. We have also warned that the adjective 'classical' is vague. In more specific terms, we can state that the treatment is classical if either relativistic or quantum corrections can be neglected.

We have also exploited a kind of rule of thumb to decide whether we are in the classical scenario or the Compton scenario. The use of the photon energy and of the electron mass suggest that whenever

$$\hbar\omega \ll m_e c^2$$
$$T_e \ll m_e c^2 \quad (1.67)$$

the diffusion problem can be formulated in classical terms. In the following discussion, we make these criteria into more substantive statements.

In the following, we use the four-vector formalism to describe the kinematics of the process shown in figure 1.9. Accordingly, we can word the kinematics of the process as given below: a photon with four-momentum $\Pi_\mu \equiv (\varepsilon, c\vec{\kappa})$ (note that the vector $\vec{\kappa}$ has the dimensions of momentum) undergoes a collision with an electron at rest $P_\mu \equiv (m_0 c^2, 0)$. The electron recoils, acquiring a momentum $P'_\mu \equiv (E', c\vec{p}')$ at the expense of the incoming photon. The outgoing photon is eventually labeled as $\Pi'_\mu \equiv (\varepsilon', c\vec{\kappa}')$. The conclusion, albeit naïve, is that we expect that the outgoing photon to have a longer wavelength. Assembling the mathematical version of this discussion is straightforward:

1. Momentum conservation

$$\vec{\kappa} = \vec{\kappa}' + \vec{p}'. \quad (1.68)$$

2. Energy conservation

$$\varepsilon + m_0 c^2 = \varepsilon' + E'. \quad (1.69)$$

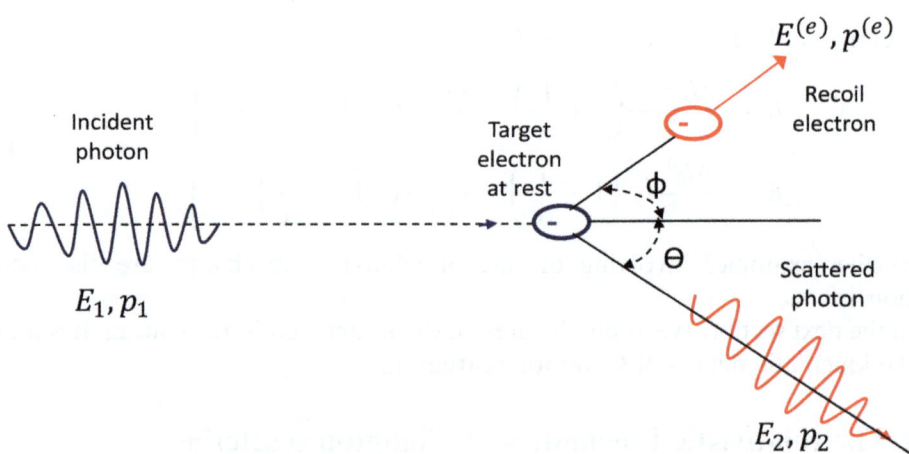

Figure 1.9. The kinematics of Compton scattering.

It is evident that upon solving for the electron output variables, we obtain:

$$P'_\mu \equiv (\varepsilon - \varepsilon' + m_0 c^2, c(\vec{\kappa} - \vec{\kappa}')). \quad (1.70)$$

The invariance of the electron momentum yields:

$$P'_\mu P'^{\mu'} \equiv (\varepsilon - \varepsilon' + m_0 c^2)^2 - c^2 |\vec{\kappa} - \vec{\kappa}'|^2 = m_0^2 c^4. \quad (1.71)$$

Furthermore, taking into account that

$$\vec{\kappa} \cdot \vec{\kappa}' = |\vec{\kappa}||\vec{\kappa}'| \cos \theta, \quad (1.72)$$

we find:

$$(\varepsilon - \varepsilon')^2 + m_0^2 c^4 + 2(\varepsilon - \varepsilon') m_0 c^2 - c^2(|\vec{\kappa}|^2 + |\vec{\kappa}'|^2) + 2c^2|\vec{\kappa}||\vec{\kappa}'| \cos(\theta) = (m_0 c^2)^2 \quad (1.73)$$

which is rearranged as shown below:

$$(\varepsilon^2 - c^2 |\vec{\kappa}|^2) + (\varepsilon^2 - c^2 |\vec{\kappa}'|^2) + 2(\varepsilon - \varepsilon') m_0 c^2 - 2\varepsilon\varepsilon' + 2c^2|\vec{\kappa}||\vec{\kappa}'| \cos(\theta) = 0. \quad (1.74)$$

Since the photon is massless and $|\vec{\kappa}| = \varepsilon/c$, the previous expression simplifies to:

$$(\varepsilon - \varepsilon') m_0 c^2 = \varepsilon\varepsilon'(1 - \cos(\theta)). \quad (1.75)$$

Using $\varepsilon = hc/\lambda$ for the photon energy, we eventually end up with:

$$(\lambda' - \lambda) m_0 c^2 = \frac{h}{m_0 c}(1 - \cos(\theta)). \quad (1.76)$$

The wavelength shift induced by the Compton process is therefore given by:

$$\frac{1}{2}\frac{\delta\lambda}{\lambda} = \frac{\varepsilon'}{m_0 c^2}\left(\sin\left(\frac{\theta}{2}\right)\right)^2 \quad (1.77)$$

$$\delta\lambda = \lambda' - \lambda.$$

The output photon energy is eventually written as:

$$\varepsilon' = \frac{\varepsilon}{1 + \frac{\varepsilon}{m_0 c^2}(1 - \cos(\theta))}. \quad (1.78)$$

Regarding the electron recoil energy, it is worth noting that the following relationship holds:

$$E' = \frac{\delta\lambda}{\lambda'}\varepsilon + m_0 c^2 \quad (1.79)$$

which is fairly interesting, as it indeed provides us with a tool for determining the electron's rest energy (for more details see the bibliography at the end of the chapter). The experimental measurements of the recoil energy and of the Compton relative wavelength shift can indeed be exploited (along with equation (1.79)) to determine the electron's mass.

It is, however, worth underlining that equation (1.79) is a direct measure of the electron's recoil kinetic energy:

$$T' = \frac{\delta\lambda}{\lambda'}\frac{\varepsilon}{m_0 c^2} \quad (1.80)$$

$$T' = \gamma' - 1.$$

Further and more specific comments on the experimental procedure and on the handling of the relevant data can be found in the bibliography at the end of the chapter.

1.5 The kinematics of inverse Compton scattering

In the previous section we have assumed that the scattering process occurs through a mechanism that involves a photon hitting an electron at rest.

This is not an absolute statement. According to the postulates of special relativity, all inertial frames are equivalent; therefore, we can 'view' the diffusion in any other reference frame linked to the laboratory frame (where we performed the calculations in our previous discussion) using a Lorentz transformation (for further comments, see the next section).

Let us now consider the case in which the electron is not at rest but is moving at a relativistic velocity.

The elements of the discussion and the strategy we follow to determine the characteristics of the outgoing photon are the same as before.

The photon and electron kinematic variables are respectively labeled with:

$$\begin{aligned}
\Pi_\mu &\equiv (\varepsilon, c\vec{\kappa}) \\
P_\mu &\equiv (E, c\vec{p}) \\
\varepsilon &= h\nu \\
\vec{\kappa} &= \frac{h}{2\pi}\vec{k} \\
E &= m_0\gamma c^2 \\
\vec{p} &= m_0\gamma c\beta.
\end{aligned} \quad (1.81)$$

The geometry of the scattering is fixed by the following conditions (see figure 1.10)

$$\begin{aligned}
\vec{\kappa} \cdot \vec{\kappa}' &= |\vec{\kappa}||\vec{\kappa}'|\cos(\theta) \\
\vec{\kappa}' \cdot \vec{p} &= |\vec{\kappa}'||\vec{p}|\cos(\phi_2) \\
\vec{\kappa} \cdot \vec{p} &= |\vec{\kappa}||\vec{p}|\cos(\phi_1).
\end{aligned} \quad (1.82)$$

If we just follow the same steps as those previously outlined, taking into account that:

$$P_\mu P^\mu = P'_\mu P'^\mu = m_0^2 c^4 \quad (1.83)$$

and solving for the final electron momentum, we end up with:

$$\varepsilon\varepsilon'(1 - \cos(\theta)) + \varepsilon'\gamma m_0 c^2(1 - \beta\cos(\phi_2)) = \varepsilon\gamma m_0 c^2(1 - \beta\cos(\phi_1)). \quad (1.84)$$

Thus, we eventually get the following for the output photon energy (for the details of the calculations, see section 1.8):

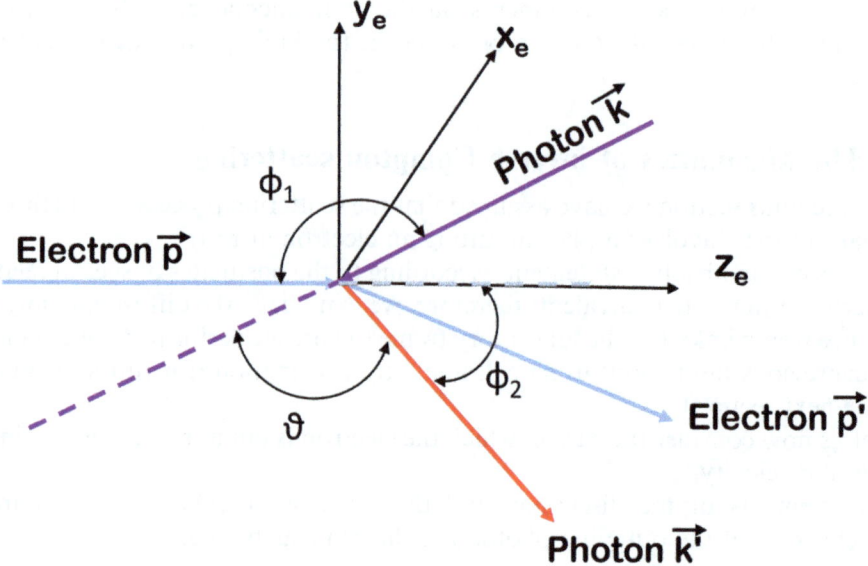

Figure 1.10. The kinematic and geometric variables of Compton backscattering.

$$\varepsilon' = \frac{1 - \beta \cos(\phi_1)}{1 - \beta \cos(\phi_2) + \frac{\varepsilon}{\gamma m_0 c^2}(1 - \cos(\theta))} \varepsilon \qquad (1.85)$$

and for the associated wavelength, we get:

$$\lambda' = \frac{1 - \beta \cos(\phi_2) + \frac{h\nu}{\gamma m_0 c^2} 1 - \cos(\theta)}{1 - \beta \cos(\phi_1)} \lambda. \qquad (1.86)$$

The Thomson scattering condition is:

$$\frac{h\nu}{\gamma m_0 c^2} \ll 1 \qquad (1.87)$$

which yields:

$$\lambda' \simeq \frac{1 - \beta \cos(\phi_2)}{1 - \beta \cos(\phi_1)} \lambda. \qquad (1.88)$$

for (1.86). In the case of CBS geometry ($\phi_2 = 0$, $\phi_1 = \pi$) and ultrarelativistic electron motion:

$$\beta \simeq 1 - \frac{1}{2\gamma^2} \qquad (1.89)$$

and the output wavelength is given by:

$$\lambda' \simeq \frac{\lambda}{4\gamma^2}. \qquad (1.90)$$

Regarding the frequency, we obtain:

$$\nu' \simeq 4\gamma^2 \nu \qquad (1.91)$$

which is an amazing result!

The output photon energy is increased by a factor of $4\gamma^2$ with respect to that of the incoming photon, which occurs at the expense of the electron energy.

This conclusion is the pivotal element of the following discussion. Just to give a preliminary idea of the possibilities offered by this process, we illustrate in figure 1.11 the output photon energy versus the electron energy for an initial photon wavelength in the near-infrared region. The consequence of the scattering is that low-energy photons (in the range of few eV) are upshifted to the hard x-ray region. In the forthcoming section, we reformulate the theory of electromagnetism from the perspective of special relativity.

1.6 The relativistic Doppler shift

Before combining electromagnetism with special relativity, we consider it useful to deal with the important, albeit not yet mentioned, notion of the **relativistic Doppler shift**, namely the frequency/wavelength shift associated with the radiation emitted or observed by a source in motion.

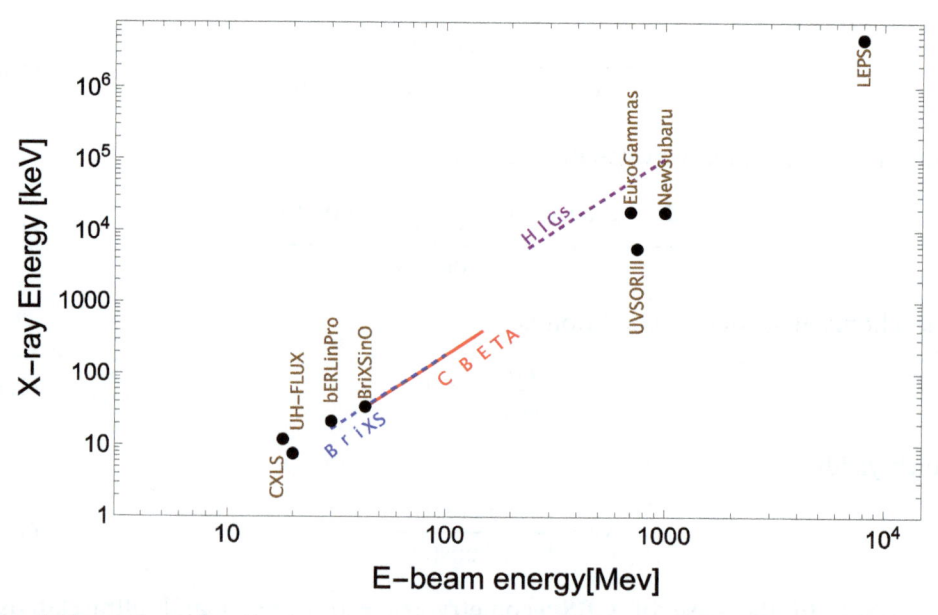

Figure 1.11. An illustration of output photon energy (keV) versus electron energy (MeV) for some of the existing/proposed CBS sources. (BriXS-Bright compact x-ray Source, BriXSino—Compat version of BriXS, bERLinPro—Berlin Energy Recovery Linac Prototype, CLS—Classical Light Scattering, CBETA—Cornell-BNL ERL Test Accelerator, UVSORIII—Ultra Violet Source III, EuroGammas—Gamma Rays based on Compton back-scattering, NewSubaru—New Short Undulator Beam Accelerated Radiation Unit, LEPS—Laser Electron Photon Experiment).

We let

$$w(x_0, \vec{x}) = w_0 Re\left[e^{ik_0 x_0 - \vec{k}\cdot\vec{x}}\right],$$
$$x_\mu \equiv (x_0, \vec{x}) = (ct, \vec{x}), \text{ and}$$
$$k_\mu \equiv \left(\frac{2\pi}{\lambda}, \vec{k}\right) \tag{1.92}$$

denote a wave propagating in an unprimed frame K. The term in the square bracket of the first equation is the complex wave amplitude, whose argument is provided by the four-vector product:

$$k_\mu x^\mu = k_0 x_0 - \vec{k}\cdot\vec{x} = \omega t - \vec{k}\cdot\vec{x}$$
$$\omega = \frac{2\pi c}{\lambda} \equiv \text{angular frequency of the wave}$$
$$\vec{k} \equiv \text{wave vector} \tag{1.93}$$
$$T = \frac{2\pi}{\omega} = \frac{1}{\nu} \equiv \text{wave period.}$$

Since we have used a four-vector formalism, we may also wonder how the wave should be described in another inertial frame. We seek a Lorentz transformation of k_μ with respect to a frame K' moving at velocity ν with respect to K. Accordingly, the Lorentz transformations yield:

$$k_0' = \gamma(k_0 - \beta k_\|)$$
$$k_\|' = \gamma(-\beta k_0 + k_\|) \qquad (1.94)$$
$$k_\perp' = k_\perp$$

where $k_\perp$ and $k_\|$ are the components of the wave vector orthogonal and parallel to the frame's direction of travel. Choosing (for simplicity) $\vec{k} = \vec{k}_\|$, the first of equations (1.94) is therefore written

$$k_0' = \gamma(k_0 - \beta k) \qquad (1.95)$$

and, on account of equations (1.93), we end up with:

$$ck_0' = \omega_0' = \gamma(1-\beta)\omega = \sqrt{\frac{1-\beta}{1+\beta}}\,\omega. \qquad (1.96)$$

As the velocity of light is the same in all inertial frames, namely $\lambda\nu = \lambda'\nu'$, we find, for the associated wavelengths:

$$\lambda' = \sqrt{\frac{1+\beta}{1-\beta}}\,\lambda. \qquad (1.97)$$

This means that, for a luminous source receding from the observer, the emitted radiation appears with a longer wavelength and a smaller frequency.

Let us now apply the previous result to ICS and note that the electron approaching the photon 'sees' the photon wavelength as being shortened according to:

$$\lambda' = \sqrt{\frac{1-\beta}{1+\beta}}\,\lambda_i. \qquad (1.98)$$

After scattering, the wave undergoes a further Doppler shift in the lab frame; therefore, in conclusion, we find:

$$\lambda' = \sqrt{\frac{1-\beta}{1+\beta}}\,\lambda_i' = \frac{1-\beta}{1+\beta}\,\lambda_i \qquad (1.99)$$

which, in the ultrarelativistic limit, reduces to equation (1.90).

This is the reason that CBS is also referred as a ***double Doppler shift*** process.

1.7 Maxwell's equations and special relativity

Maxwell's equations (MEs) were formulated before the introduction of special relativity. They are not invariant under the GTs, and were shown by J J Larmor to be such under an extension of GT, later recognized as the Lorentz transformations. Accordingly, MEs written in the Heaviside vector notation

$$\vec{\nabla} \cdot \vec{E} = \frac{\rho}{\varepsilon_0}, \qquad \vec{\nabla} \times \vec{B} - \frac{1}{c^2}\frac{\partial}{\partial t}\vec{E} = \mu_0 \vec{J}$$

$$\vec{\nabla} \cdot \vec{B} = 0, \qquad \vec{\nabla} \times \vec{E} + \frac{\partial}{\partial t}\vec{B} = 0 \qquad (1.100)$$

$$\vec{\nabla} \equiv \left(\frac{\partial}{\partial x}, \frac{\partial}{\partial y}, \frac{\partial}{\partial z}\right)$$

should implicitly contain the formalism of special relativity.

In equations (1.100) ρ and $\vec{J}$ are the charge and current densities, respectively, and $\vec{\nabla}$, $\vec{\nabla}\cdot$, and $\vec{\nabla}\times$ are the gradient, divergence, and curl operators, respectively, satisfying the algebraic rules:

$$\begin{aligned}
\vec{\nabla} \cdot \vec{\nabla} = \vec{\nabla}^2 &= \frac{\partial^2}{\partial x^2} + \frac{\partial^2}{\partial y^2} + \frac{\partial^2}{\partial z^2}, \\
\vec{\nabla} \times (\vec{\nabla} \times \vec{C}) &= -\vec{\nabla}^2 \vec{C} + \vec{\nabla}(\vec{\nabla} \cdot \vec{C}) \\
\vec{\nabla} \cdot \vec{\nabla} \times \vec{C} &= 0.
\end{aligned} \qquad (1.101)$$

If we apply the operator $\vec{\nabla}\cdot$ to the second equation of (1.100)

$$\vec{\nabla} \cdot \left(\vec{\nabla} \times \vec{B} - \frac{1}{c^2}\frac{\partial}{\partial t}\vec{E}\right) = \mu_0 \vec{\nabla} \cdot \vec{J} \qquad (1.102)$$

we find, after applying the rules in equation (1.101), that:

$$-\frac{1}{c^2}\frac{\partial}{\partial t}\vec{\nabla} \cdot \vec{E} = \mu_0 \vec{\nabla} \cdot \vec{J}. \qquad (1.103)$$

On account of the identity $c^2 = \frac{1}{\varepsilon_0 \mu_0}$ and the first equation of (1.100), we end up with the continuity equation:

$$\frac{\partial}{\partial t}\rho + \vec{\nabla} \cdot \vec{J} = 0. \qquad (1.104)$$

We now introduce the following four-vectors:

$$\begin{aligned}
\partial_\mu &\equiv \partial_0, \vec{\nabla}, \qquad \partial_0 = \frac{1}{c}\frac{\partial}{\partial t} \\
J_\mu &\equiv (c\rho, -\vec{J}).
\end{aligned} \qquad (1.105)$$

A straightforward observation is useful to clarify the physical meaning of the four-vector J_μ. As the charge is a relativistic invariant, the density of a system of charges at rest can be denoted by ρ_0, allowing us to conclude that the associated four-current is

$$J_\mu = c\rho_0 u_\mu \qquad (1.106)$$

which resembles the definition of the four-momentum vector given in equation (1.30). We note that the continuity equation (1.104) is formally equivalent to the four-vector product

$$\partial_\mu J^\mu = \frac{\partial \rho}{\partial t} + \vec{\nabla} \cdot \vec{J} = 0. \tag{1.107}$$

The 'norm' of the four-vector ∂_μ can be written as:

$$\partial_\mu \partial^\mu = \frac{1}{c^2} \frac{\partial^2}{\partial t^2} - \vec{\nabla}^2,$$
$$\vec{\nabla}^2 = \frac{\partial^2}{\partial x^2} + \frac{\partial^2}{\partial y^2} + \frac{\partial^2}{\partial z^2}, \tag{1.108}$$

and is recognized as the d'Alembert operator. Two auxiliary potentials (the scalar Φ and the vector $\vec{A}$) are exploited to define the electric and magnetic fields, namely:

$$\vec{E} = -\frac{\partial}{\partial t}\vec{A} - \vec{\nabla}\Phi$$
$$\vec{B} = \vec{\nabla} \times \vec{A}. \tag{1.109}$$

The equations specifying the auxiliary potentials are obtained by substituting the last two equations into the first two equations of (1.100):

$$-\vec{\nabla}^2 \Phi - \frac{\rho}{\varepsilon_0} = \frac{\partial}{\partial t}\vec{\nabla} \cdot \vec{A}$$
$$-\vec{\nabla}^2 \vec{A} + \frac{1}{c^2}\frac{\partial^2}{\partial t^2}\vec{A} - \mu_0 \vec{J} = -\vec{\nabla}\left[(\vec{\nabla} \cdot \vec{A}) + \frac{1}{c^2}\frac{\partial \Phi}{\partial t}\right]. \tag{1.110}$$

The equation for the vector potential reduces to a homogeneous d'Alembert equation if we set

$$\frac{1}{c^2}\frac{\partial \Phi}{\partial t} + (\vec{\nabla} \cdot \vec{A}) = 0. \tag{1.111}$$

Using this last condition in the equation defining the vector potential, we eventually find:

$$\frac{1}{c^2}\frac{\partial^2 \Phi}{\partial t^2} - \vec{\nabla}^2 \Phi = \frac{\rho}{\varepsilon_0}$$
$$\frac{1}{c^2}\frac{\partial^2}{\partial t^2}\vec{A} - \vec{\nabla}^2 \vec{A} = \mu_0 \vec{J}. \tag{1.112}$$

If we interpret Φ and $\vec{A}$ in terms of the four-vector formalism and set

$$A_\mu \equiv \left(\frac{\Phi}{c}, -\vec{A}\right) \tag{1.113}$$

we can cast equation (1.111) in the form:

$$\partial_\mu A^\mu = 0 \tag{1.114}$$

which is known as the Lorentz **gauge**, whose physical meaning will be discussed in the following. We note that equation (1.114) is invariant for the Lorentz transformations and therefore holds in any inertial frame (specifically, if ∂'_μ and A'_μ denote the transformed operator and potential, respectively, condition (1.114) can be written as $\partial'_\mu A'^\mu = 0$). The four-vector (1.113) undergoes the standard Lorentz transformation:

$$\begin{pmatrix} \frac{\Phi'}{c} \\ A'_x \end{pmatrix} = \begin{pmatrix} \gamma\left(\frac{\Phi}{c} - vA_x\right) \\ \gamma\left(-\frac{v}{c}\Phi + A_x\right) \end{pmatrix}. \tag{1.115}$$

And the Lorentz transformations for the electric field and the magnetic field can now be viewed as a consequence of equations (1.109) and (1.115). Before outlining the specific computational steps, we note that in nonrelativistic electromagnetism, the magnetic field is viewed by a moving charge as an electric field orthogonal to the direction of motion, namely:

$$\vec{E}' \propto \vec{v} \times \vec{B}. \tag{1.116}$$

The relativistic generalization can be guessed from the ordinary Lorentz transformation shown in figure 1.6 at the beginning of section 1.3; thus

$$\begin{aligned} \vec{E}'_\perp &= \gamma(\vec{E}_\perp + \vec{v} \times \vec{B}) \\ \vec{B}'_\perp &= \gamma\left(\vec{B}_\perp - \frac{\vec{v} \times \vec{E}}{c^2}\right), \\ \vec{E}'_\parallel &= \vec{E}_\parallel, \qquad \vec{B}'_\parallel = \vec{B}_\parallel \end{aligned} \tag{1.117}$$

To be less vague and avoid any misunderstanding, we are obliged to appear pedantic. In equation (1.117), the orthogonal and parallel components are defined with respect to the direction of motion. If we assume that the motion takes place in the x-direction, we find e.g.

$$\vec{E}'_y = \gamma(\vec{E}_y - v B_z). \tag{1.118}$$

It is evident that the definitions (1.108) of the electric and magnetic fields, in terms of the vector and scalar potentials, are invariant under Lorentz transformations, therefore:

$$\begin{aligned} \vec{E}' &= -\frac{\partial}{\partial t'}\vec{A}' - \vec{\nabla}'\Phi' \\ \vec{B}' &= \vec{\nabla}' \times \vec{A}' \end{aligned} \tag{1.119}$$

where (see section 1.8)

$$\begin{aligned}\frac{\partial}{\partial t'} &= \frac{\partial}{\partial t} + v\frac{\partial}{\partial x}, \\ \frac{\partial}{\partial x'} &= \frac{\partial}{\partial x} - \frac{v}{c^2}\frac{\partial}{\partial t}.\end{aligned} \qquad (1.120)$$

The correctness of transformations (1.117) can now easily be checked by plugging transformations (1.115) into (1.118), which gives

$$\vec{E}'_y = \gamma\left[-\frac{\partial}{\partial t}A_y - \frac{\partial}{\partial y}\Phi - v\left(\frac{\partial}{\partial x}A_y - \frac{\partial}{\partial y}A_x\right)\right] = \\ = \gamma\left[-\left(\frac{\partial}{\partial t} + v\frac{\partial}{\partial x}\right)A_y - \frac{\partial}{\partial y}\Phi + vA_x\right] = \vec{E}'_y. \qquad (1.121)$$

The last identity is obtained as a consequence of the first equation of (1.119).

The check for the other components is done by following the same procedure.

A less intuitive procedure is based on the introduction of the electromagnetic tensor, defined as:

$$F^{\mu\nu} = \frac{\partial A^\nu}{\partial x_\mu} - \frac{\partial A^\mu}{\partial x_\nu} \qquad (1.122)$$

$$\mu, \nu = 0, 1, 2, 3$$

which can be expressed as a 4 × 4 matrix whose entries can be constructed by noting, for example, that:

$$\begin{aligned}F^{\mu\mu} &= \frac{\partial A^\mu}{\partial x_\mu} - \frac{\partial A^\mu}{\partial x_\mu} = 0 \\ F^{01} &= \frac{\partial A^0}{\partial x_1} - \frac{\partial A^1}{\partial x_0} = \frac{E_x}{c} \\ F^{10} &= -\frac{E_x}{c} \\ F^{12} &= \frac{\partial A^1}{\partial x_2} - \frac{\partial A^2}{\partial x_1} = -[\vec{\nabla} \times \vec{A}]_z = -B_z.\end{aligned} \qquad (1.123)$$

Accordingly, we obtain (see section 1.8)

$$F^{\mu\nu} = \begin{pmatrix} 0 & -\frac{E_x}{c} & -\frac{E_y}{c} & -\frac{E_z}{c} \\ \frac{E_x}{c} & 0 & -B_z & B_y \\ \frac{E_y}{c} & B_z & 0 & -B_x \\ \frac{E_z}{c} & -B_y & B_x & 0 \end{pmatrix}. \qquad (1.124)$$

The Lorentz transformations of the tensor $F^{\mu\nu}$ are governed by:

$$F'^{\mu\nu} = \hat{L}(v) F^{\mu\nu} \hat{L}(v)^T. \tag{1.125}$$

Further details of the topic described in this section can be found in section 1.8.

The material discussed so far constitutes the **_theoretical minimum_** required to understand the remaining parts of this book.

1.8 Comments and exercises

In the concluding part of this chapter, we add comments and exercises regarding the concepts developed in the previous sections. Most of our derivations are based on simplified physical and mathematical arguments. The following parts are intended to partially remedy this deficiency and to propose further elements of analysis in physical and mathematical terms.

1.8.1 Electron–photon interactions: a qualitative phenomenology

We have specified the physical constants that characterize Thomson scattering. Within this context, the electron's classical radius plays a pivotal role. It is therefore important to clarify its physical meaning, specified by the identity given below:

$$\frac{e^2}{4\pi\varepsilon_0} \frac{1}{r_0} = m_o c^2. \tag{1.126}$$

This states that r_0 represents the radial distance from an electron, where the electrostatic energy acquired by a test charge with magnitude $|e|$ equals the electron's rest mass energy. We can improve our understanding of condition (1.126) by setting

$$\int_{r_0}^{\infty} |e| E(r) \, dr = m_o c^2$$

$$E(r) = \frac{e}{4\pi\varepsilon_0} \frac{1}{r^2}. \tag{1.127}$$

The relevant physical meaning is transparent: it confirms that the energy required to bring a charge $|e|$ to infinity when subjected to an electric field $E(r)$ is equivalent to the electron's rest mass energy. This last observation is suggestive of further speculations, which will be more carefully treated in the second part of this book. In the early 1930s, Sauter suggested the possibility of observing the creation of electron–positron pairs from the vacuum by applying a strong electric field to a given region of space. This idea was pursued by other authors and, about two decades later, the problem was rigorously formulated by Schwinger within the context of quantum electrodynamics. A simple argument that allows us to quantify the Sauter–Schwinger electric field (E_{ss}) that allows this process is given below. The physical mechanism that allows the production of an electron–positron pair is due to the work done by E_{ss} on the charges, which must be equivalent to the mass of the produced particle pair. The distance over which the work is done is estimated using the Heisenberg principle, yielding:

$$\Delta x \simeq \frac{\hbar}{\Delta p} = \frac{\hbar}{m_0 c} = \lambdabar_c. \qquad (1.128)$$

We thus obtain:

$$eE_{ss}\lambdabar_e = m_0 c^2$$
$$E_{ss} = \frac{m_0 c^2}{e\,\lambdabar_e} \simeq 1.323 \times 10^{18}\ \text{V m}^{-1}. \qquad (1.129)$$

In figure 1.12 we show a pictorial view of this process. It can be interpreted as tunneling through the energy gap required for particle creation; it takes place when the field transfers an energy to the pairs that is of the order of their rest masses. Even though the second part of this book discusses (with more suitable tools) the process of pair creation, we invite the reader to find a reasonable argument to explain why the probability of its occurrence can be expressed as:

$$P_{(e^+ e^-)} \propto \exp\left(-\pi \frac{E_{ss}}{E}\right). \qquad (1.130)$$

Furthermore, it is evident from 1.130 that for intensities exceeding E_{ss}, pair production becomes more significant, and the presence of these pairs may influence the field at a macroscopic scale. For this reason, nonlinear effects break the linearity of MEs. It is instructive to understand how the fields $E(r_0)$ and E_{ss} compare. We invite the reader to prove that:

$$\frac{E(r_0)}{E_{ss}} = \frac{\lambdabar_e}{r_0} = \alpha^{-1}$$
$$\alpha = \frac{e^2}{4\pi\varepsilon_0 \hbar c} \simeq \frac{1}{137}. \qquad (1.131)$$

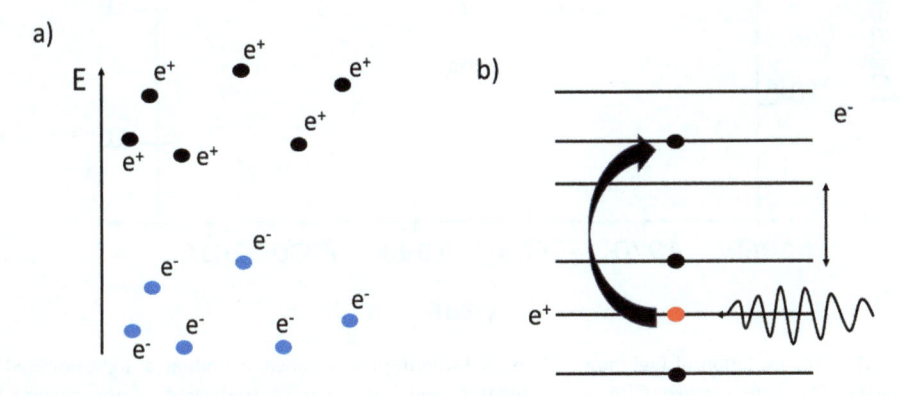

Figure 1.12. (a) The action of an intense electric field on vacuum and the creation of pairs. (b) The transition from negative to positive electron states.

Show also that, corresponding to E_{ss}, the following magnetic field can be defined:

$$B_{ss} = \frac{m_0 c}{\lambda_e} \simeq 4.41 \times 10^9 \text{ T}. \qquad (1.132)$$

It is interesting to speculate about the possibility of reaching the Sauter–Schwinger limit using an intense laser field. To this aim, we note that the intensity of a a hypothetical electromagnetic wave bearing an electric field equivalent to E_{ss} is given by:

$$\varepsilon_0 c E_{ss}^2 = \frac{1}{4\pi}\left(\frac{c}{\alpha \lambda_e^3}\right) m_0 c^2 \simeq 4.463 \times 10^{29} \text{ W cm}^{-2}. \qquad (1.133)$$

In figure 1.13 we have reported the evolution (over time) of the available laser source intensities. To date, the largest intensity $\simeq 10^{23}$ W cm^{-2} has been achieved at the Institute for Basic Science in South Korea, and we are still far (six orders of

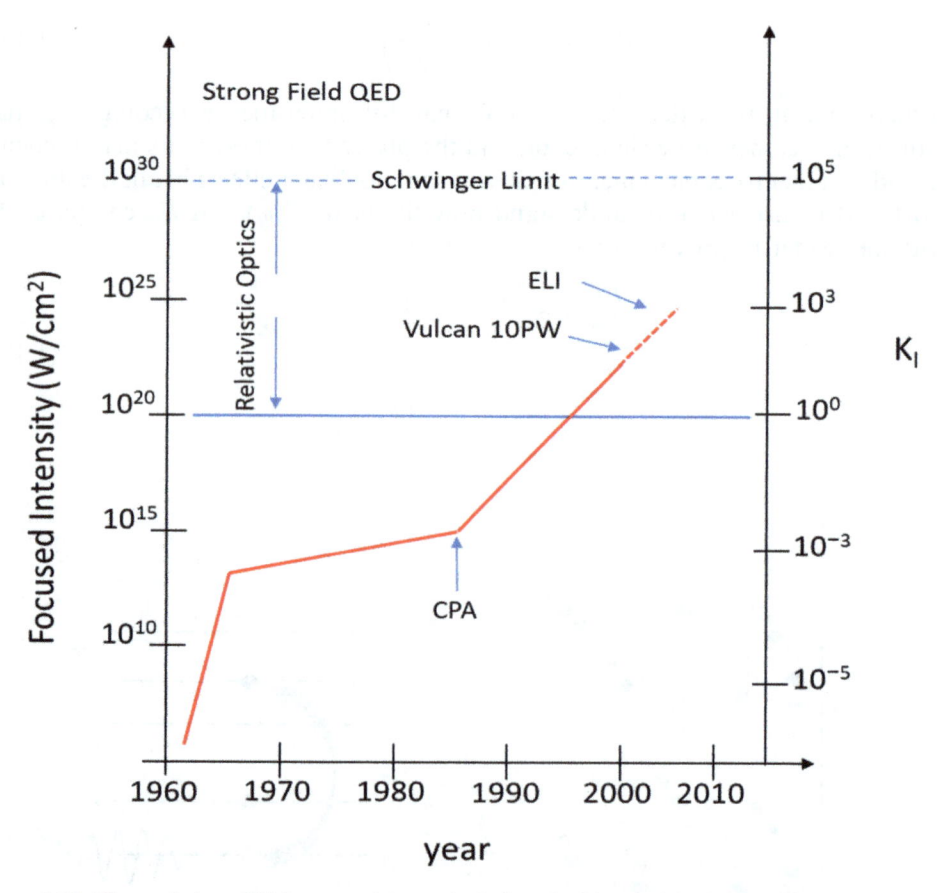

Figure 1.13. The evolution of high-powered lasers, including the historical evolution of light-powered laser technology. The knee marked CPA is the breakthrough due to the introduction of the chirped pulse amplification technique. Reprinted figure with permission from Mourou G 2019 Nobel Lecture: Extreme light physics and application *Rev. Mod. Phys.* **91** 030501, copyright (2019) by the American Physical Society.

magnitude) below the Sauter–Schwinger threshold. However, the existing powerful lasers in the region above 10^{20} W cm^{-2} offer tools of great interest that can be used to explore new physics in electron Compton scattering. We have anticipated aspects of our later discussions to see some implications that can be reached through the use of the elementary discussions developed in this introductory chapter.

1.8.2 Generalities related to electromagnetic fields

In the derivation of the Thomson cross section, we mentioned the Poynting vector, which is the physical quantity that accounts for the transport of the radiation flux.

This statement should be specified by exploiting the proper mathematical tools.

The first point to be underscored is that MEs implicitly contain electromagnetic wave propagation. In order to substantiate this last point, we emphasize once more that MEs in vacuum and in the absence of charges and currents read:

$$\vec{\nabla} \cdot \vec{E} = 0, \qquad \vec{\nabla} \cdot \vec{B} = 0$$
$$\vec{\nabla} \times \vec{E} = -\frac{\partial}{\partial t}\vec{B}, \qquad \vec{\nabla} \times \vec{B} = \varepsilon_0 \mu_0 \frac{\partial}{\partial t}\vec{E} \qquad (1.134)$$
$$\varepsilon_0 \mu_0 = \frac{1}{c^2}, \qquad \vec{E} \equiv \vec{E}(\vec{r}, t), \vec{B} \equiv \vec{B}(\vec{r}, t).$$

Repeating, for convenience's sake, the vector calculus exercise outlined in section 1.7, we find the following wave equation for the electric field:

$$\vec{\nabla} \times \vec{\nabla} \times \vec{E} = -\frac{\partial}{\partial t}\vec{\nabla} \times \vec{B} \to \vec{\nabla} \times \vec{\nabla} \times \vec{E} = -\varepsilon_0 \mu_0 \frac{\partial^2}{\partial t^2}\vec{E}$$
$$\vec{\nabla} \times \vec{\nabla} \times \vec{E} = -\vec{\nabla}^2 \vec{E} + \vec{\nabla} \cdot \vec{E} = -\vec{\nabla}^2 \vec{E} \to \qquad (1.135)$$
$$\to \vec{\nabla}^2 \vec{E} - \frac{1}{c^2}\frac{\partial^2}{\partial t^2}\vec{E} = 0$$

and similarly, for the vector $\vec{B}$,

$$\vec{\nabla}^2 \vec{B} - \frac{1}{c^2}\frac{\partial^2}{\partial t^2}\vec{B} = 0. \qquad (1.136)$$

We make the assumption that the waves propagate in the z-direction (see figure 1.14) and write the solutions (see below for further comments) of the previous wave equations as:

$$\vec{E}(\vec{r}, t) = \vec{E}_0(x, y, z)e^{i(kz-\omega t+\delta)}$$
$$\vec{B}(\vec{r}, t) = \vec{B}_0(x, y, z)e^{i(kz-\omega t+\delta)}$$
$$\vec{E}_0(x, y, z) = \vec{E}_0(x, y, z)e^{i\delta}, \vec{B}_0(x, y, z) = \vec{B}_0(x, y, z)e^{i\delta} \qquad (1.137)$$
$$k = \frac{2\pi}{\lambda}, \quad \omega = \frac{2\pi c}{\lambda}.$$

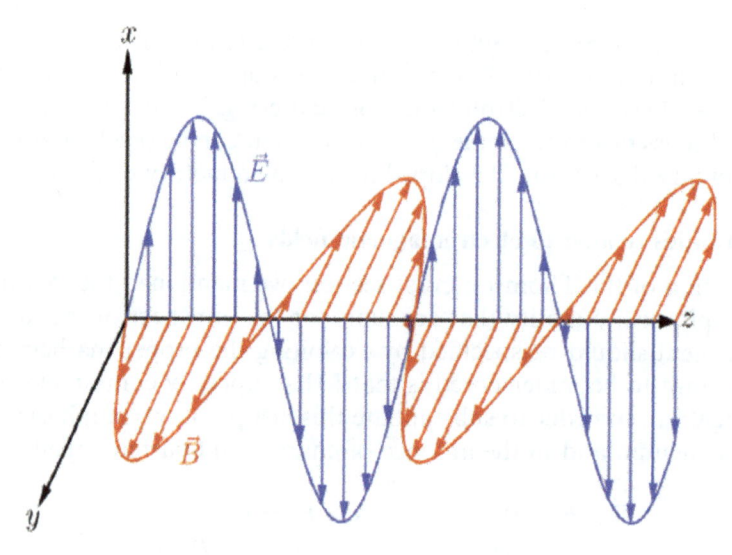

Figure 1.14. Plane wave propagation.

The mnemonic for the various declinations of k and ω in generic wave propagation is reported in figure 1.15. We have introduced a complex vector whose real parts yield the corresponding 'physical' quantities, namely:

$$\vec{E}(\vec{r}, t) = Re[\vec{\tilde{E}}(\vec{r}, t)]$$
$$\vec{B}(\vec{r}, t) = Re[\vec{\tilde{B}}(\vec{r}, t)]. \quad (1.138)$$

It is worth highlighting that equations (1.137) assume that both electric and magnetic fields oscillate with the same phase. (Is it physically possible to have different phases?)

MEs impose restrictions on the field distributions (specifically on their dependence on space coordinates), which depend on the conditions of the specific problem under study. To increase our confidence in the previous statements, the following exercises are suggested.

Exercise 1.1. Use the first line in equation (1.134) to prove that:
1. The electric and magnetic fields have no dependence on the transverse coordinates (x, y).
2. No field component (electric/magnetic) is allowed in the direction of propagation.

Solution:
Use the divergence operator in Cartesian coordinates

$$\vec{\nabla} = \hat{x}\frac{\partial}{\partial x} + \hat{y}\frac{\partial}{\partial y} + \hat{z}\frac{\partial}{\partial z}$$

$$v_\varphi = \frac{\omega}{k} \quad v_\varphi \quad v_\varphi = \frac{\omega}{k}$$

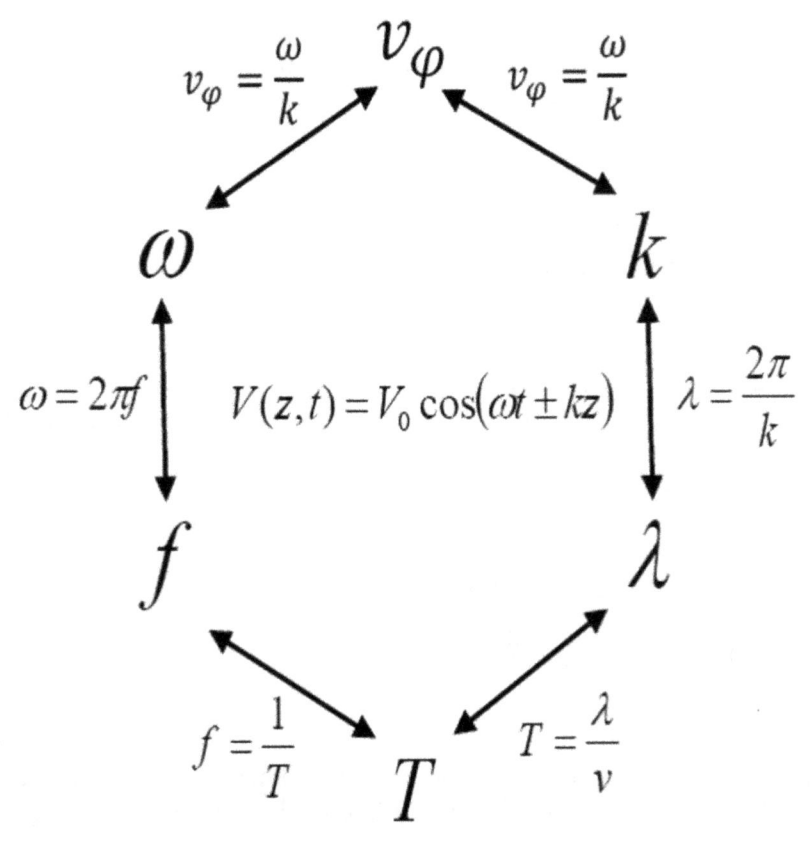

$$\omega \qquad k$$

$$\omega = 2\pi f \quad V(z,t) = V_0 \cos(\omega t \pm kz) \quad \lambda = \frac{2\pi}{k}$$

$$f \qquad \lambda$$

$$f = \frac{1}{T} \qquad T \qquad T = \frac{\lambda}{v}$$

Figure 1.15. A mnemonic ring for the quantities specifying wave propagation. Here, $v_\varphi = \frac{\omega}{k}$ denotes the phase velocity, which for propagation in vacuum is equivalent to the group velocity $v_g = \frac{\partial \omega}{\partial k}$.

and write (consider for brevity the electric field only)

$$\vec{\tilde{E}}_0(x,y,z) = [\hat{x}E_{0,x} + \hat{y}E_{0,y} + \hat{z}E_{0,z}]e^{i\delta}.$$

Using the first of MEs for the electric field in the absence of charge, we obtain

$$Re[\vec{\nabla}\cdot\vec{E}(\vec{r},t)] = 0 \rightarrow \left(\hat{x}\frac{\partial}{\partial x} + \hat{y}\frac{\partial}{\partial y} + \hat{z}\frac{\partial}{\partial z}\right) \cdot \left\{[\hat{x}E_{0,x} + \hat{y}E_{0,y} + \hat{z}E_{0,z}]e^{i\,kz-\omega t+\delta}\right\} = 0.$$

Furthermore, from the orthogonality of the unit vectors $\hat{x}$, $\hat{y}$, and $\hat{z}$, we find the conditions

$$\frac{\partial}{\partial x}(E_{0,x}e^{i\,(kz-\omega t+\delta)}) = 0$$

$$\frac{\partial}{\partial y}(E_{0,y}e^{i\,(kz-\omega t+\delta)}) = 0$$

$$\frac{\partial}{\partial z}(E_{0,z}e^{i\,(kz-\omega t+\delta)}) = 0.$$

The phase δ does not depend on space; therefore, the first two identities in the previous equations ensure that the electric field component $E_{0,x}$ does not explicitly depend on x and that $E_{0,y}$ is independent of y. The third condition, written in the form

$$ikE_{0,z}e^{i(kz-\omega t+\delta)} = 0$$

is fulfilled if $E_{0,z} = 0$. The same conclusions (as conveniently checked by the reader) can be obtained for the magnetic field components.

Exercise 1.2. Use the second line in equation (1.134) to prove that for a monochromatic plane wave in free space, the following conditions are true for the electric and magnetic fields:
  1. They oscillate in phase.
  2. They are mutually orthogonal and perpendicular to the propagation direction.
  3. They are linked by the identity

$$\vec{B} = \frac{1}{c}\hat{z} \times \vec{E}. \tag{1.139}$$

Solution:
We obtain the solution using the procedure outlined in the previous comment. Regarding the left-hand side of the second line in (1.134), we explicitly get

$$\vec{\nabla} \times \vec{E} = -\frac{\partial}{\partial t}\vec{B} \rightarrow$$

$$\rightarrow \begin{vmatrix} \hat{x} & \hat{y} & \hat{z} \\ \frac{\partial}{\partial x} & \frac{\partial}{\partial y} & \frac{\partial}{\partial z} \\ \tilde{E}_x & \tilde{E}_y & 0 \end{vmatrix} = -\frac{\partial}{\partial t}(\tilde{B}_x\hat{x} + \tilde{B}_y\hat{y}) \rightarrow \begin{array}{l} \frac{\partial}{\partial z}\tilde{E}_y = \frac{\partial}{\partial t}\tilde{B}_x \\ \frac{\partial}{\partial z}\tilde{E}_x = -\frac{\partial}{\partial t}\tilde{B}_y. \end{array}$$

It is evident that the last two lines of the previous identities can be explicitly written as:

$$E_{0,y}\frac{\partial}{\partial z}e^{i(kz-\omega t+\delta)} = B_{0,x}\frac{\partial}{\partial t}e^{i(kz-\omega t+\delta)}$$

$$E_{0,x}\frac{\partial}{\partial z}e^{i(kz-\omega t+\delta)} = -B_{0,y}\frac{\partial}{\partial t}e^{i(kz-\omega t+\delta)}$$

which contain all the answers to the questions in exercise 1.8.2. We indeed find that in order to be consistent, the phase δ appearing in the arguments of the electric and magnetic oscillating terms 'must' be the same. Accordingly, electric and magnetic fields oscillate in phase.

Furthermore, by keeping the derivatives of both sides, we obtain:

$$kE_{0,y} = -\omega B_{0,x} \rightarrow B_{0,x} = -\frac{k}{\omega}E_{0,y},$$

$$kE_{0,x} = \omega B_{0,y} \rightarrow B_{0,y} = \frac{k}{\omega}E_{0,x}$$

which are used to state that the fields are mutually orthogonal and that since

$$\frac{k}{\omega} = \frac{1}{c}|\hat{k}|, \; \hat{k} = \frac{\vec{k}}{|\vec{k}|}$$

we can eventually write

$$\vec{B} = \frac{1}{c}\hat{k} \times \vec{E}$$

which reduces to equation (1.139) if, as in the present case, $\hat{k} \| \hat{z}$.

Exercise 1.3. Specialize the previous results to the case of a plane wave polarized in the $\hat{x}$ direction.
Hint:
write $\vec{E} = E\,\vec{x}$ and prove that $\vec{B} = \frac{1}{c}E\hat{y}$.

Exercise 1.4. Derive the electric and magnetic fields associated with a monochromatic wave propagating in a generic direction $\hat{k}$.
Solution:
Introduce the vector triple specified by (see figure 1.16)

$$\hat{t} \equiv (\hat{n}, \hat{n} \times \hat{k}, \hat{k})$$
$$\hat{n} \cdot \hat{k} = 0.$$

The electric field is supposed to be polarized along $\hat{n}$. Therefore, since the propagation is characterized by the same properties, regardless of the propagation direction, the following substitutions are sufficient:

$$\hat{x} \rightarrow \hat{n}, \; \hat{y} \rightarrow \hat{n} \times \hat{k}, \; \hat{z} \rightarrow \hat{k}$$
$$\vec{k} \equiv k_x\hat{x} + k_y\hat{y} + k_z\hat{z} \equiv \text{wave} - \text{vector}$$
$$kz \rightarrow \vec{k} \cdot \vec{r} = k_x x + k_y y + k_z z, \; \omega \equiv |\vec{k}|c.$$

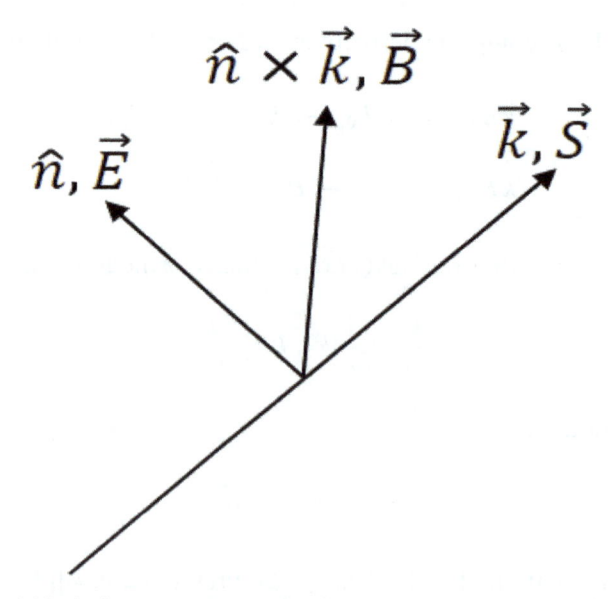

Figure 1.16. A plane wave propagation field and its Poynting vector direction.

This gives:

$$\vec{E} = E\,\vec{n},\ \vec{B} = \frac{1}{c}\hat{k} \times \vec{E} = \frac{1}{c}E\,(\hat{k} \times \hat{n})$$

$$\hat{n} \cdot \hat{k} = 0 \rightarrow \vec{B} \cdot \vec{E} = 0$$

$$\vec{\Lambda}(\vec{r}, t) = Re\left[\begin{pmatrix}\vec{E_0} \\ \vec{B_0}\end{pmatrix} e^{i\vec{k}\cdot\vec{r} - \omega t + \delta}\right] =$$

$$= \begin{pmatrix}\vec{E_0} \\ \vec{B_0}\end{pmatrix} \cos(\vec{k}\cdot\vec{r} - \omega t + \delta) \equiv \text{instantaneous electric and magnetic fields.}$$

The relationship $\hat{n} \cdot \hat{k} = 0$ is usually referred to as the transversality condition. Figure 1.16 shows the geometric framework of electromagnetic wave propagation in vacuum. In the following, we point out that such a simple view no longer holds under different conditions (propagation in plasmas or intense waves determining vacuum nonlinearities).

Exercise 1.5. The instantaneous energy density transported by a monochromatic plane wave is

$$u_{EM}(r, t) = u_E(\vec{r}, t) + u_M(\vec{r}, t)$$

$$u_E(\vec{r}, t) = \frac{1}{2}\varepsilon_0 E(\vec{r}, t)^2,\ u_M(\vec{r}, t) = \frac{1}{2\mu_0}E^2(\vec{r}, t)^2 \qquad (1.140)$$

$$E(\vec{r}, t)^2 = \vec{E}(\vec{r}, t) \cdot \vec{E}(\vec{r}, t).$$

Show that the relevant units are [J m^{-3}] and that

$$u_E(\vec{r}, t) = u_M(\vec{r}, t) = \varepsilon_0 E(\vec{r}, t)^2 = \varepsilon_0 E_0^2 \cos(\vec{k} \cdot \vec{r} - \omega t + \delta)^2. \quad (1.141)$$

Exercise 1.6. The vacuum field impedance associated with an electromagnetic plane wave is defined as

$$\vec{Z}(\vec{r}, t) = \vec{E}(\vec{r}, t) \times \frac{1}{\vec{H}(\vec{r}, t)},$$

$$\vec{H}(\vec{r}, t) = \frac{1}{\mu_0} \vec{B}(\vec{r}, t).$$

Show that

$$\vec{Z}(\vec{r}, t) = \sqrt{\frac{\mu_0}{\varepsilon_0}} \hat{k},$$

$$\sqrt{\frac{\mu_0}{\varepsilon_0}} \simeq 377 \Omega. \quad (1.142)$$

Solution:
Even though the inverse of a vector may appear an ill-defined concept (indeed, it is), we get a quick answer by noticing that, given a generic vector $\vec{v}$, we have:[3]

$$\frac{1}{\vec{v}} = \frac{\vec{v}}{\vec{v} \cdot \vec{v}} = \frac{\vec{v}}{|\vec{v}|^2}$$

which helps us to get

$$\vec{Z}(\vec{r}, t) = \frac{\mu_0}{c} \vec{E}(\vec{r}, t) \times (E(\vec{r}, t) \times (\hat{n} \times \hat{k})) \frac{1}{E(\vec{r}, t)^2} =$$
$$= \frac{\mu_0}{c} \hat{k} = \sqrt{\frac{\mu_0}{\varepsilon_0}} \hat{k}. \quad (1.143)$$

The previous result is extremely meaningful from a physical point of view and goes beyond the specific aspects of the problems we are studying. Equation (1.143) states that the vacuum impedance is a vector directed in the propagation direction and has a constant amplitude. The last property is justified by the fact that the vacuum 'must' be invariant under any transformation (translations, rotations, Lorentz...). The vector nature is slightly more complicated and should be interpreted within the context of electromagnetic field quantization (see below and the second part of this book).

---

[3] The identity is not correctly defined. It indeed assumes that when we 'multiply' the numerator and denominator by the same quantity, the original fraction is left unchanged. The statement is true for scalars but doubtful for vectors. In this case we have either internal (scalar) or external (vector) products, therefore we make the assumption that $\frac{1}{\vec{v}} = \frac{1 \cdot \vec{v}}{\vec{v} \cdot \vec{v}}$ and $1 \cdot \vec{v} = \vec{v}$.

Regarding the numerical value reported in the second equation of (1.142), we note that:

$$\varepsilon_0 \simeq 8.85 \times 10^{-12} \text{ Farad (F) m}^{-1}$$
$$\mu_0 \simeq 4\pi\, 10^{-7} \text{ H m}^{-1}$$
$$Z_0 = \sqrt{\frac{\mu_0}{\varepsilon_0}} \simeq 10^2 \sqrt{\frac{40\pi}{8.85}} \simeq 376.7\, \Omega.$$

Prove also that $\left[\frac{H}{\text{Farad (F)}}\right] = [\Omega^2]$ using the identities $U_L = LI^2$, $U_C = \frac{Q^2}{C}$, $[RC] = [T]$, where L, I are inductance and current, respectively, Q, C are charge and capacitance, respectively, and R is resistance.

Exercise 1.7. Derive the instantaneous Poynting vector for a plane wave moving in an arbitrary direction $\vec{k}$.
Solution:
The definition of the Poynting vector is

$$\vec{S}(\vec{r}, t) = \frac{1}{\mu_0}\vec{E}(\vec{r}, t) \times \vec{B}(\vec{r}, t) \tag{1.144}$$

which for a plane wave yields

$$\vec{S}(\vec{r}, t) = \frac{1}{\mu_0 c} E_0^2 \cos^2(\vec{k}\cdot\vec{r} - \omega t)\hat{k}. \tag{1.145}$$

It is therefore easily inferred that:

$$\vec{S}(\vec{r}, t) = \frac{1}{\mu_0 c} E_0^2 \cos^2 - (\vec{k}\cdot\vec{r} - \omega t)\hat{k} = \varepsilon_0 c E_0^2 \cos^2(\vec{k}\cdot\vec{r} - \omega t)\hat{k}. \tag{1.146}$$

It is accordingly important to underline that the physical units of the Poynting vector are $[\vec{S}] = [\text{W m}^{-2}]$, namely watts per square meter, or those of an *intensity* or *flux density*.

The definition in equation (1.144) is just a mathematical statement. The role of $\vec{S}$ in electromagnetism is of the utmost importance; it is therefore necessary to add further comments on the relevant meaning.

In figure 1.17, we have indicated a region of space delimited by a certain spherical volume V, containing a free charge and an electromagnetic field whose instantaneous internal stored energy is (for brevity, we have neglected the arguments $(\vec{r}, t)$ of the fields):

$$U_{em}(t) = \frac{1}{2}\int_v \left(\varepsilon_0 E^2 + \frac{1}{\mu_0}B^2\right)dv \rightarrow [\text{J}].$$

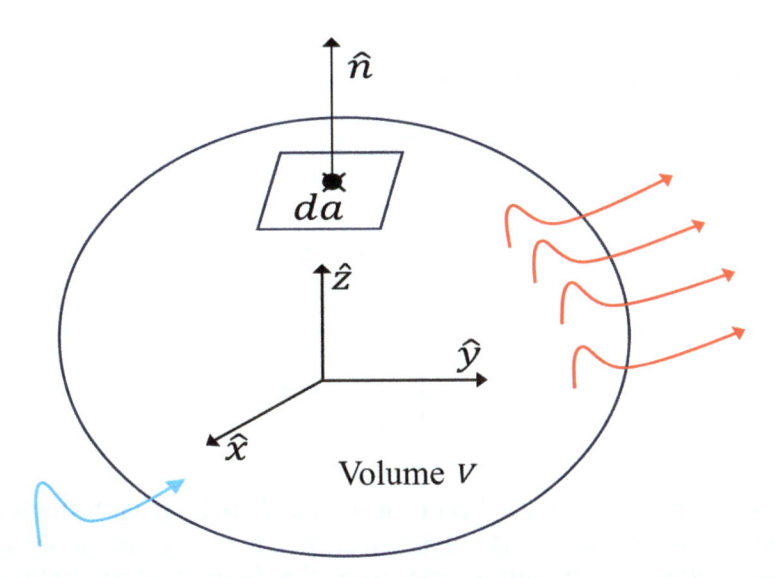

Figure 1.17. The geometry of the Poynting theorem. Here, $d\vec{a}$ represents the oriented surface element.

The **Poynting theorem** states that: *the instantaneous work $W(t)$ done by the fields on the charge corresponds to the decrease of the instantaneous energy minus the instantaneous energy flowing outside the boundary surface.* We have, accordingly:

$$\frac{d}{dt}W(t) = -\frac{d}{dt}U_{em}(t) - \Phi_{out}(t)$$

$$\Phi_{out}(t) = \oint \vec{S}(\vec{r}, t) \cdot d\vec{a}.$$

The Poynting theorem contains different subtleties, which we briefly mention here:
1. The work done on the charge increases the charge energy (kinetic/potential).
2. The power $\frac{d}{dt}W(t)$ consists of two contributions, namely the mechanical and electromagnetic parts, so that the theorem can be reformulated in the following differential form:

$$\frac{\partial}{\partial t}[u_{mech}(\vec{r}, t) + u_{em}(\vec{r}, t)] = -\vec{\nabla} \cdot \vec{S}(\vec{r}, t).$$

The version shown above resembles the mass continuity equation.

We invite the reader to reconcile the Poynting theorem with the phenomenology of the heat dissipation of a current inside a wire.

Exercise 1.8. Derive the Poynting theorem in the absence of charges inside the enclosed volume region.
Solution:

$$-\frac{\partial}{\partial t}u_{EM}(r, t) = -\varepsilon_0 \vec{E} \cdot \frac{\partial \vec{E}}{\partial t} - \frac{1}{\mu_0}\vec{B} \cdot \frac{\partial \vec{B}}{\partial t} = \frac{1}{\mu_0}\left(-\vec{E} \cdot (\vec{\nabla} \times \vec{B}) + \vec{B} \cdot (\vec{\nabla} \times \vec{E})\right). \quad (1.147)$$

Then use the vector calculus theorem

$$\vec{\nabla}\cdot(\vec{A}\times\vec{C}) = \vec{C}\cdot(\vec{\nabla}\times\vec{A}) - \vec{A}\cdot(\vec{\nabla}\times\vec{A})$$

set

$$-\frac{\partial}{\partial t}u_{EM}(r,t) = \vec{\nabla}\cdot\vec{S}$$

and conclude that

$$\vec{\nabla}\cdot\vec{S} = \vec{\nabla}\cdot\left(\frac{1}{\mu_0}\vec{E}\times\vec{B}\right) \tag{1.148}$$

which proves the theorem.

The proof we have just outlined is due to Heaviside (who published this clean and transparent derivation after the Poynting paper, which was based on a much heavier mathematical formalism). It must be emphasized that from equation (1.148), we derive

$$\vec{S} = \frac{1}{\mu_0}\vec{E}\times\vec{B} + \vec{G}$$

where $\vec{G}$ is a divergence-free vector ($\vec{\nabla}\cdot\vec{G} = 0$). For the relevant physical meaning, the reader is referred to the literature listed at the end of the chapter.

Exercise 1.9. Use the notions we have just outlined to infer the existence of a 'linear momentum density' associated with the Poynting vector.
Solution:
Use a purely dimensional argument and note that

$$\left[\frac{\vec{S}}{c^2}\right] = \left[\frac{E}{TL^2L^2}\right] = \left[\frac{\frac{ML^2T^2}{T^3}}{L^2L^2}\right] = \left[\frac{MLT^{-1}}{L^3}\right].$$

We can therefore 'guess' that the linear momentum density for a plane wave is

$$\vec{P}_{em}(\vec{r},t) = \frac{\varepsilon_0}{c}E_0^2\cos^2(\vec{k}\cdot\vec{r} - \omega t)\hat{k} = \frac{1}{c}u_{em}(r,t)\hat{k}$$

and the relevant units are kg m^{-2} s^{-1}.

Exercise 1.10. Use the result of the previous exercise to derive the 'angular momentum density' and infer the existence of a 'linear momentum density' associated with a plane wave.

Hint:
Using the definition of angular momentum, we get

$$\vec{L}_{em}(\vec{r}, t) = \vec{r} \times \vec{P}_{em}(\vec{r}, t) = \frac{1}{c} u_{em}(r, t) \vec{r} \times \hat{k}.$$

It is important to underline that the wave angular momentum depends on the choice of the origin defining the vector $\vec{r}$.

Exercise 1.11. We have stressed that Thomson scattering consists of two processes of absorption and emission with an instantaneous delay. Is there any argument capable of quantifying the elapsed time?
Solution:
In our 'classical' description of Thomson scattering, we treated the problem using a macroscopic description, which accounted for the scattering process by exploiting an electromagnetic plane wave, the associated fields, and the Poynting vector.

At a more elementary scale the wave is an ensemble of photons, which are first absorbed and then emitted by the scattering center.

In figure 1.18 we have used a Feynman diagram to describe the process (see the second part of the book for an adequate explanation). The transition between the absorption and emission processes is mediated by the virtual electron state.

The most natural tool with which to derive the order of magnitude of the elapsed time between the two process is the indetermination principle, namely:

$$\Delta E \, \Delta t \simeq \frac{\hbar}{2}.$$

The energy uncertainty is associated with the photon energy, thus we obtain

$$\Delta t \simeq \frac{\hbar}{2\hbar\omega} = \frac{T}{4\pi}$$

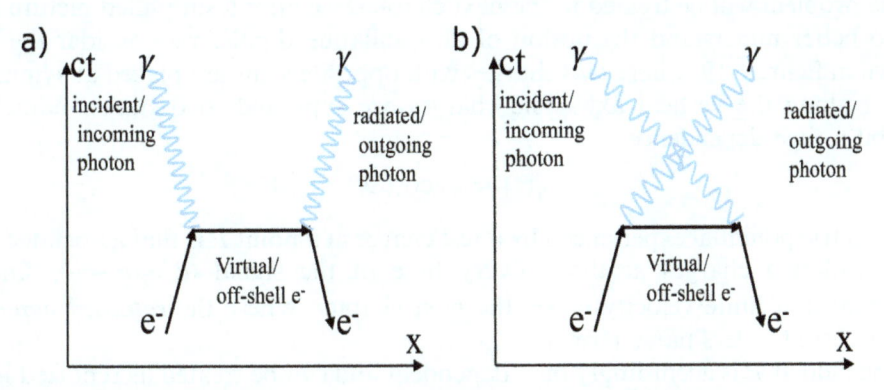

Figure 1.18. A Feynman space-time diagram of Compton scattering, which consists of two elementary processes of absorption (of an incoming photon) and emission (of an outgoing photon) at different space-time positions. The relevant physical and mathematical details will be discussed in the second volume.

for the associated time uncertainty, which is significantly smaller than the wave (photon) oscillation period.

Exercise 1.12. In the derivation of the Thomson cross section, we have mentioned only the electric field, and we have underlined that the diffusion can be ascribed to an electric dipole transition. So, what is the role of the magnetic field (carried by the photon) in terms of its effect on the charged particle?
Solution:
The force acting on a charged particle is the Lorentz force, which consists of an electric part and a magnetic part, namely:

$$\frac{d\vec{p}}{dt} = q\,(\vec{E} + \vec{v} \times \vec{B}).$$

Accordingly, the interaction of a plane wave with an electron can be written as:

$$\frac{d\vec{p}}{dt} = -|e|E_o \cos(kz - \omega t)\left(\hat{x} + \frac{\vec{v}}{c} \times \hat{z}\right).$$

We have assumed that the electron is initially at rest. The process can therefore be viewed as two consecutive steps:
1. The electric field induces a motion component in the $\hat{x}$ direction.
2. The magnetic component induces a further motion component in the direction $\hat{x} \times \hat{z}$.

It is therefore evident that this further term contributes to the Larmor formula with a magnetic dipole component driven by the magnetic field $B_0 = 1/cE_o \ll E_0$. However, the corresponding contribution is far smaller than its corresponding electric counterpart.

In the main body of the chapter and in this section, we have often mentioned dipole radiation, but it has been left an undefined concept. Even though this aspect of the problem will be treated in the next chapter, we offer a simplified picture here.

To better understand the notion of an oscillating dipole, we consider the view shown in figure 1.19, where two charges with opposite signs are placed at symmetric points along the vertical axis. The charges are supposed to oscillate, namely to exhibit a time dependence

$$q(t) = q_0 \cos(\omega t).$$

The electric potential experienced by a test charge at a point P is that associated with the oscillating charges at the delivery time of the signal ($t' = t - \frac{r}{c}$), since it propagates at finite velocity c (see the next chapter, where the **retarded potentials** are discussed in less naïve terms).

The field at P is accordingly time dependent and can be treated as reported in the exercises suggested below, which are of particular importance and should be followed carefully.

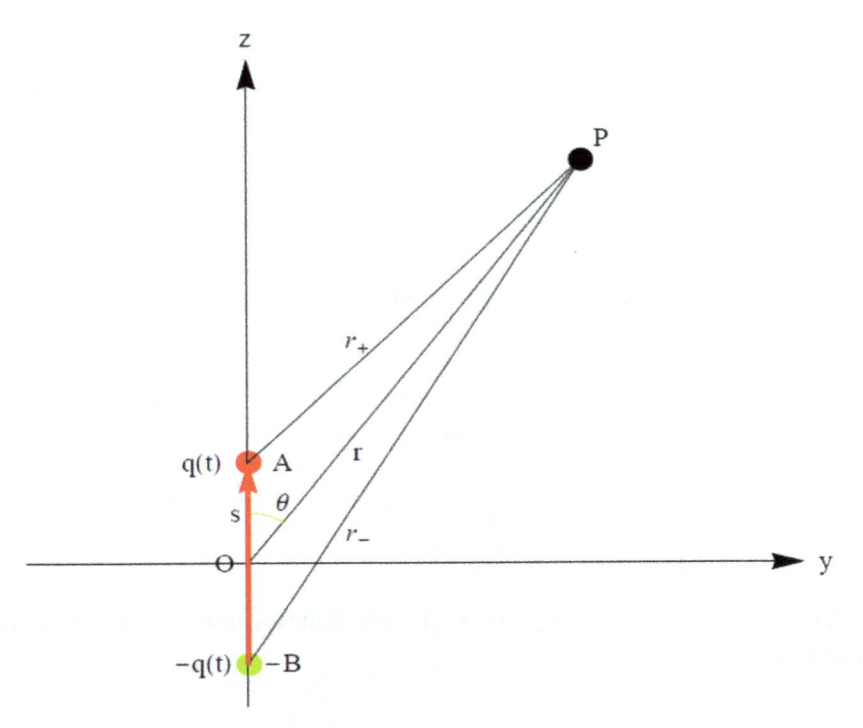

Figure 1.19. Electric dipole geometry.

Exercise 1.13. Show that the electric potential associated with an electric dipole $E1$ at a generic point P is

$$V^{El}(r, t) \simeq -\frac{\omega\, d}{4\pi\varepsilon_0 c}\frac{\cos\theta}{r}\sin(\omega t - kr) \quad \text{for } s \ll \lambda \ll r,$$

$$k = \frac{\omega}{c}, \quad d = qs.$$

Hint:
Note that

$$V^{El}(r, t) = \frac{q_0}{4\pi\varepsilon_0}\left[\frac{\cos\left(\omega(t - \frac{r_+}{c})\right)}{r_+} - \frac{\cos\left(\omega(t - \frac{r_-}{c})\right)}{r_-}\right]$$

$$r_{\pm} = \sqrt{r^2 + \left(\frac{s}{2}\right)^2 \pm sr\cos(\theta)}.$$

Expand at the order $o(\frac{s}{r})$ and find

$$V^{El}(r, t) \simeq -\frac{q_0 s \cos\theta}{4\pi\varepsilon_0 r}\left[k\sin(\omega t - kr) - \frac{\cos(\omega t - kr)}{r}\right].$$

Neglect the last term in the square brackets and finally find the desired result.

It is evident that, if, instead of the charges, the relevant distance were to vary harmonically, the potential experienced at point P would remain the same.

Exercise 1.14. Show that the vector potential associated with an oscillating dipole is

$$\vec{A}^{El}(r, t) \simeq -\frac{\mu_0 d}{4\pi}\frac{\omega}{r} \sin(\omega t - kr)\hat{z}. \tag{1.149}$$

Hint:
Taking into account that an oscillating charge q is arranged as shown in figure 1.19 and the relevant vector potential is at P, we can write

$$\vec{A}^{El}(r, t) = -\frac{\mu_0\, q\omega}{4\pi} \int_{-d/2}^{d/2} \frac{\sin(\omega t - kr)}{\rho} dz\, \hat{z}$$

$$\rho = \sqrt{r^2 - 2rz\cos\theta + z^2}.$$

Since $|z| \leqslant \frac{s}{2} \ll r \to \rho \simeq r$, equation (1.149) follows after integrating on the variable z.

Exercise 1.15. Write the formulas for an electric field and a magnetic field generated by an oscillating dipole in spherical coordinates.
Hint:
The use of equations (1.109) yields

$$\vec{E}^{El}(r, t) = -\frac{\partial}{\partial t}\vec{A}^{El}(r, t) - \vec{\nabla} V^{El}(r, t)$$

$$\vec{H}^{El}(r, t) = \vec{\nabla} \times \vec{A}^{El}(r, t).$$

Use equation (1.149); then, after changing to spherical coordinates (see figure 1.20), we get

$$\vec{E}^{El}(r, t) = -\frac{\mu_0 \omega^2 d}{4\pi}\left(\frac{\sin(\theta)}{r}\right)\cos(\omega t - kr)\hat{\theta}$$

$$\vec{B}^{El}(r, t) = \frac{1}{c}\hat{r} \times \vec{E}^{El}(r, t) \tag{1.150}$$

(easier said than done!).

The derivation of the previous identities requires a certain algebraic effort (which can be found in many textbooks; we recommend the lectures by Professor Steven Errede, Department of Physics, University of Illinois at Urbana-Champaign, Illinois, where the derivation has been done in detail and depth).

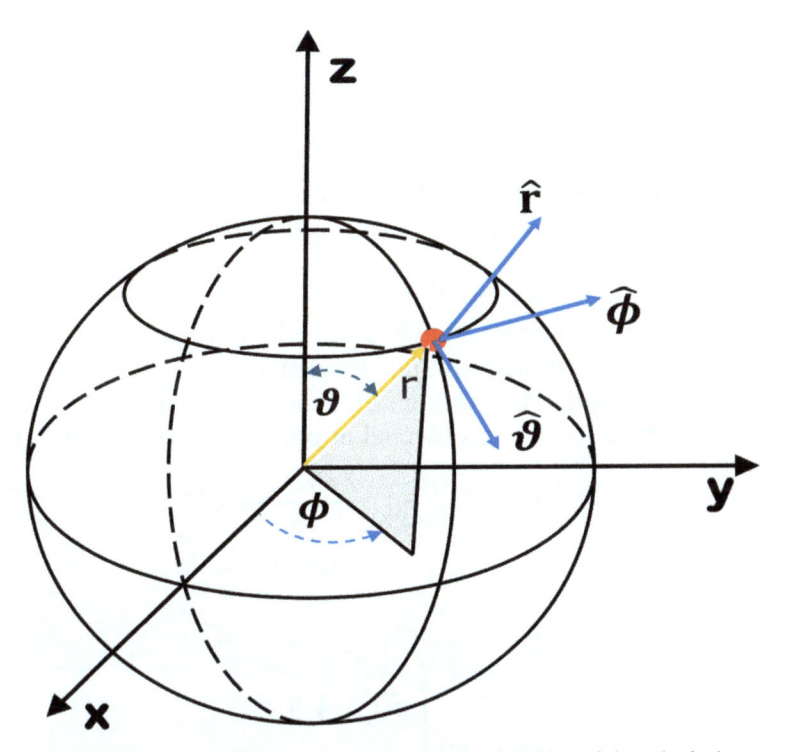

Figure 1.20. The geometry of spherical coordinates and the definition of the spherical vector triple.

It is, however, worth noting that:
1. Equations (1.150) hold in the far-field region ($s \ll \lambda \ll r$).
2. Both fields exhibit the same dependence on the distance r.
3. The electric and magnetic fields are in phase.
4. They exhibit the same angular dependence on the angle θ.

Exercise 1.16. Derive the Poynting vector associated with the electric dipole fields.
Hint: Use the standard definition of the Poynting vector and find

$$\vec{S}^{E1} = \frac{\mu_0}{c}\left[\frac{\omega^2 d}{4\pi}\left(\frac{\sin(\theta)}{r}\right)\cos(\omega t - kr)\right]^2 \hat{r} \qquad (1.151)$$

which holds because

$$\hat{r} \times \hat{\theta} = \hat{\varphi}$$
$$\hat{\theta} \times \hat{\varphi} = \hat{r}$$
$$\hat{\theta} \times (\hat{r} \times \hat{\theta}).$$

Exercise 1.17. Discuss the behavior of the retarded fields associated with an oscillating magnetic dipole.
Solution:
With reference to figure 1.21, the definition of a magnetic dipole follows from elementary physical notions, namely

$$\vec{m} = I\vec{S}_C$$

where I is the current flowing through a closed loop C with a radius b, and $\vec{S}_C = \pi b^2 \hat{z}$ is the associated oriented area directed along z.

If the current is assumed to be harmonically time dependent, i.e. $I(t) = I_0 \cos(\omega t)$, we expect that the retarded vector observed at point P can be written as:

$$\vec{A}^{M1}(r, t) = \frac{\mu_0}{4\pi} \int_C \frac{I \cos[\omega \, t_r]}{r} d\vec{l}'$$

$$r = r - r'(t_r), \qquad t_r = t - \frac{r}{c}.$$

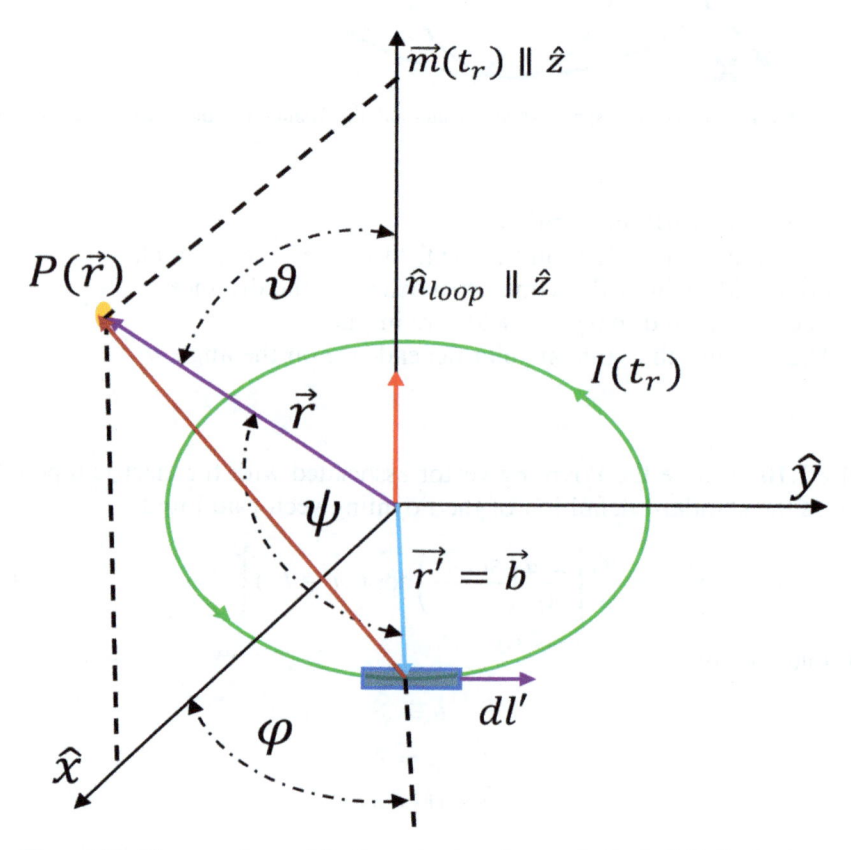

Figure 1.21. The geometry used for the evaluation of magnetic dipole field distributions.

By following a procedure analogous to that of **exercise 1.15** (see Errede's lectures for the details), we obtain the following for the vector potential in the far-field limit ($b \ll \lambda \ll r$) and in polar coordinates:

$$\vec{A}^{M1}(r, t) = -\frac{\mu_0 \omega \, m}{4\pi c} \frac{\sin(\theta)}{r} \sin\left[\omega t - \frac{r}{c}\right] \hat{\varphi}.$$

In the absence of charge densities, there is no scalar potential contribution.

Exercise 1.18. Derive the electric and magnetic fields associated with the magnetic dipole moment.
Hint:
The calculation is slightly simpler than the case discussed in **exercise 1.14**. We find indeed that

$$\vec{E}^{M1}(r, t) \approx \frac{\mu_0 \omega^2 \, |\vec{m}|}{4\pi c} \frac{\sin(\theta)}{r} \cos\left[\omega t - \frac{r}{c}\right] \hat{\varphi}$$

$$\vec{B}^{M1}(r, t) = \frac{1}{c} \hat{r} \times \vec{E}^{M1}(r, t). \qquad (1.152)$$

The comments outlined above for $\vec{E}^{E1}$, $\vec{B}^{E1}$ (points (a)–(d)) hold for equation (1.152); furthermore, it should also be noted that $\vec{E}^{M1}(r, t) \perp \vec{E}^{E1}(r, t)$; $\vec{B}^{M1}(r, t) \perp \vec{B}^{E1}(r, t)$.

Exercise 1.19. Find the Poynting vector for the M1 radiation.
Solution:
By analogy to equation (1.151), we find (see figure 1.22 for a correct understanding of vector products):

$$\vec{S}^{M1} = \frac{\mu_0}{c^3} \left[\frac{\omega^2 m}{4\pi} \left(\frac{\sin(\theta)}{r} \right) \cos(\omega t - kr) \right]^2 \hat{r}.$$

A few concluding comments are in order
1. Both the M1 and E1 Poynting vectors exhibit a $\sin(\theta)^2$ dependence and are invariant along φ. This justifies the characteristic doughnut shape shown in figure 1.5.
2. The relative intensities associated with $M1/E1$ are fixed by the ratio

$$\frac{|\vec{S}^{M1}|}{|\vec{S}^{E1}|} = \frac{1}{c^2} \left(\frac{m}{d}\right)^2 = \frac{\pi^2}{c^2} \frac{I_0^2 b^4}{e^2 s^2}.$$

Since $I_0 = e\omega$, we find:

$$\frac{|\vec{S}^{M1}|}{|\vec{S}^{E1}|} = \frac{\pi^2}{c^2} \frac{\omega^2 b^4}{s^2}.$$

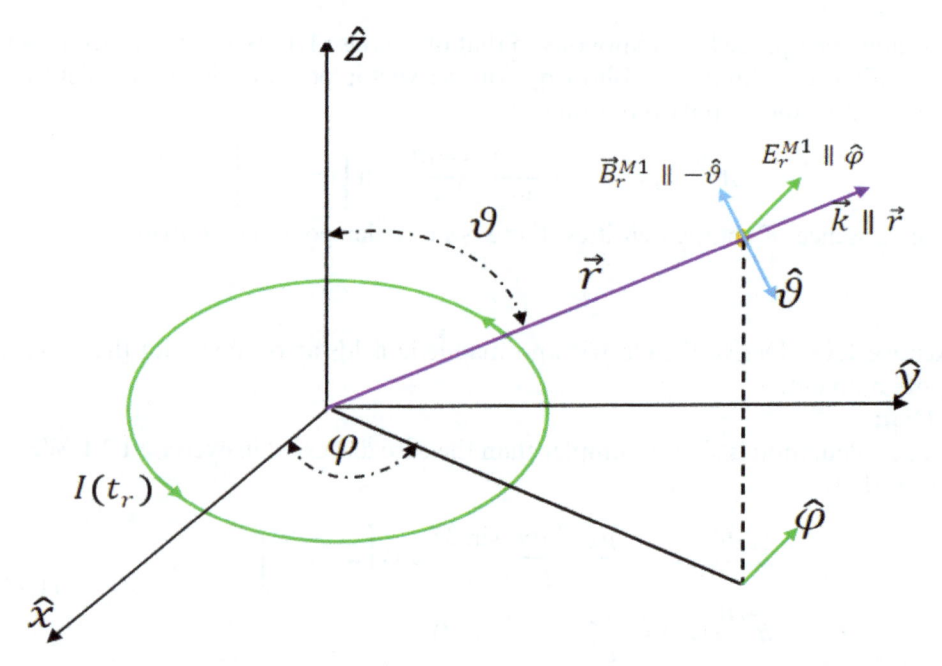

Figure 1.22. Electric and magnetic field orientations associated with a magnetic dipole.

Furthermore, keeping $\pi b = s$, we end up with

$$\frac{|\vec{S}^{M1}|}{|\vec{S}^{E1}|} = \frac{1}{c^2}\omega\, b^2 \ll 1.$$

The notions summarized in this section are essential for an understanding of the forthcoming chapters.

1.8.3 Special relativity, notation, Maxwell's equations, and gauge invariance

Exercise 1.20. Prove the following theorem: in vacuum, a free electron can neither emit a photon nor absorb radiation.
Solution:
This 'theorem' is just a consequence of the conservation of momentum and energy.
Start with:

$$p_{e,\mu} \equiv (E_e, \vec{p_e})$$
$$p_{f,\mu} \equiv (\hbar\omega, \hbar\vec{k})$$

(the labels 'e' and 'f' denote electrons and photons, respectively). Consider the process $e^- \to e^- + \gamma$ and apply the four-momentum conservation identity:

$$\vec{P_e} = \vec{P'_e} + \vec{P_f} \to \frac{m\vec{v}}{\sqrt{1-\left(\frac{v}{c}\right)^2}} = \frac{m\vec{v'}}{\sqrt{1-\left(\frac{v'}{c}\right)^2}} + \hbar\vec{k}$$

$$E_e = E'_e + E_f \to \frac{mc^2}{\sqrt{1-\left(\frac{v}{c}\right)^2}} = \frac{mc^2}{\sqrt{1-\left(\frac{v'}{c}\right)^2}} + \hbar\omega.$$

Assume that $\vec{v'} \cdot \vec{k} = |\vec{v'}||\vec{k}|\cos(\alpha)$ and find

$$\frac{v^2}{1-\left(\frac{v}{c}\right)^2} = \frac{v'^2}{1-\left(\frac{v'}{c}\right)^2} + \left(\frac{\hbar k}{m}\right)^2 + 2\frac{v'\left(\frac{\hbar k}{m}\right)\cos(\alpha)}{\sqrt{1-\left(\frac{v'}{c}\right)^2}},$$

$$\frac{1}{1-\left(\frac{v}{c}\right)^2} = \frac{1}{1-\left(\frac{v'}{c}\right)^2} + \left(\frac{\hbar\omega}{mc^2}\right)^2 + 2\frac{\frac{\hbar\omega}{mc^2}}{\sqrt{1-\left(\frac{v'}{c}\right)^2}}.$$

Use the second equation to get

$$\frac{\left(\frac{v}{c}\right)^2}{1-\left(\frac{v}{c}\right)^2} = \frac{\left(\frac{v'}{c}\right)^2}{1-\left(\frac{v'}{c}\right)^2} + \left(\frac{\hbar\omega}{mc^2}\right)^2 + 2\frac{\frac{\hbar\omega}{mc^2}}{\sqrt{1-\left(\frac{v'}{c}\right)^2}}.$$

Combine it with the first and find

$$\left(\frac{\hbar\omega}{mc}\right)^2 + 2\frac{\frac{\hbar\omega}{m}}{\sqrt{1-\left(\frac{v'}{c}\right)^2}} = +\left(\frac{\hbar k}{m}\right)^2 + 2\frac{v'\left(\frac{\hbar k}{m}\right)\cos(\alpha)}{\sqrt{1-\left(\frac{v'}{c}\right)^2}}.$$

Solving for $\cos(\alpha)$, we eventually end up with

$$\cos(\alpha) = \frac{c}{2v'}\left(\frac{\omega}{kc}\right)\left[\left[\left(\frac{\hbar\omega}{mc}\right) - \left(\frac{kc}{\omega}\right)\right]\sqrt{1-\left(\frac{v'}{c}\right)^2} + 2\frac{\omega}{kc}\right].$$

So far, we have not made the assumption that $\omega = kc$, thus we find $\cos(\alpha) = \frac{c}{v'} > 1$.

This rules out any possibility of the emission or absorption of photons by an electron in vacuum.

It is evident that things change if $\omega \neq kc$, which may happen in a waveguide or in a dispersive medium. We invite the reader to discuss these possibilities.

Exercise 1.21. Write the frequency of the Compton backscattered photon as

$$\omega' = \frac{1+\beta}{1-\beta+\sqrt{1-\beta^2}\frac{\hbar\omega}{m_0 c^2}}\omega. \tag{1.153}$$

1. Derive the limit $\omega'|_{\gamma\gg1}$ and show that it is independent. Comment on the physical content of the obtained result.
2. Keep the limit $\beta \to 0$ and show that it yields the previously discussed Compton shift.

Solution:
The second part of the exercise does not need any comment.
Regarding the first part, we note that the last equation can be rewritten as

$$\omega'|_{\gamma\gg1} = \frac{\xi^2}{1+\xi\frac{\hbar\omega}{m_0 c^2}}\omega|_{\gamma\gg1} \simeq \frac{\sqrt{\xi|_{\gamma\gg1}}\,m_0 c^2}{\hbar}$$

$$\xi = \sqrt{\frac{1+\beta}{1-\beta}}. \tag{1.154}$$

Bearing in mind that $\xi|_{\gamma\gg1} \simeq \sqrt{2}\,\gamma$, the final result is

$$\omega'|_{\gamma\gg1} \simeq \frac{2E}{\hbar}. \tag{1.155}$$

This apparently trivial conclusion gives us an idea of the role of quantum recoil. With increasing electron energy, the acquired recoil increases and therefore the output photon energy does not diverge.

Exercise 1.22. An atom that emits radiation at frequency ν' in its rest frame is moving at high velocity. Derive the frequency observed in the laboratory frame as a function of the angle θ determined by the atom's velocity and the direction of observation (see figure 1.23).
Solution:
The solution of the problem requires just a few wise assumptions:
1. We assume that the direction of the atom's propagation coincides with the $\hat{z}$ axis.
2. The photon's direction of momentum forms an angle θ (in the lab frame) with the direction of motion.
3. The atom's recoil associated with the photon emission is neglected.

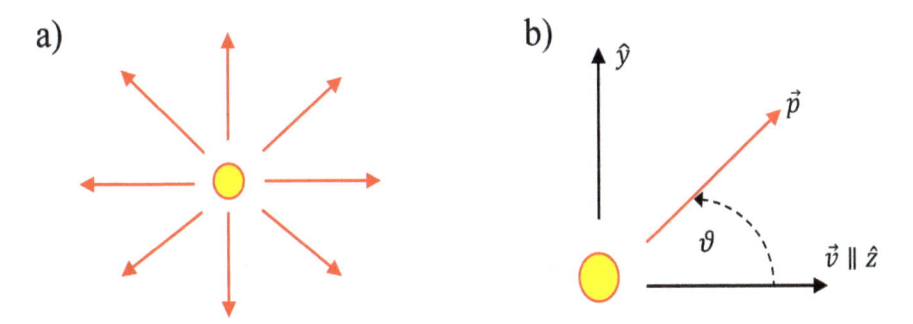

Figure 1.23. The emission process: (a) in an atom's rest frame, (b) in the laboratory frame.

According to the Lorentz transformations (see section (1.2)), we can link the rest and lab frame variables as follows:

$$p^\mu \equiv \begin{pmatrix} \frac{h\nu}{c} \\ p_x \\ p_y \\ p_z \end{pmatrix} = \frac{h\nu}{c}\begin{pmatrix} 1 \\ 0 \\ \sin(\theta) \\ \cos(\theta) \end{pmatrix}$$

$$p^\mu = \begin{pmatrix} \gamma & 0 & 0 & \gamma\beta \\ 0 & 1 & 0 & 0 \\ 0 & 0 & 1 & 0 \\ \gamma\beta & 0 & 0 & \gamma \end{pmatrix}\begin{pmatrix} \frac{h\nu'}{c} \\ p'_x \\ p'_y \\ p'_z \end{pmatrix} = \frac{h\nu'}{c}\begin{pmatrix} \gamma & 0 & 0 & \gamma\beta \\ 0 & 1 & 0 & 0 \\ 0 & 0 & 1 & 0 \\ \gamma\beta & 0 & 0 & \gamma \end{pmatrix}\begin{pmatrix} 1 \\ 0 \\ \sin(\theta') \\ \cos(\theta') \end{pmatrix} = \frac{h\nu'}{c}\begin{pmatrix} \gamma(1+\beta\cos(\theta')) \\ 0 \\ \sin(\theta') \\ \gamma(\beta+\cos(\theta')) \end{pmatrix}. \quad (1.156)$$

Comparing the previous equations, we obtain

$$\nu = \gamma(1+\beta\cos(\theta'))\nu',$$
$$\sin(\theta) = \frac{\sin(\theta')}{\gamma(1+\beta\cos(\theta'))}, \quad \sin(\theta') = \frac{\sin(\theta)}{\gamma(1-\beta\cos(\theta))},$$
$$\cos(\theta) = \frac{\cos(\theta')+\beta}{1+\beta\cos(\theta')} \quad \cos(\theta') = \frac{\cos(\theta)-\beta}{1-\beta\cos(\theta)} \quad \nu = \gamma\left[1+\beta\frac{\cos(\theta)-\beta}{1-\beta\cos(\theta)}\right]\nu'. \quad (1.157)$$

The physical consequences of the previous identities are worth underlining:
1. At large velocities ($\beta \to 1$), the photons are emitted in the forward direction.
2. For $\theta = 0$, it is also that case that $\theta' = 0$.
3. For $\theta = \pi$, it is also that case that $\theta' = \pi$.

We finally invite the reader to comment on the meaning of figure 1.24.

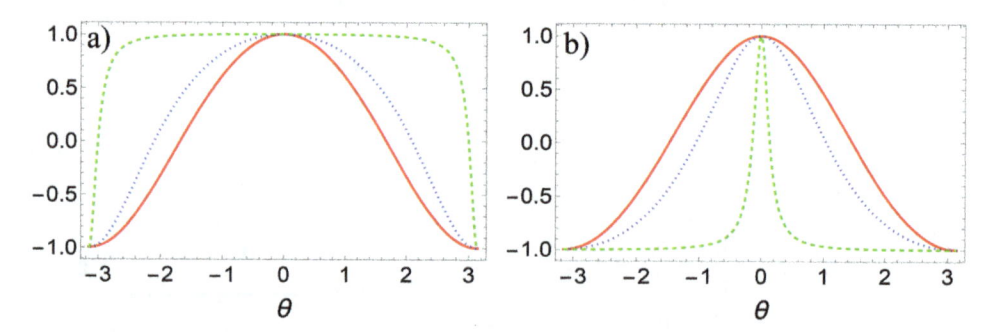

Figure 1.24. (a) $\cos(\theta')$ versus θ and (b) $\cos(\theta)$ versus θ' for different values of β (0.99 dashed, 0.5 dotted, 0.1 continuous).

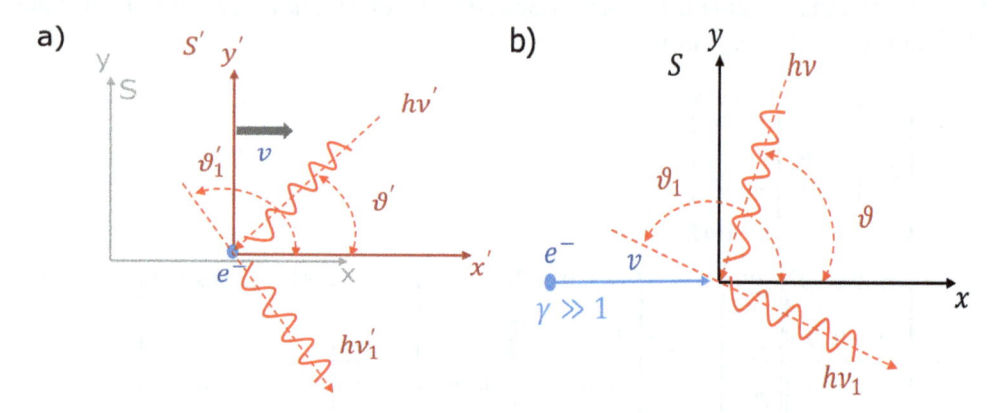

Figure 1.25. CBS (a) in an electron's rest frame and (b) in the lab frame.

Exercise 1.23. Consider what is shown in figure 1.25 and comment on the relevant physical meaning.

Solution:

Figure 1.25(a) shows the kinematics of CBS in both the electron's rest (primed) frame and in the laboratory (unprimed) frame. The incoming photon seen in the primed frame undergoes a Doppler shift.

If the photon energy remaining in this frame is much smaller than the electron's rest mass, the scattering process is ordinary Thomson diffusion. Accordingly, the incoming and outgoing photons have the same frequency and hence the same energy.

When transforming back to the rest frame, we should consider a further Doppler contribution. The photon energy in the unprimed frame is therefore given by

$$\varepsilon_{\text{out}} = \gamma^2 (1 - \beta \cos(\theta))(1 - \beta \cos(\theta')). \tag{1.158}$$

Noting that due to (see the previous exercise)

$$\cos(\theta') = \frac{\cos(\theta) - \beta}{1 - \beta \cos(\theta)} \qquad (1.159)$$

we obtain

$$\theta' = \pi \rightarrow \theta = \pi. \qquad (1.160)$$

In the case of backscattering, we eventually end up with

$$\varepsilon_{\text{out}} \simeq 4\gamma^2 \varepsilon. \qquad (1.161)$$

Exercise 1.24. Prove the non-invariance of Maxwell's equations under GT.
Solution:
The discussion of this problem is highly conceptual and necessarily elaborated from the formal point of view. We refer to a simplified point of view which is outlined below:

1. We recall that the GTs, in their simplified form, read as follows:

$$t' = t,$$
$$\vec{r}' = \vec{r} - \vec{v}_0 t$$
$$\vec{r} \equiv (x^1, x^2, x^3).$$

Keeping the derivative of the vector $\vec{r}'$ with respect to time, we obtain the velocity composition law:

$$\vec{v}' = \vec{v} - \vec{v}_0.$$

Furthermore, GTs do not allow any length contraction or time dilatation.

2. The transformation of the associated partial derivatives can be written as:[4]

$$\frac{\partial}{\partial t'} = \frac{\partial}{\partial t}\frac{\partial t}{\partial t'} + \frac{\partial}{\partial x^\alpha}\frac{\partial x^\alpha}{\partial t'} =$$
$$= \frac{\partial}{\partial t} + \vec{v} \cdot \vec{\nabla} \qquad (1.162)$$
$$\vec{\nabla}' = \vec{\nabla}.$$

3. Note that if we require the invariance of the Lorentz force

$$\vec{F} = q(\vec{E} + \vec{v} \times \vec{B})$$

under GT, we obtain (using also equation (1.162))

---

[4] We have used the contra-variant notation with $\frac{\partial}{\partial x^\alpha}\frac{\partial x^\alpha}{\partial t'} \rightarrow \sum_{\alpha=1}^{3}\frac{\partial}{\partial x^\alpha}\frac{\partial x^\alpha}{\partial t'}$.

$$\vec{F} = \vec{F}' \rightarrow \vec{E} + \vec{v} \times \vec{B} = \vec{E}' + \vec{v}' \times \vec{B}' \rightarrow$$
$$\rightarrow \vec{E} + \vec{v} \times \vec{B} = \vec{E}' + (\vec{v} - \vec{v}_0) \times \vec{B}' \rightarrow$$
$$\rightarrow \vec{E}' + \vec{v} \times (\vec{B}' - \vec{B}) = \vec{E} + \vec{v}_0 \times \vec{B}'.$$

The simplest solutions of the last equation are the (nonrelativistic) transformations of the electric and magnetic fields:

$$\begin{aligned}\vec{B}'(\vec{r}', t') &= \vec{B}(\vec{r}, t), \\ \vec{E}'(\vec{r}', t') &= \vec{E}(\vec{r}, t) + \vec{v}_0 \times \vec{B}(\vec{r}, t)\end{aligned} \quad (1.163)$$

which ensure the invariance of the Lorentz force in two reference systems moving with relatively constant velocities.

4. It is evident that, as a consequence of equations (1.162) and (1.163), the magnetic fields are solenoidal $\vec{\nabla}' \cdot \vec{B}' = \vec{\nabla} \cdot \vec{B} = 0$ in both systems.
5. Consider Gauss' law $\vec{\nabla} \cdot \vec{E} = \frac{\rho}{\varepsilon_0}$ and apply the transformations (1.162) and (1.163), which yields

$$\vec{\nabla}' \cdot \vec{E}'(\vec{r}', t') = \vec{\nabla}' \cdot \vec{E}(\vec{r}, t) - \vec{v}_0 \cdot \vec{\nabla} \times \vec{B}(\vec{r}, t).$$

The last term $\vec{v}_0 \cdot (\vec{\nabla} \times \vec{B}(\vec{r}, t))$ does not vanish under all circumstances, therefore the Galilean invariance (of Gauss' law) is broken.

We invite the reader to use the procedure we have outlined to prove that Faraday's law is invariant under GT, whereas this statement does not hold for Ampère's law.

1.8.4 Special relativity: four-vectors and their associated matrix formalism

In this subsection we briefly review the covariant/contra-variant formalism, which is the most common way to treat vectors in four-dimensional space-time. Here, we give the main definitions; the concepts are discussed more thoroughly in section 1.8.

The four-dimensional vectors specifying space and time coordinates are defined as:

$$\begin{aligned}x^\mu &= \begin{pmatrix} x^0 & x^1 & x^2 & x^3 \end{pmatrix}, \\ x^0 &= ct \\ \mu &= 0, 1, 2, 3\end{aligned} \quad (1.164)$$

in which components one to three are recognized as the ordinary space vectors, namely:

$$(x^1 \ x^2 \ x^3) \rightarrow \vec{r} \equiv (x, y, z). \quad (1.165)$$

The above four-vector is said to be contra-variant, while its covariant counterpart reads

$$x_\mu = (x_0\ x_1\ x_2\ x_3), \qquad \mu = 0, 1, 2, 3$$
$$x_0 = x^0$$
$$x_1 = -x^1 \qquad (1.166)$$
$$x_2 = -x^2,$$
$$x_3 = -x^3.$$

The following rule should be kept in mind: *the transition from covariant to contravariant forms, and vice versa, namely the raising or the lowering of the index μ ($x_\mu \to x^\mu / x^\mu \to x_\mu$) occurs through a change of sign of the ordinary space components.*

Any row contra/covariant four-vector is complemented by a column companion, i.e.:

For the generalized Bessel functions, see the following:

$$(x^0\ x^1\ x^2\ x^3) \to \begin{pmatrix} x^0 \\ x^1 \\ x^2 \\ x^3 \end{pmatrix}$$
$$(x_0\ x_1\ x_2\ x_3) \to \begin{pmatrix} x_0 \\ x_1 \\ x_2 \\ x_3 \end{pmatrix}. \qquad (1.167)$$

The Lorentz transformations can therefore be written as:

$$x'_\mu = \hat{L}(\beta) \begin{pmatrix} ct \\ x \\ y \\ z \end{pmatrix}$$
$$\hat{L}(\beta) = \begin{pmatrix} \gamma & -\gamma\beta & 0 & 0 \\ -\gamma\beta & \gamma & 0 & 0 \\ 0 & 0 & 1 & 0 \\ 0 & 0 & 0 & 1 \end{pmatrix}. \qquad (1.168)$$

The 4 × 4 matrix on the right-hand side of equation (1.168) is the Lorentz matrix. If, for simplicity, we consider the 2 × 2 part only, we find:

$$\hat{L} = \begin{pmatrix} \gamma & -\gamma\beta \\ -\gamma\beta & \gamma \end{pmatrix} \qquad (1.169)$$

whose determinant is unitary:

$$|\hat{L}| = \beta^2 - \gamma\beta^2 = 1. \qquad (1.170)$$

The inverse transformation is written as:

$$x_\mu = \hat{L}^{-1} x'_\mu$$
$$\hat{L}(v)^{-1} = \begin{pmatrix} \gamma & \gamma\beta \\ \gamma\beta & \gamma \end{pmatrix} = \hat{L}(-v). \tag{1.171}$$

The norm of the vectors x^μ/x_μ is defined analogously to that of the ordinary spatial case (for further comments, see section 1.8) and reads

$$x_\mu x^\mu = \begin{pmatrix} ct \\ x \\ y \\ z \end{pmatrix}(ct\ -x\ -y\ -z) = ct^2 - |\vec{r}|^2$$
$$\vec{r} = (x, y, z)$$
$$|\vec{r}|^2 = x^2 + y^2 + z^2. \tag{1.172}$$

It is easily shown (see section 1.8) that the norm (1.172) is conserved by the Lorentz transformation, namely:

$$x'_\mu x'^\mu = x_\mu x^\mu. \tag{1.173}$$

Before closing this subsection, we exploit the previously developed concepts to derive important relationships involving the angular transformations. This is a consequence of the fact that the scalar product is also a Lorentz invariant, i.e.:

$$(x_1)'_\mu (x_2)'^\mu = (x_1)_\mu (x_2)^\mu. \tag{1.174}$$

Assuming that

$$\vec{r}_1 \cdot \vec{r}_2 = |\vec{r}_1||\vec{r}_2| \cos(\theta)$$
$$\vec{r}'_1 \cdot \vec{r}'_2 = |\vec{r}'_1||\vec{r}'_2| \cos(\theta') \tag{1.175}$$

we eventually find the following relationship between the angles θ, θ':

$$\cos(\theta') = -\frac{c^2(t_1 t_2 - t'_1 t'_2) - |\vec{r}_1||\vec{r}_2|\cos(\theta)}{|\vec{r}'_1||\vec{r}'_2|} \tag{1.176}$$

which is a general identity that is valid for transformations with different velocities and different orientations between them.

A simpler derivation valid for a less general case is discussed below. The velocity composition between the two inertial frames can be directly inferred from the Lorentz transformations, which yield:

$$\frac{dx'}{dt'} = \frac{(dx - v dt)}{dt - \frac{v dx}{c^2}} = \frac{\frac{dx}{dt} - v}{1 - \frac{v \frac{dx}{dt}}{c^2}}. \tag{1.177}$$

Assuming that in the unprimed system the velocity is $dx/dt = u$, we end up with:

$$u' = \frac{u-v}{1-\frac{vu}{c^2}}$$
$$u = \frac{u'+v}{1+\frac{vu'}{c^2}}.$$
(1.178)

It is evident that if $u = c$ then $u' = c$; in other words, the velocity of light is invariant under Lorentz transformation.[5]

It should be underscored that the velocity composition holds if the velocities u and v are parallel. For a more general discussion, see section 1.8.

Consider now a photon traveling along the x component with velocity $u' = c \cos(\theta')$. From the second of equations (1.178), we find (see section 1.8 for further comments):

$$c \cos(\theta) = \frac{c \cos(\theta') + v}{1 + \beta \cos(\theta')}$$
(1.179)

and eventually:

$$c \cos(\theta') = \frac{c \cos\theta - v}{1 - \beta \cos(\theta)}.$$
(1.180)

Exercise 1.25. Introduce the metric tensor

$$g^{\mu\nu} = g_{\mu\nu} = \begin{pmatrix} 1 & 0 & 0 & 0 \\ 0 & -1 & 0 & 0 \\ 0 & 0 & -1 & 0 \\ 0 & 0 & 0 & -1 \end{pmatrix}$$

and show that:

$$g_{\alpha\gamma} g^{\gamma\beta} = \delta_\alpha^\beta$$

$$\delta_\alpha^\beta = \begin{cases} 1, & \alpha = \beta \\ 0, & \alpha \neq \beta \end{cases}.$$

Hint:
Use the Einstein convention and write:

$$g_{\alpha\gamma} g^{\gamma\beta} = \sum_{\alpha=0}^{4} g_{\alpha\gamma} g^{\gamma\beta} = g_{\alpha 0} g^{0\beta} + g_{\alpha 1} g^{1\beta} + g_{\alpha 2} g^{2\beta} + g_{\alpha 3} g^{3\beta}.$$

---

[5] It should be noted that the invariance of the velocity of light is not a consequence of the Lorentz transformations but a consequence of the assumption of postulates 1 and 2.

Exercise 1.26. Show that the metric tensor is the element linking contra-variant and covariant vectors, as given below:
$$x_\mu = g_{\mu\nu} x^\nu, \quad x^\mu = g^{\mu\nu} x_\nu.$$

Hint:
Define the column vector
$$x_\nu = \begin{pmatrix} x^0 \\ x^1 \\ x^2 \\ x^3 \end{pmatrix}$$

$$g_{\mu\nu} x^\nu = \begin{pmatrix} 1 & 0 & 0 & 0 \\ 0 & -1 & 0 & 0 \\ 0 & 0 & -1 & 0 \\ 0 & 0 & 0 & -1 \end{pmatrix} \begin{pmatrix} x^0 \\ x^1 \\ x^2 \\ x^3 \end{pmatrix} = \begin{pmatrix} x^0 \\ -x^1 \\ -x^2 \\ -x^3 \end{pmatrix} = x_\mu.$$

Exercise 1.27. Write the Lorentz transformations in terms of the matrix
$$\Lambda^\mu_\nu = \begin{pmatrix} \gamma & -\gamma\beta & 0 & 0 \\ -\gamma\beta & \gamma & 0 & 0 \\ 0 & 0 & 1 & 0 \\ 0 & 0 & 0 & 1 \end{pmatrix}.$$

Hint:
Note that
$$\begin{pmatrix} \gamma & -\gamma\beta & 0 & 0 \\ -\gamma\beta & \gamma & 0 & 0 \\ 0 & 0 & 1 & 0 \\ 0 & 0 & 0 & 1 \end{pmatrix} \begin{pmatrix} x^0 \\ x^1 \\ x^2 \\ x^3 \end{pmatrix} = \begin{pmatrix} \gamma(x^0 - \beta x^1) \\ \gamma(-\beta x^0 + x^1) \\ x^2 \\ x^3 \end{pmatrix} \rightarrow x'^\mu = \Lambda^\mu_\nu x^\nu. \quad (1.181)$$

The right-hand side of the last equation provides the 'synthetic' form of the Lorentz transformation in contra-variant form.

Exercise 1.28. Write the corresponding Lorentz transformation for motion in the $x^2 = y$ direction.
Hint:
Let β_y denote the reduced velocity along y. We find
$$x'^\mu = {}_y\Lambda^\mu_\nu x^\nu = \begin{pmatrix} \gamma & 0 & -\gamma\beta_y & 0 \\ 0 & 1 & 0 & 0 \\ -\gamma\beta_y & 0 & \gamma & 0 \\ 0 & 0 & 0 & 1 \end{pmatrix} \begin{pmatrix} ct \\ x \\ y \\ z \end{pmatrix}.$$

Exercise 1.29. Write the corresponding Lorentz transformation for a generic motion with components β_x, β_y, β_z.
Hint:

$$x'^{\mu} = \left({}_x\Lambda^{\mu}_{\tau} \cdot {}_y\Lambda^{\tau}_{\lambda} \cdot {}_z\Lambda^{\lambda}_{\sigma}\right) x^{\sigma} =$$

$$= \begin{pmatrix} \gamma & -\gamma\beta_x & 0 & 0 \\ -\gamma\beta_x & \gamma & 0 & 0 \\ 0 & 0 & 1 & 0 \\ 0 & 0 & 0 & 1 \end{pmatrix} \begin{pmatrix} \gamma & 0 & -\gamma\beta_y & 0 \\ 0 & 1 & 0 & 0 \\ -\gamma\beta_y & 0 & \gamma & 0 \\ 0 & 0 & 0 & 1 \end{pmatrix} \begin{pmatrix} \gamma & 0 & 0 & -\gamma\beta_z \\ 0 & 1 & 0 & 0 \\ 0 & 0 & 1 & 0 \\ -\gamma\beta_z & 0 & 0 & \beta \end{pmatrix} \begin{pmatrix} ct \\ x \\ y \\ z \end{pmatrix}. \quad (1.182)$$

Furthermore, it is evident that the previous matrices yield:

$$\vec{r} = x\hat{x} + y\hat{y} + z\hat{z},$$
$$\vec{\beta} = \beta_x\hat{x} + \beta_y\hat{y} + \beta_z\hat{z}$$
$$\gamma = \frac{1}{\sqrt{1-|\vec{\beta}|^2}}, \quad |\vec{\beta}|^2 = \beta_x^2 + \beta_y^2 + \beta_z^2. \quad (1.183)$$

Exercise 1.30. Use the notions elaborated so far ((1.182) and (1.183)) to prove that a compact and elegant way of writing the Lorentz transformation for a generic direction $\vec{r}$ is

$$x'^{\mu} = \begin{pmatrix} \gamma & -\gamma\vec{\beta}\cdot \\ -\gamma\vec{\beta} & 1 + \frac{(\gamma-1)\vec{\beta}}{|\vec{\beta}|^2}(\vec{\beta}\cdot) \end{pmatrix} \begin{pmatrix} ct \\ \vec{r} \end{pmatrix} = \begin{pmatrix} \gamma(ct - \vec{\beta}\cdot\vec{r}) \\ \vec{r} + \frac{(\gamma-1)(\vec{\beta}\cdot\vec{r})\vec{\beta}}{|\vec{\beta}|^2} - \gamma\vec{\beta}ct \end{pmatrix} \quad (1.184)$$

and derive equation (1.181) as a particular case.
Hint:
The first question is just matter of tedious (but worthwhile) algebra that is required to obtain the right-hand side of equation (1.182) in a vector form.

It is also instructive to write the second component of the two-column vector in equation (1.184) as:

$$\vec{r}' - \vec{r} = (\hat{u} \cdot \vec{M})\hat{u}$$
$$\hat{u} = \frac{\vec{\beta}}{|\vec{\beta}|}, \quad \vec{M} = \gamma(\vec{r} - \vec{\beta}\,ct) - \vec{r}$$

and draw the vector composition of the space vector Lorentz transformation.

Regarding the second question, note that:

$$\vec{r} + \frac{(\gamma-1)}{|\vec{\beta}|^2}(\vec{\beta}\cdot\vec{r})\vec{\beta} - \gamma\vec{\beta}ct \to x[1+(\gamma-1)] - \gamma\beta ct = \ldots.$$

Consider, for simplicity,

$$\Lambda^\mu_\nu(\beta) = \begin{pmatrix} \gamma & -\gamma\beta \\ -\gamma\beta & \gamma \end{pmatrix} \tag{1.185}$$

and show that:

$$[\Lambda^\mu_\nu(\beta)]^{-1} = \Lambda^\mu_\nu(-\beta).$$

Apply the following identity from elementary matrix algebra:

$$\hat{A} = \begin{pmatrix} a & b \\ c & d \end{pmatrix} \to \hat{A}^{-1} = \frac{1}{|\hat{A}|}\begin{pmatrix} d & -b \\ -c & a \end{pmatrix}$$

$$|\hat{A}| = \det(\hat{A}) = ad - bc$$

and eventually find:

$$[\Lambda^\mu_\nu(\beta)]^{-1} = \begin{pmatrix} \gamma & \gamma\beta \\ \gamma\beta & \gamma \end{pmatrix} = \Lambda^\mu_\nu(-\beta).$$

Exercise 1.31. Use the identity (4D Minkowski space representation)

$$x^\mu \equiv \begin{pmatrix} ct \\ ix \\ iy \\ iz \end{pmatrix}, \qquad x_\mu \equiv \begin{pmatrix} ct & ix & iy & iz \end{pmatrix}$$

to show that

$$x_\mu x^\mu = (ct)^2 - x^2 - y^2 - z^2$$

and derive the relevant Lorentz transformations.
Solution:
The answer to the first question does not require any conceptual effort, and the same is true for the second. The Lorentz transformation can indeed be written as (we limit ourselves to the 2 × 2 case for simplicity):

$$x'^\mu = \begin{pmatrix} \gamma & -i\gamma\beta \\ i\gamma\beta & \gamma \end{pmatrix}\begin{pmatrix} ct \\ ix \end{pmatrix} = \begin{pmatrix} \gamma(ct - \beta x) \\ i\gamma(x + \beta ct) \end{pmatrix}.$$

It is also easy to check that:

$$x'_\mu x'^\mu = (ct\ ix)\begin{pmatrix} \gamma & i\gamma\beta \\ -i\gamma\beta & \gamma \end{pmatrix}\begin{pmatrix} \gamma & -i\gamma\beta \\ i\gamma\beta & \gamma \end{pmatrix}\begin{pmatrix} ct \\ ix \end{pmatrix} =$$

$$= (ct\ ix)\begin{pmatrix} \gamma^2(1-\beta^2) & 0 \\ 0 & \gamma^2(1+\beta^2) \end{pmatrix}\begin{pmatrix} ct \\ ix \end{pmatrix} = (ct\ ix)\begin{pmatrix} \gamma^2(1-\beta^2)ct \\ i\gamma^2(1-\beta^2)x \end{pmatrix} =$$

$$= \gamma^2(1-\beta^2)(ct)^2 - \gamma^2(1-\beta^2)x^2 = (ct)^2 - x^2 = x_\mu x^\mu.$$

The results outlined in this exercise are conceptually important. The point of view introduced by Minkowski (mathematically) unifies space and time. Within this context the Lorentz matrix can indeed be viewed as a rotation matrix, if we make the following correspondence:

$$\gamma \Rightarrow \cos(\theta)$$
$$\beta \Rightarrow -i\tan(\theta)$$
$$\gamma\beta \Rightarrow -i\sin(\theta).$$

It is important to note that the Minkowski metric is (formally) Euclidean, in the sense that the norm in 4D space is defined as:

$$||x|| = \sqrt{\sum_{\alpha=0}^{3} x^{\alpha 2}}.$$

The negative sign in the spatial coordinates is just due to the presence of the imaginary unit in their definition. In the contra-/covariant formalism, the metric is non-Euclidean, and using the vector formalism we find:

$$||x||^2 = (ct\ x)\begin{pmatrix} 1 & 0 \\ 0 & -1 \end{pmatrix}\begin{pmatrix} ct \\ x \end{pmatrix} = (ct\ x)\begin{pmatrix} ct \\ -x \end{pmatrix} = ct^2 - x^2.$$

Exercise 1.32. Show that the use of the identity

$$\gamma^2 - (\gamma\beta)^2 = 1$$

allows the following (formal) identity:

$$\gamma = \cosh(\phi)$$
$$\beta = \tanh(\phi)$$

and that the matrix in equation (1.185) can be written as:

$$\Lambda^\mu_\nu(\beta) = \begin{pmatrix} \cosh(\phi) & -\sinh(\phi) \\ -\sinh(\phi) & \cosh(\phi) \end{pmatrix}.$$

Solution:
It is evident that the kinematic invariant exhibits an analogy with the fundamental identity of the hyperbolic trigonometry

$$\cosh(\phi)^2 - \sinh(\phi)^2 = 1.$$

This analogy yields

$$\gamma = \cosh(\phi)$$
$$\gamma\beta = \sinh(\phi)$$

and $\beta = \tanh(\phi)$. Without entering into further details (see the bibliography at the end of the chapter), we note that it is possible to introduce a further kinematic variable $\phi = \tanh^{-1}(\beta)$, known as rapidity. According to the previous discussion, the Lorentz transformation can be viewed as hyperbolic rotation in space-time. We believe that the reader should more carefully consider the conceptual basis of these last considerations and weigh up the (formal and substantial) analogy between the relativistic velocity composition and the additive property of the hyperbolic tangent, namely:

$$\tanh(\alpha + \beta) = \frac{\tanh(\alpha) + \tanh(\beta)}{1 + \tanh(\alpha)\tanh(\beta)}.$$

1.8.5 Comments on the gauge invariance of classical electromagnetism

We have introduced contra-variant and covariant derivatives, defined as

$$\partial_\mu = \frac{\partial}{\partial x^\mu} = \left(\frac{\partial}{\partial(ct)}, \vec{\nabla}\right) \text{ covariant}$$

$$\partial^\mu = \frac{\partial}{\partial x_\mu} \equiv \left(\frac{\partial}{\partial(ct)}, -\vec{\nabla}\right) \text{ contra} - \text{variant}.$$

Accordingly, one gets:

$$\partial_\mu \partial^\mu = \frac{1}{c^2}\frac{\partial^2}{\partial t^2} - |\vec{\nabla}|^2.$$

We have seen that the underlying formalism is extremely useful for writing MEs in a very compact form. The exercises that follow allow the reader to gain further confidence with the associated technicalities.

Exercise 1.33. In the main body of the chapter, we have written MEs using the electromagnetic tensor. The following identity (to be proved) yields the definition of the electric and magnetic fields in terms of the tensor $F^{\mu\nu}$ in conjunction with specific values of the indices ($\mu\nu$):

$$F^{\mu\nu} = \partial^\mu A^\nu - \partial^\nu A^\mu = \begin{cases} \vec{E} = -\left(\frac{\partial}{\partial t}\vec{A} + \vec{\nabla}\Phi\right) \\ \vec{B} = \vec{\nabla} \times \vec{A} \end{cases}$$

where:

$$A^\mu \equiv \left(\frac{\Phi}{c}, \vec{A}\right).$$

Hint:
The computation is straightforward. It can indeed be noted that

$$F^{01} = \partial^0 A^1 - \partial^1 A^0 = \frac{1}{c}\frac{\partial}{\partial t}A_x + \frac{1}{c}\frac{\partial}{\partial x}\Phi = -\frac{E_x}{c},$$

$$F^{12} = \partial^1 A^2 - \partial^2 A^1 = -\frac{\partial A_y}{\partial x} + \frac{\partial A_x}{\partial y} = -\left(\vec{\nabla}\times\vec{A}\right)_z = -B_z.$$

Exercise 1.34. Use the properties of the $F^{\mu\nu}$ tensor to prove that the homogeneous (i.e. in the absence of charge and current) ME reads $\partial_\mu F^{\mu\nu} = 0$.

Hint:
Note that the electromagnetic tensor is given by:

$$F^{\mu\nu} = \begin{pmatrix} 0 & -\frac{E_x}{c} & -\frac{E_y}{c} & -\frac{E_z}{c} \\ \frac{E_x}{c} & 0 & -B_z & B_y \\ \frac{E_y}{c} & B_z & 0 & -B_x \\ \frac{E_z}{c} & -B_y & B_x & 0 \end{pmatrix}.$$

Therefore, we find:

$$\partial_\alpha F^{\alpha 0} = \frac{\partial}{\partial x^0}F^{00} + \frac{\partial}{\partial x^1}F^{10} + \frac{\partial}{\partial x^2}F^{20} + \frac{\partial}{\partial x^3}F^{30} \to \vec{\nabla}\cdot\vec{E} = 0$$

$$\partial_\alpha F^{\alpha 1} = \frac{\partial}{\partial x^0}F^{01} + \frac{\partial}{\partial x^1}F^{11} + \frac{\partial}{\partial x^2}F^{21} + \frac{\partial}{\partial x^3}F^{31} = -\frac{1}{c^2}\frac{\partial}{\partial t}E_x + \left(\frac{\partial}{\partial y}B_z - \frac{\partial}{\partial z}B_y\right) \to$$

$$\to \vec{\nabla}\times\vec{B} = \frac{1}{c^2}\frac{\partial\vec{E}}{\partial t}.$$

Exercise 1.35. Use the properties of the $F^{\mu\nu}$ tensor to prove that the inhomogeneous (in the presence of charge and current) MEs read:

$$\partial_\mu F^{\mu\nu} = \mu_0 J^\nu$$

$$J^\mu \equiv (c\rho, \vec{J}).$$

Hint:
Use the same procedure as before and find, for example, that

$$\partial_\alpha F^{\alpha 0} = \frac{\partial}{\partial x^0} F^{00} + \frac{\partial}{\partial x^1} F^{10} + \frac{\partial}{\partial x^2} F^{20} + \frac{\partial}{\partial x^3} F^{30} = \mu_0 J^0 \rightarrow$$
$$\rightarrow \vec{\nabla} \cdot \vec{E} = \frac{\rho}{\varepsilon_0}$$
$$\partial_\alpha F^{\alpha 1} = \frac{\partial}{\partial x^0} F^{01} + \frac{\partial}{\partial x^1} F^{11} + \frac{\partial}{\partial x^2} F^{21} + \frac{\partial}{\partial x^3} F^{31} = -\frac{1}{c^2}\frac{\partial}{\partial t} E_x + \left(\frac{\partial}{\partial y} B_z - \frac{\partial}{\partial z} B_y\right) = \mu_0 J_x \rightarrow$$
$$\rightarrow \vec{\nabla} \times \vec{B} = \frac{1}{c^2}\frac{\partial \vec{E}}{\partial t} + \mu_0 \vec{J}.$$

Exercise 1.36. Show that the four-vector $\mu_0 J^\mu$ can be written as:

$$\mu_0 J^\mu \equiv \left(\frac{1}{c}\vec{\nabla}\cdot\vec{E},\ \vec{\nabla}\times\vec{B} - \frac{1}{c^2}\frac{\partial \vec{E}}{\partial t}\right)$$

and comment on the physical meaning of the Lorentz invariant $J_\mu J^\mu$.
Hint:
The first question is straightforward and does not require comment.
The second should be explained with some care. It should be noted that the current $\vec{J}$ density is linked to the charge density ρ:

$$\rho = \frac{Q}{V}, \qquad Q \equiv \text{charge}, \qquad V \equiv \text{volume}$$
$$\vec{J} \equiv \rho \vec{v}, \qquad \vec{v} \equiv \text{velocity of the charge}.$$

Therefore, we find:
$$J_\mu J^\mu = \rho c^2 1 - \beta^2.$$

Exercise 1.37. Show that the identity
$$\partial_\mu J^\mu = 0$$
is equivalent to the charge continuity equation.
Hint:
By simply applying the definition of the covariant derivative, we obtain:
$$\frac{\partial}{\partial t}\rho + \vec{\nabla}\cdot\vec{J} = 0.$$

Exercise 1.38. Show that any transformation of the type
$$A^\nu \rightarrow A^\nu + \frac{\partial}{\partial x_\nu}\lambda \tag{1.186}$$
leaves the definition of the electromagnetic tensor 'invariant.'

Hint:
It can easily be confirmed that the extra term induced by the previous transformation is identically vanishing. We find indeed that

$$F^{\mu\nu} \to F^{\mu\nu} + \Lambda^{\mu\nu}$$

$$\Lambda^{\mu\nu} = \frac{\partial^2}{\partial x_\mu \partial x_\nu}\lambda - \frac{\partial^2}{\partial x_\nu \partial x_\mu}\lambda = 0.$$

Exercise 1.39. Explain the importance of the Lorentz gauge condition

$$\partial_\nu A^\nu = 0. \tag{1.187}$$

Solution:
Before commenting on the importance of the previous condition, we note that in terms of vector and scalar fields, equation (1.187) reads:

$$\vec{\nabla} \cdot \vec{A} = -\frac{1}{c^2}\frac{\partial \Phi}{\partial t} \tag{1.188}$$

which is a kind of continuity equation (losses from the vector potential are gains for its scalar counterpart).

Relationship (1.187) is ensured by the fact that, according to equation (1.186), it is always possible to find a gauge transformation such that:

$$\partial_\nu A'^\nu = \partial_\nu A^\nu + \partial_\nu \partial^\nu \lambda = 0 \tag{1.189}$$

which is always true if

$$\partial_\nu \partial^\nu \lambda = 0.$$

The advantage of the Lorentz gauge is that it simplifies the equation

$$\partial_\nu F^{\mu,\nu} = \partial_\nu (\partial^\mu A^\nu - \partial^\nu A^\mu) = \mu_0 J^\mu.$$

Since $\partial_\nu \partial^\mu = \partial^\mu \partial_\nu$, we find:

$$\partial_\nu (\partial^\mu A^\nu - \partial^\nu A^\mu) = \partial^\mu (\partial_\nu A^\nu) - \partial_\nu \partial^\nu A^\mu$$
$$\partial_\nu \partial^\nu A^\mu = -\mu_0 J^\mu.$$

It should now be underlined that from the last identity, it follows that

$$\partial_\nu \partial^\nu (\partial_\mu A^\mu) = -\mu_0 \partial_\mu J^\mu.$$

Put in these terms, the charge conservation equation follows from Lorentz invariance.

Exercise 1.40. Explain why the 'instantaneous Coulomb gauge' $\vec{\nabla} \cdot \vec{A} = 0$ should be handled with care when dealing with relativistic electromagnetic problems.
Hint:
Note that the condition $\vec{\nabla} \cdot \vec{A} = 0$ is not a relativistic invariant.

Exercise 1.41. Discuss the electromagnetic gauge conditions and show that these ensure that the photon is a massless particle.
  Hint:
  The answer to this problem is not straightforward. We invite the reader to think about this statement, which goes well beyond the physics of electromagnetic fields.

Exercise 1.42. Guess the form of the MEs in the hypothesis of massive photons, and show that in this case, the gauge condition does not hold.
  Solution:
  The method we choose is based on simple assumptions and follows the lines proposed by Alexandru Proca during the first half of the last century (see the bibliography). We assume the existence of a photon mass m_γ without mentioning any mechanism for the 'generation' of a massive photon field.
  The assumption of a massive field implies that:
  1. The definition of the Poynting vector should be modified to include the mass term.
  2. The same holds for the energy density.

Regarding the last point, we note that the energy density should read:

$$W = \frac{1}{2}\left(\varepsilon_0 |\vec{E}|^2 + \frac{|\vec{B}|^2}{\mu_0} + \text{massive terms}\right). \tag{1.190}$$

It is important to underscore that the new terms should be defined in terms of the electromagnetic constants ε_0, μ_0, the vector and scalar potentials, and a characteristic term including the photon mass. We assume that the last term in equation (1.190) can be split into electric and magnetic contributions. To define the first using what we have at hand, we note that the quantities

$$\Gamma_\gamma = \frac{\Phi}{\lambda_\gamma}$$
$$\vec{\Lambda}_\gamma = \frac{\vec{A}}{\lambda_\gamma} \tag{1.191}$$

(where λ_γ is a characteristic length) have the dimensions of an electric field and a magnetic field, respectively. Having assumed the existence of a photon mass, the most natural definition of λ_γ is through the associated Compton wavelength, namely:

$$\lambda_\gamma = \frac{\hbar}{m_\gamma c}. \tag{1.192}$$

Accordingly, the electromagnetic energy density is

$$W = \frac{1}{2}\left(\varepsilon_0 |\vec{E}|^2 + \frac{|\vec{B}|^2}{\mu_0} + \varepsilon_0 \Gamma_\gamma^2 + \frac{|\vec{\Lambda}_\gamma|^2}{\mu_0}\right) \quad (1.193)$$

and the Poynting vector is

$$\vec{S} = \frac{1}{\mu_0}(\vec{E} \times \vec{B} + \Gamma\vec{\Lambda}). \quad (1.194)$$

Exercise 1.43. We invite the reader to derive the Proca version of MEs.
Solution:
Use the electromagnetic Lagrangian written as:

$$L = -\frac{1}{4}F_{\mu\nu}F^{\mu\nu} - J_\mu A^\mu + \frac{1}{2\lambda_\gamma^2}A_\mu A^\mu \quad (1.195)$$

and show (use the formalism outlined in the course of the chapter) that the massive contributions are modified as follows:

$$\vec{\nabla} \cdot \vec{E} = \frac{\rho}{\varepsilon_0} - \frac{\Phi}{\lambda_\gamma^2},$$

$$\vec{\nabla} \times \vec{B} = \mu_0 \vec{J} + \frac{1}{c^2}\frac{\partial \vec{E}}{\partial t} - \frac{\vec{A}}{\lambda_\gamma^2} \quad (1.196)$$

$$\vec{\nabla} \times \vec{E} = -\frac{\partial \vec{B}}{\partial t}$$

$$\vec{\nabla} \cdot \vec{B} = 0.$$

According to the previous identity, Gauss' law and Ampère's law are modified by a nonzero photon mass.

Exercise 1.44. Note that the definition of electric and magnetic fields in terms of scalar and vector potentials is left unchanged by the introduction of the photon mass term. Use equations (1.196) to derive the wave equations for electric and magnetic fields.
Solution:
Manipulating and combining the third and second equations (as already explained) gives:

$$\vec{\nabla}^2 \vec{E} - \vec{\nabla}(\vec{\nabla} \cdot \vec{E}) = \mu_0 \frac{\partial}{\partial t}\vec{J} + \frac{1}{c^2}\frac{\partial^2}{\partial t^2}\vec{E} - \frac{1}{\lambda_\gamma^2}\frac{\partial}{\partial t}\vec{A} \rightarrow$$

$$\rightarrow \vec{\nabla}^2 \vec{E} - \frac{1}{c^2}\frac{\partial^2}{\partial t^2}\vec{E} = \frac{1}{\varepsilon_0}\vec{\nabla}\rho + \mu_0\frac{\partial}{\partial t}\vec{J} + \frac{1}{\lambda_\gamma^2}\left(-\vec{\nabla}\Phi - \frac{\partial}{\partial t}\vec{A}\right). \quad (1.197)$$

In the absence of charges and currents, the previous identity reduces to

$$\vec{\nabla}^2 \vec{E} - \frac{1}{c^2}\frac{\partial^2}{\partial t^2}\vec{E} = \left(\frac{m_\gamma c}{\hbar}\right)^2 \vec{E} \qquad (1.198)$$

which is the so-called Klein–Gordon equation for a massive field.

Exercise 1.45. Derive the continuity equation from equation (1.196):
Hint:
Use the second Proca equation to find:

$$0 = \mu_0 \vec{\nabla}\cdot\vec{J} + \frac{1}{c^2}\frac{\partial}{\partial t}\vec{\nabla}\cdot\vec{E} - \frac{\vec{\nabla}\cdot\vec{A}}{\lambda_\gamma^2} \to$$

$$\to \mu_0 \vec{\nabla}\cdot\vec{J} + \frac{1}{c^2\varepsilon_0}\frac{\partial}{\partial t}\rho = \frac{1}{\lambda_\gamma^2}\left(\frac{\partial}{\partial t}\Phi + \vec{\nabla}\cdot\vec{A}\right). \qquad (1.199)$$

The last term on the right is $\partial_\mu A^\mu$ and can therefore be written as:

$$\mu_0 \partial_\mu J^\mu = \frac{1}{\lambda_\gamma^2}\partial_\mu A^\mu. \qquad (1.200)$$

Therefore, the conclusion is that to ensure the validity of the continuity equation, it is necessary to choose $\partial_\mu A^\mu = 0$. Such a condition (namely the **Lorentz gauge**) is no longer a free choice but a mandatory step. Accordingly, if $m_\gamma \neq 0$, then the gauge invariance is broken.

A different argument, which implies that Proca's theory is not gauge invariant, is given below. The definitions of the Poynting vector and of electromagnetic energy density depend on the measurability of either the electric or magnetic field and of the potentials. This means that the latter are to be unambiguously determined. This statement is equivalent to assessing the non-gauge invariance of the theory.

Exercise 1.46. Find a physical argument relating electromagnetic gauge invariance, charge, and energy conservation.
Solution:
The answer is not straightforward; we report here a very transparent example given by Wigner in 1949 (see the bibliography). To support the argument, we underscore that the physical content of the gauge principle can be rephrased by saying that **no physical phenomenon depends on the absolute value of the potential.** Suppose that the creation of charge is allowed in some process involving a closed system which is embedded in a Faraday cage. The latter is charged and the system is operated to create the charge via a mechanism requiring a certain amount of energy E. The closed system, hence the charge, is taken out of the cage and is pulled away from it along with a certain potential line with an associated scalar potential Φ, thus

gaining a certain quantity of energy W. The process is then reversed and the charge is annihilated, gaining the same amount of energy employed to create it. The closed system is eventually accommodated inside the cage. In conclusion, the energy balance is:

$$E_T = E_Q + W + E_{-Q} = W \qquad (1.201)$$
$$W = Q\Phi$$

where $E_{\pm Q}$ is the energy required to create/annihilate the charge and $E_Q = -E_{-Q}$. The conclusions are that:
  a) The conservation of energy is violated.
  b) The amount of energy gained depends on the absolute value of the potential.

Accordingly, the assumption that the charge is not conserved implies non-conservation of the energy and invalidates the gauge invariance.

Exercise 1.47. The equations for the propagation of an electromagnetic field in a rectangular waveguide read:

$$\left(\frac{1}{c^2}\frac{\partial^2}{\partial t^2} - \frac{\partial^2}{\partial z^2} + Q^2\right)\psi = 0, \qquad Q = \pi\sqrt{\left(\frac{n}{a}\right)^2 + \left(\frac{l}{b}\right)^2}. \qquad (1.202)$$

Show that they can be associated with the propagation of a massive field.
Hint:
Note that the last term can be written as:

$$Q = \frac{m_\gamma c}{\hbar}. \qquad (1.203)$$

Furthermore, note that the waveguide's photon mass can be written as:

$$m_\gamma = \frac{\hbar \omega_c}{c^2} \qquad (1.204)$$

where $\omega_c = \pi c\sqrt{1/a^2 + 1/b^2}$ is the cutoff frequency.

References and further reading

For classical electrodynamics and Thomson scattering, see the following:

Mourou G 2019 Nobel Lecture: Extreme light physics and application *Rev. Mod. Phys.* **91** 030501

Maxwell J C 2010 *A Treatise on Electricity and Magnetism* **1** (Cambridge: Cambridge University Press) doi: https://dx.doi.org/10.1017/CBO9780511709333

Thomson J J 2009 *Elements of the Mathematical Theory of Electricity and Magnetism* 4th edn (Cambridge: Cambridge University Press) doi: https://dx.doi.org/10.1017/CBO9780511694141

Jackson J D 1999 *Classical Electrodynamics* 3rd edn (New York: Wiley) http://cdsweb.cern.ch/record/490457

Griffiths D J 2017 *Introduction to Electrodynamics* 4th edn (Cambridge: Cambridge University Press) doi: https://dx.doi.org/10.1017/9781108333511

Purcell E M and Morin D J 2013 *Electricity and Magnetism* 3rd edn (Cambridge: Cambridge University Press) doi: https://dx.doi.org/10.1017/CBO9781139012973

Panofsky W K H and Phillips M 1962 *Classical Electricity and Magnetism* **vol XIV** 2nd edn (Reading, MA: Addison-Wesley Publishing Company) p 494

Thomson J J n.d. *'Geometric Proof' of the Larmor Formula* https://blog.cupcakephysics.com/electromagnetism/2014/11/23/thomsons-derivation-of-the-larmor-formula.html

Smoot G F 2001 *Physics 139 Relativity Notes* http://aether.lbl.gov/www/classes/p139/homework/homework.html

Thomas J n.d. *Larmor Power* https://brilliant.org/wiki/larmor-power/

Singal A K 2015 Compatibility of Larmor's formula with radiation reaction for an accelerated charge arXiv:1411.4456

Longair M S 2011 *High Energy Astrophysics* 3rd edn (Cambridge: Cambridge University Press) doi: https://dx.doi.org/10.1017/CBO9780511778346

Dattoli G and Nguyen F 2018 Free electron laser and fundamental physics *Prog. Part. Nucl. Phys.* **99** 1–28 doi: https://dx.doi.org/10.1016/j.ppnp.2018.01.004

For further discussion on equations 1.79, 1.80 see:

Ng E 2011 The Kinematics and Electron Cross sections of Compton Scattering https://web.stanford.edu/~edwin98/compton-scattering-paper.pdf M.I.T. Junior Lab Staff, 'Compton Scattering,'

For synchrotron, undulator, and FEL radiation, see the following:

Krinsky S, Perlman M L and Watson R E 1979 *Characteristics of Synchrotron Radiation and of Its Sources BNL-27678; TRN: 80-009587* Brookhaven National Lab doi: https://dx.doi.org/10.2172/5395804

Colson W 1981 The nonlinear wave equation for higher harmonics in free-electron lasers *IEEE J. Quant. Electron.* **17** 1417–27

Colson W B 1990 Classical free electron laser theory *Laser Handbook, Volume 6: Free Electron Lasers* ed Colson W B and Pellegrini C and Renieri A (Amsterdam: North-Holland) 5

Dattoli G and Renieri A 1985 Experimental and theoretical aspects of the free-electron laser *Laser Handbook* **4** ed Stitch M L and Bass M S (Amsterdam: North-Holland Physics) doi: https://dx.doi.org/10.1016/B978-0-444-86927-2.50005-X

Kim K J 1989 Characteristics of synchrotron radiation *AIP Conf Proc.* **184** 565–632

Dattoli G, Renieri A and Torre A 1993 *Lectures on the Theory of Free Electron Laser and Related Topics* (Singapore: World Scientific) doi: https://dx.doi.org/10.1142/1334

Saldin E, Schneidmiller E V and Yurkov M V 2000 *The Physics of Free Electron Lasers* Advanced Texts in Physics (Berlin, Heidelberg: Springer) (reissued in softcover 2010) doi: https://dx.doi.org/10.1007/978-3-662-04066-9

Ciocci F *et al* 2000 *Insertion Devices for Synchrotron Radiation and Free Electron Laser* vol 6 (Series on Synchrotron Radiation Techniques and Applications.) (Singapore: World Scientific) doi: https://dx.doi.org/10.1142/4066

Errede S 2007 *Physics 435 Lecture Notes, Handouts, Etc.* https://hep.physics.illinois.edu/home/serrede/P435/P435_Lectures.html

Dattoli G *et al* 2017 *Charged Beam Dynamics, Particle Accelerators and Free Electron Lasers* (IOP Plasma Physics Series) (Bristol: IOP Publishing) doi: https://dx.doi.org/10.1088/978-0-7503-1239-4

For Compton and CBS, see the following:

Compton A H 1923 A quantum theory of the scattering of x-rays by light elements *Phys. Rev.* **21** 483–502

Compton A H 1923 The spectrum of scattered X-rays *Phys. Rev.* **22** 409–13

Heitler W 1936 *The Quantum Theory of Radiation* vol 5 1st edn (International Series of Monographs on Physics) (Oxford: Oxford University Press)

Arutyunian F R and Tumanian V A 1963 The Compton effect on relativistic electrons and the possibility of obtaining high energy beams *Phys. Lett.* **4** 176–8

Milburn R H 1963 Electron scattering by an intense polarized photon field *Phys. Rev. Lett.* **10** 75–7

Fiocco G and Thompson E 1963 Thomson scattering of optical radiation from an electron beam *Phys. Rev. Lett.* **10** 89–91

Goldman I I 1964 Intensity effects in compton scattering *Phys. Lett.* **8** 103–6 ISSN: 0031-9163 doi: https://dx.doi.org/10.1016/0031-9163(64)90728-0

Nikishov A I and Ritus V I 1964 Quantum processes in the field of a plane electromagnetic wave and in a constant field 1 *Sov. Phys. JETP* **19** 529–41 http://jetp.ras.ru/cgi-bin/e/index/e/19/2/p529?a=list

Nikishov A I and Ritus V I 1964 Quantum processes in the field of a plane electromagnetic wave and in a constant field 2 *Sov. Phys. JETP* **19** 1191–9 http://jetp.ras.ru/cgi-bin/e/index/e/19/2/p529?a=list

Brown L S and Kibble T W B 1964 Interaction of intense laser beams with electrons *Phys. Rev.* **133** A705–19

Kibble T W B 1965 Frequency shift in high-intensity Compton scattering *Phys. Rev.* **138** B740–53

Kibble T W B 1966 Refraction of electron beams by intense electromagnetic waves *Phys. Rev. Lett.* **16** 1054–6

Kibble T W B 1966 Mutual refraction of electrons and photons *Phys. Rev.* **150** 1060–9

Reiss H R and Eberly J H 1966 Green's function in intense-field electrodynamics *Phys. Rev.* **151** 1058–66

Sarachik E S and Schappert G T 1970 Classical theory of the scattering of intense laser radiation by free electrons *Phys. Rev. D* **1** 2738–53

For an overview of Compton sources and related topics, see the following:

Hajima R 2016 Status and perspectives of Compton sources *Phys. Procedia* **84** 35–9

Krafft G A and Priebe G 2010 Compton sources of electromagnetic radiation *Rev. Accel. Sci. Technol.* **03** 147–63

Weller H R *et al* 2009 Research opportunities at the upgraded HIγS facility *Prog. Part. Nucl. Phys.* **62** 257–303 https://www.sciencedirect.com/science/article/pii/S0146641008000434

Graves W S 2014 Compact x-ray source based on burst-mode inverse Compton scattering at 100 kHz *Phys. Rev. ST Accel. Beams* **17** 120701

For special relativity, see the following:

Morin D 2017 *Special Relativity for the Enthusiastic beginner* CreateSpace. see also http://www.people.fas.harvard.edu/ædjmorin/chap11.pdf

Tsamparlis M 2019 *Special Relativity: An Introduction with 200 Problems and Solutions* (Cham: Springer) doi: https://dx.doi.org/10.1007/978-3-030-27347-7

Artioli M, Babusci D, and Artioli M 2012 Appunti di Relatività Ristretta e di Cinematica Relativistica. INFN-12-18/LNF, https://www.openaccessrepository.it/record/21050

For a further discussion of electromagnetic field gauge invariance, see the previously mentioned lectures by S Errede.

The original articles by Proca can be found in:

Proca G A 1988 *Alexandre Proca. Oeuvre Scientifique Publiée* (Rome, Italy: S.I.A.G.)

For a historical perspective, see the following:

Wigner E P 1949 Invariance in physical theory *Proc. Am. Phil. Soc.* **93** 521–6 http://www.jstor.org/stable/3143140

Poenaru D N and Calboreanu A 2006 Alexandru Proca (1897–1955) and his equation of the massive vector boson field *Europhys. News* **37** 24–6

For more recent contributions on the subject, see the following:

Robles P and Claro F 2012 Can there be massive photons? A pedagogical glance at the origin of mass *Eur. J. Phys.* **33** 1217

Hu B J and Ruan C L 1998 Propagation properties of a plasma waveguide in an external magnetic field *J. Phys. D* **31** 2151

Alshannaq S S and Rojas R G 2008 *XXIX URSI, General Assembly On the Modelling of Metallic Waveguides from DC to Light* (Chicago, IL: Union Radio-Scientifique Internationale) 1–4 https://www.ursi.org/proceedings/procGA08/papers/DBp2.pdf

Hoeneisen B 2006 Trying to understand mass arXiv:hep-th/0609080

Wang Z Y and Xiong C D 2007 Photons inside a waveguide as massive particles arXiv:0708.3519

IOP Publishing

Backscattering Sources, Volume 1
Theoretical framework and Thomson backscattering sources
Alessandro Curcio, Giuseppe Dattoli and Emanuele Di Palma

Chapter 2

Thomson backscattering radiation

2.1 Compton scattering and Thomson scattering

In the previous chapter we have underlined the elemental nature of Thomson scattering, which is the scattering of a photon during the fundamental process of photon–electron collision. The archetypal role of this interaction stems from the nature of the elements involved, which are the constitutive blocks of electromagnetism itself. We have also mentioned that Thomson scattering is the classical limit of the more general photon–electron collision process called Compton scattering. Even though sharp boundaries between one physical domain and another are very often arbitrary and hide the complexity of the related problems, we have underlined that the 'classical' nature of the Thomson process can be captured by saying that it occurs when the recoil acquired by the electron is negligible compared to its momentum before the collision. Conversely, when the effects of momentum recoil are important, the classical description provided by Thomson scattering is not sufficiently precise. Quantum mechanical corrections or full *ab initio* quantum computations are therefore necessary for a satisfactory description of the process. In Compton scattering, momentum recoil is a quantum effect in the sense that it derives from a binary photon–electron collision, where the photon must be treated as a particle instead of a wave. We have carefully noted that in Thomson scattering, the electron interacts with an electromagnetic wave, namely a collection of 'soft' photons, carrying a momentum that is not 'hard' enough to transfer a significant momentum to the charge in the direction of wave propagation. The recoil due the transfer of momentum from the photon (associated with the Poynting vector directed along the propagation axis) is the boundary that marks the transition from the classical regime to the quantum regime. The involvement of momentum transfer is not extraneous to the classical process; indeed, it takes place, but not along the wave propagation axis. In the Compton backscattering (CBS) process, on-axis recoil is of crucial importance, since it can induce phase retardation (the Doppler effect), with dramatic consequences for the observed radiation spectrum.

In this section we want to qualitatively demonstrate the above statements using the elementary kinematics and electrodynamics outlined in the previous chapter. The aim is to reveal the most relevant physics of Thomson scattering and CBS and the intimate connection between the two processes. We have seen in chapter 1 that according to the kinematics of the photon–electron collision, $\phi_1 = \pi$ and $\theta = \pi - \phi_2$, and the backscattered photon energy for small angles of observation ϕ_2 is:

$$\hbar \omega_s = \frac{1 + \beta_0}{1 - \beta_0 \cos\theta + \dfrac{\hbar \omega_0}{\gamma_0 m_0 c^2}(1 + \cos\theta)} \hbar \omega_0 \simeq \frac{4\gamma_0^2 \hbar \omega_0}{1 + \gamma_0^2 \theta^2 + \dfrac{4\gamma_0 \hbar \omega_0}{m_0 c^2}} \qquad (2.1)$$

where we have redefined $\phi_2 = \theta$ to follow the notation used later in this book, in which the polar observation angle is denoted by θ in spherical coordinates. When the initial energy of the photon $\hbar \omega_0$ is much smaller than $m_0 c^2/4\gamma_0$, the collision is essentially elastic for the electron, corresponding to the classical case of relativistic Thomson scattering, where the last term in the denominator of equation (2.1) is negligible. Although already done, we repeat the same proof. We start by defining the electron recoil $\vec{R}$:

$$\vec{R} = \vec{p}^{(1)} - \vec{p}^{(2)} = \hbar \vec{k}^{(2)} - \hbar \vec{k}^{(1)} \simeq \hbar \vec{k}^{(2)} \qquad (2.2)$$

where, in this equation, we have exploited the conservation of momentum and the fact that the scattered photon is significantly upshifted compared to the incoming photon. We can establish a correspondence between the recoil and the Compton term in the denominator of equation (2.1). The recoil can be calculated from the energy of the on-axis backscattered photon:

$$R \simeq \hbar k^{(2)} \simeq 4 \frac{\gamma_0^2 \hbar \omega_0}{c}. \qquad (2.3)$$

Therefore, the Compton term in the denominator of equation (2.1) becomes:

$$\frac{4\gamma_0 \hbar \omega_0}{m_0 c^2} \simeq \frac{R}{p^{(1)}}. \qquad (2.4)$$

Equation (2.4) states that the Compton term is important only when the electron recoil after collision cannot be negligible compared to the initial value of the electron momentum. This is also depicted in figure 2.1, where one of the two interaction diagrams of Compton/Thomson scattering is displayed (this is enough for the arguments of this section, but a more adequate explanation in terms of Feynman diagrams and their associated formalism will be found in the companion volume describing hard Compton effects). The only description that naturally includes the recoil is one that has a quantum nature, since this description implies the particle nature of the photon itself. The electromagnetic interaction energy density can be expressed as $H = j^\mu A_\mu$. This result will be derived later in this book. However, here we want to give a naïve explanation of the diagrams in figure 2.1 based on the

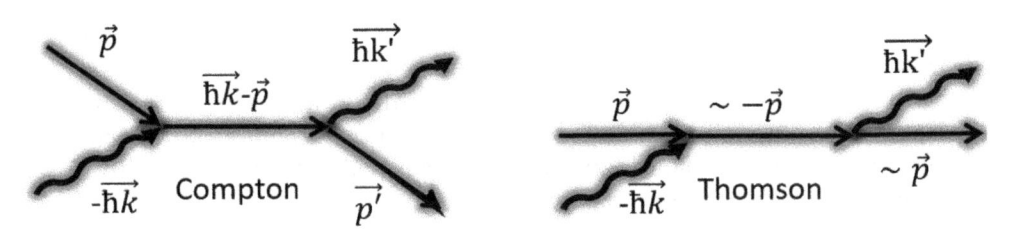

Figure 2.1. Feynman diagrams expressing the difference between Compton scattering (a) and Thomson scattering (b). In case (b), the electron dynamics remain practically unperturbed after the interaction process.

arguments we presented so far, namely those of classical kinematics and electrodynamics. The reader who feels uncomfortable with the arguments used below can safely skip this part without any risk of losing information required to understand the mechanisms relevant to x-ray CBS sources. At the vertex of each diagram one can imagine that an electromagnetic interaction occurs between the electron and the photon fields described by H. Three lines constitute the interaction vertex: two for the electron and one for the photon. The latter is described by the field A^μ with momentum $\hbar \vec{k}$. One of the electron lines, associated with $p^{(e|1)}$, describes the electron before the interaction; the other, associated with $p^{(e|2)}$, describes the electron after collision. The two electron lines are hidden in the field that we have called current density, i.e. j^μ. In fact, in classical electrodynamics, $j^\mu = c\rho_0 u^\mu$, where the charge density ρ_0 and the velocity u^μ are considered constant during the radiation process. The relation between classical and quantum mechanics is given by the statement that $\rho_0 = -e|\psi|^2/V$, where ψ is the electron wave function and V the considered volume of space. In classical electrodynamics, one evidently has $|\psi|^2 = \psi^{*(1)}\psi^{(1)}$, where $\psi^{(1)}$ is the initial wave function before collision and $|\psi|^2$ is the probability of finding the electron at some point in space-time. Through the connection we have created between the classical and quantum descriptions of the current density, it is finally possible to explicitly note the two electron fields $\psi^{*(1)}$ and $\psi^{(1)}$. For the classical case of Thomson scattering, the second line is associated with the field $\psi^{*(1)}$, which carries the same amount of momentum as $\psi^{(1)}$ (but a different sign). From Maxwell's equations, we know that the radiation field is proportional to the current density, i.e. $A^\mu \propto j^\mu$. In the classical case, there is no connection between the radiation field and the electron recoil because the interaction potential H is proportional to $|\psi|^2 = \psi^{*(1)}\psi^{(1)}$, which does not contain any information about the recoil. Instead, in quantum electrodynamics, $\rho_0 \neq const$. The current density is now a matrix element between a final and an initial state of the electron. Therefore, we can state $j^\mu \to j_{21}^\mu \propto \psi^{*(2)}\psi^{(1)}$ (in this naïve derivation, we have not mentioned the spinorial nature of the wave function and the presence of Dirac matrices in the definition of the electromagnetic current). Accordingly, expressing the electron as a plane wave, we write the associated wave function as:

$$\psi^{(1)} \propto e^{-i\hbar k^{\mu(1)}} \to \psi^{*(2)}\psi^{(1)} \propto e^{-i(k^{\mu(1)} - k^{\mu(2)})r_\mu} \equiv e^{-iR^\mu r_\mu}. \tag{2.5}$$

It is evident that the electron recoil naturally occurs in equation (2.5). In quantum electrodynamics the radiation field is $A^\mu \propto j_{21}^\mu$, which is related to the electron recoil, in agreement with the relativistic kinematics of the electron–photon collision process we derived earlier. We will treat all these aspects rigorously and quantitatively later in this work. However, here we have highlighted the fact that only a quantum description can naturally include the electron recoil after a collision with a photon (Compton scattering), while Thomson scattering is realized when the electron recoil is negligible, which is a purely classical condition.

2.2 Electron dynamics under intense wave excitation

In this section we use classical means to study the dynamics of relativistic electrons under the effect of an intense laser wave. The tools we employ in our study are therefore Maxwell's equations and the relativistic version of Newton's equations of motion.

2.2.1 Electron motion in an intense plane wave

The discussion of chapter 1 aimed to assess the qualitative aspects of the free electron–laser interaction. The deeper study we are going to outline reveals a wealth of phenomenology, including high-intensity field effects such as harmonic generation.

Let us consider an electromagnetic wave characterized by its vector potential $\vec{A}$. In the Coulomb gauge chosen here, $\Phi = 0$, and the initial photon field is completely determined by $\vec{A}$. Moreover, we assume that $\vec{A}$ is purely transverse, i.e. the only nonzero components of the vector potential are A_x and A_y. The electron motions for the three dimensions of space are given by:

$$\begin{aligned}
\frac{dp_x}{dt} &= -e\left(-\frac{\partial A_x}{\partial t} - v_z \partial_z A_x\right) \\
\frac{dp_y}{dt} &= -e\left(-\frac{\partial A_y}{\partial t} - v_z \partial_z A_y\right) \\
\frac{dp_z}{dt} &= -ev_x \partial_z A_x + v_y \partial_z A_y \\
m_0 c^2 \frac{d\gamma}{dt} &= e\frac{\partial A_x}{\partial t} v_x + e\frac{\partial A_y}{\partial t} v_y.
\end{aligned} \qquad (2.6)$$

The plane wave impinging on the electron can be described by:

$$\vec{A}(\vec{r}, t) = \vec{A}_0(\vec{r}, t) e^{-i\omega_0 t - i\vec{k}\cdot\vec{r}} \qquad (2.7)$$

where the sign of the photon momentum $\hbar \vec{k}$ is negative, since we consider a head-on collision between the electron and the photon. Given the planar symmetry of the problem, we try to assume that all the physical variables depend solely on the phase coordinate $\zeta = z + ct$. First, we notice that the space and time derivatives of the electromagnetic field are related to each other:

$$\partial_t \vec{A} = c\partial_\zeta \vec{A} = c\partial_z \vec{A}. \qquad (2.8)$$

Furthermore, we set $\vec{k} = \omega_0 \hat{z}/c$. By combining the last two equations of (2.6) via equation (2.8), it is possible to derive a relation between the axial momentum of the electron and its energy:

$$m_0 c^2 (\gamma(\zeta) - \gamma_0) = -c(p_z(\zeta) - p_z(0)). \tag{2.9}$$

Therefore, the axial momentum is determined by:

$$p_z(\zeta) = p_z(0) - m_0 c(\gamma(\zeta) - \gamma_0). \tag{2.10}$$

The transverse momenta are easily found if we recognize that $d_t = \partial_t + v_z \partial_z$, thus:

$$p_{x,y}(t) = p_{x,y}(0) + eA_{x,y}(t). \tag{2.11}$$

At this point it is worth giving a more explicit expression for the axial momentum as a function of the electromagnetic field, as we have done for the transverse momenta. We start from the definition of the Lorentz factor in terms of the components of the momentum vector:

$$m_0^2 c^4 \gamma^2(\zeta) = m_0^2 c^4 + p_x^2 c^2 + p_y^2 c^2 + p_z^2 c^2. \tag{2.12}$$

We then express the electron energy by means of equation (2.9):

$$m_0^2 c^4 \gamma^2(\zeta) = m_0^2 c^4 \gamma_0^2 - 2m_0 c^3 \gamma_0 p_z(\zeta) - p_z(0) + p_z^2(\zeta)c^2 - 2p_z(\zeta)p_z(0)c^2 + p_z^2(0)c^2. \tag{2.13}$$

Finally, we combine equations (2.9) and (2.12) to obtain the desired expression for the axial momentum component:

$$p_z(\zeta) = p_z(0) - \frac{e\vec{p}_{\perp 0} \cdot \vec{A}}{(\gamma_0 m_0 c + p_z(0))} - \frac{e^2 |A|^2}{2(\gamma_0 m_0 c + p_z(0))}. \tag{2.14}$$

In deriving the electron motion in terms of coordinates, we will find the following definition useful:

$$\vec{v}(\zeta) = \frac{\vec{p}(\zeta)}{\gamma(\zeta) m_0} = \frac{\vec{p}(\zeta)}{\gamma_0 m_0 + \frac{p_z(0)}{c} - \frac{p_z(\zeta)}{c}}. \tag{2.15}$$

Furthermore, it is also found that:

$$1 + \frac{v_z}{c} = \frac{\gamma_0 m_0 c + p_z(0) - p_z}{\gamma_0 m_0 c + p_z(0) - p_z} + \frac{p_z}{\gamma_0 m_0 c + p_z(0) - p_z} = \frac{\gamma_0 m_0 c + p_z(0)}{\gamma_0 m_0 c + p_z(0) - p_z(\zeta)}. \tag{2.16}$$

By combining equations (2.15) and (2.16), it is possible to find the following interesting identity:

$$\frac{\vec{v}}{1 + \frac{v_z}{c}} = \frac{\vec{p}c}{\gamma_0 m_0 c + p_z(0) - p_z} \frac{\gamma_0 m_0 c + p_z(0) - p_z}{\gamma_0 m_0 c + p_z(0)} = \frac{\vec{p}c}{\gamma_0 m_0 c + p_z(0)}. \tag{2.17}$$

Equation (2.17) turns to be extremely useful for the purpose of finding a closed expression for the electron coordinates. We take into account that, on account of the assumptions mentioned above, the time derivative reads:

$$\frac{d}{dt} = \frac{d\zeta}{dt}\frac{d}{d\zeta} = c\left(1 + \frac{v_z}{c}\right)\frac{d}{d\zeta}. \quad (2.18)$$

Therefore, equation (2.17) can be explicitly written as the derivative of the electron space vector with respect to the phase coordinate ζ:

$$\frac{\vec{v}}{1 + \frac{v_z}{c}} = \frac{1}{1 + \frac{v_z}{c}}\frac{d\vec{r}}{dt} = \frac{1}{1 + \frac{v_z}{c}}c\left(1 + \frac{v_z}{c}\right)\frac{d\vec{r}}{d\zeta} = c\frac{d\vec{r}}{d\zeta} \quad (2.19)$$

In conclusion, combining equations (2.17) and (2.19) allows us to exactly solve the electron motion as follows:

$$x(\zeta) = x_0 + \frac{p_x(0)}{\gamma_0 m_0 c + p_z(0)}\zeta + \frac{e}{\gamma_0 m_0 c + p_z(0)}\int_0^\zeta d\zeta' A_x(\zeta')$$

$$y(\zeta) = y_0 + \frac{p_y(0)}{\gamma_0 m_0 c + p_z(0)}\zeta + \frac{e}{\gamma_0 m_0 c + p_z(0)}\int_0^\zeta d\zeta' A_y(\zeta') \quad (2.20)$$

$$z(\zeta) = z_0 + \frac{p_z(0)}{\gamma_0 m_0 c + p_z(0)}\zeta - \frac{e\vec{p}_{\perp 0} \cdot \int_0^\zeta d\zeta' \vec{A}(\zeta')}{(\gamma_0 m_0 c + p_z(0))^2} - \frac{e^2 \int_0^\zeta d\zeta' |A(\zeta')|^2}{2(\gamma_0 m_0 c + p_z(0))^2}.$$

The last equations account for the dynamics of an electron wiggling under the action of an electromagnetic field. In the following we study the limits of very small and very large field amplitudes, in particular to study the energy exchange between the particle and the wave.

2.2.2 The limit of very small field amplitude

For the sake of simplicity, without much loss of generality for the purposes of this section, we assume that the electron's initial transverse momentum is zero, i.e. $\vec{p}_\perp = 0$. For $eA \ll \gamma_0 mc$, the coordinate z does not depend upon the external field, and a uniform motion is recovered:

$$z \simeq z_0 + \frac{p_z(0)}{\gamma_0 m_0 c + p_z(0)}(z + ct) \rightarrow z_0 + \frac{\gamma_0 m_0 c}{\gamma_0 m_0 c + p_z(0)}z = \frac{p_z(0)}{\gamma_0 m_0 c + p_z(0)}ct \rightarrow z(t) = z_0 + v_z(0)t \quad (2.21)$$

where $v_z(0) = p_z(0)/\gamma_0 m_0$. Thus, the transverse motion becomes:

$$x(t) = \frac{ec}{\gamma_0 m_0 c + p_z(0)}\int_0^t dt' A_x(t')$$

$$y(t) = \frac{ec}{\gamma_0 m_0 c + p_z(0)}\int_0^t dt' A_y(t'). \quad (2.22)$$

Moreover, for a nonrelativistic electron (or an electron starting at rest), we recover the classical solution:

$$x(t) = \frac{e}{m_0} \int_0^t dt' A_x(t')$$
$$y(t) = \frac{e}{m_0} \int_0^t dt' A_y(t'). \qquad (2.23)$$

We have thus demonstrated that, for low field amplitude, the relativistic solution of the equation of motion for a charge under the action of an electromagnetic wave reduces to a form which is similar to the nonrelativistic solution. For the particle energy, we use equation (2.9) to obtain:

$$E = \gamma m_0 c^2 = E_0 - c(p_z(\zeta) - p_z(0)) \simeq E_0 = \gamma_0 m_0 c^2 \qquad (2.24)$$

where we have used $eA \ll \gamma_0 mc$, i.e. $p_z(t) \simeq p_z(0)$ via equation (2.14). Therefore, for very small field amplitudes, there is no important energy exchange between the particle and the wave (but the energy exchange is still nonzero).

2.2.3 The limit of very large field amplitude: direct laser acceleration

For any physical problem involving high field strengths, it is very useful to define the so-called *relativistic parameter*:

$$a_0 = \frac{eA_0}{m_0 c} \qquad (2.25)$$

which was described in chapter 1 as being associated with the intensity-dependent electron mass-shift term. The name *relativistic parameter* is due to the fact the parameter is only large enough to determine relativistic electron dynamics and relativistic features of the scattered radiation when the momentum transfer eA_0 from the field to the electron is of the order of $m_0 c$. The so-called *relativistic optics* is all based on the condition $a_0 \gg 1$, where the fields are large enough to excite particles of the optical media to relativistic energies. For $a_0 \gg 1$, it is necessary to consider equations (2.20) without approximation to study the electron motion. For the particle energy, similarly to the case of small field amplitudes, we use equation (2.9) to obtain:

$$E = E_0 + \frac{e^2 |A|^2 c}{2(\gamma_0 m_0 c + p_z(0))}. \qquad (2.26)$$

Equation (2.26) indicates that the electron increments its kinetic energy along the propagation axis of the electromagnetic wave, and this net energy exchange is truly important when the field imposes relativistic oscillating energy on the charge. In the case of Thomson backscattering between ultrarelativistic electrons and and an infrared laser beam, $\gamma_0 m_0 c + p_z(0) \simeq 2\gamma_0 m_0 c$, thus:

$$E \simeq \gamma_0 m_0 c^2 + \frac{a_0^2 m_0 c^2}{2\gamma_0}. \qquad (2.27)$$

The second term on the right-hand side of equation (2.27) corresponds to an increase in electron kinetic energy due to a so-called 'ponderomotive' action of the intense wave. Comparing the second term with the first shows that such ponderomotive energy is significant for $a_0 \simeq \gamma$. For an infrared laser based on an active medium such as Ti:Sa, the laser frequency is $\omega_0 \simeq 2.35 \times 10^{15}$ Hz. For a Ti:Sa laser pulse that has an energy of 6 J, a duration of 30 fs, and a beam waist of 10 μm, the associated vector potential is $A_0 \simeq 0.01$ Vsm^{-1}. The corresponding relativistic parameter is approximately $a_0 \simeq 5$. For an an electron with an energy of 100 MeV, the second term of equation (2.27) is equivalent to $\simeq 0.07$ MeV, which is much smaller than $E_0 = 100$ MeV. In order to get $a_0 \simeq \gamma$, one would need a laser pulse energy larger than 10 kJ, which is not available for laser systems with such a short duration!

2.3 Retarded potentials

In this section we obtain a generalized expression for the electromagnetic field carried by the electron. By the word 'generalized,' we mean an expression that goes well beyond the Coulomb field and that is also valid when the electron is moving under external forces, as in the case of Thomson scattering. The wave equation for the vector potential A^μ associated with any current density j^μ was found in the previous chapter:

$$\partial_\nu \partial^\nu A^\mu = -\mu_0 j^\mu. \qquad (2.28)$$

Here, j^μ represents the moving electron and μ_0 is the magnetic permeability of vacuum. In order to solve equation (2.28), we use a Green's function. We first take the space-time Fourier transform of equation (2.28):

$$\left(-\frac{\omega^2}{c^2} + k^2\right)\widetilde{A}^\mu = -\mu_0 \widetilde{j}^\mu. \qquad (2.29)$$

Then, according to the solution method, we need to solve the following integral to find the *photon propagator* in the space-time domain, defined as:

$$\frac{1}{(2\pi)^4} \int d^4k \frac{e^{-ik^\mu r_\mu}}{-\frac{\omega^2}{c^2} + k^2}. \qquad (2.30)$$

The solution of the wave equation (2.28) is simply the convolution of the propagator at equation (2.30) with the electron current density j^μ. However, before we obtain the complete expression for the photon propagator, we work out the spatial part of the integral to interpret A^μ in terms of the so-called *retarded potential*. Indeed, we obtain:

$$\frac{1}{(2\pi)^3} \int d^3k \frac{e^{i\vec{k}\cdot\vec{r}}}{-\frac{\omega^2}{c^2} + k^2} = -\frac{1}{(2\pi)^2 r} \int_{-\infty}^{\infty} \frac{dk\,k}{k^2 - \frac{\omega^2}{c^2}} \sin(kr) = -\frac{e^{i\frac{\omega r}{c}}}{4\pi r}. \qquad (2.31)$$

The time Fourier transform of the electromagnetic field of the electron is thus found to be:

$$\tilde{A}^\mu(\vec{r}, \omega) = \mu_0 \int \frac{e^{i\frac{\omega}{c}|\vec{r}-\vec{r}'|}}{4\pi|\vec{r}-\vec{r}'|} \tilde{j}^\mu(\vec{r}', \omega) d^3r'. \tag{2.32}$$

Equation (2.32) is the Fourier transform of the *retarded potential*. In fact, the same potential expressed in the time domain becomes:

$$A^\mu(\vec{r}, t) = \frac{\mu_0}{2\pi} \int d^3r' \int d\omega \frac{e^{i\frac{\omega}{c}|\vec{r}-\vec{r}'|-i\omega t}}{4\pi|\vec{r}-\vec{r}'|} \tilde{j}^\mu(\vec{r}', \omega) = \frac{\mu_0}{4\pi} \int d^3r' \frac{j^\mu\left(\vec{r}', t - \frac{|\vec{r}-\vec{r}'|}{c}\right)}{|\vec{r}-\vec{r}'|} \tag{2.33}$$

where, in the last part of equation (2.33), we have used the definition of the inverse time Fourier transform. The reason for the name 'retarded potential' is clear from equation (2.33), since the field A^μ and the associated current density j^μ are retarded by each other, i.e. the field at the space-time point $(\vec{r}, t)$ is determined by the current at the space-time point $(\vec{r}', t')$, where $t' = t - |\vec{r} - \vec{r}'|/c$.

2.3.1 Liénard–Wiechert potentials

Equation (2.33) is actually more general than a simple expression of the electromagnetic field associated with a moving electron. It yields the general expression for an arbitrary current density. If we solve the integral for the case of a moving electron, the previous result is recognized as the Liénard–Wiechert potential, after the scientists who first obtained it. For an electron moving along a trajectory $\vec{r}_e(t)$, the associated current density is:

$$j^\mu(\vec{r}, t) = -e\left(c\delta(\vec{r} - \vec{r}_e(t)), -\frac{d\vec{r}_e}{dt}\delta(\vec{r} - \vec{r}_e(t))\right). \tag{2.34}$$

The combination of equation (2.34) and equation (2.33) yields:

$$A^\mu(\vec{r}, t) = -\frac{e\mu_0}{4\pi} \int d^3r' \frac{\left(c\delta(\vec{r}' - \vec{r}_e(t')), -\frac{d\vec{r}_e}{dt'}\delta(\vec{r}' - \vec{r}_e(t'))\right)}{|\vec{r} - \vec{r}'|}. \tag{2.35}$$

Equation (2.35) can be recast into:

$$A^\mu(\vec{r}, t) = -\frac{e\mu_0}{4\pi} \int\int d^3r' d\tau \frac{\left(c\delta(\vec{r}' - \vec{r}_e(\tau)), -\frac{d\vec{r}_e}{d\tau}\delta(\vec{r}' - \vec{r}_e(\tau))\right)}{|\vec{r} - \vec{r}'|} \delta(\tau - t'(\tau)) \tag{2.36}$$

which is solved by:

$$A^\mu(\vec{r}, t) = -\frac{e\mu_0}{4\pi} \int d\tau \frac{\left(c, -\frac{d\vec{r}_e}{d\tau}\right)}{|\vec{r} - \vec{r}_e(\tau)|} \delta\left(\tau - t + \frac{|\vec{r} - \vec{r}_e(\tau)|}{c}\right). \tag{2.37}$$

The Dirac delta of the function of time is expanded as:

$$\delta(\tau - t'(\tau)) = \frac{\delta(\tau - t')}{1 + \frac{1}{c}\frac{d}{dt'}\left(\sqrt{(\vec{r} - \vec{r}_e(t')) \cdot (\vec{r} - \vec{r}_e(t'))}\right)} = \frac{\delta(\tau - t')}{1 - \frac{1}{c}\frac{\vec{r} - \vec{r}_e}{|\vec{r} - \vec{r}_e|} \cdot \frac{d\vec{r}_e}{dt'}}. \quad (2.38)$$

Therefore, plugging equation (2.38) into equation (2.37), we obtain:

$$A^\mu(\vec{r}, t) = -\frac{e\mu_0}{4\pi} \frac{\left(c, -\frac{d\vec{r}_e}{dt'}\right)}{|\vec{r} - \vec{r}_e(t')| - \frac{1}{c}(\vec{r} - \vec{r}_e(t')) \cdot \frac{d\vec{r}_e}{dt'}}. \quad (2.39)$$

If we define $R^\mu(t') = (|\vec{r} - \vec{r}_e(t')|, -\vec{r} + \vec{r}_e(t'))$ and $v_e^\mu(t') = (c, -d\vec{r}_e/dt')$, a covariant form of the Liénard–Wiechert potentials of the electron can be found:

$$A^\mu(\vec{r}, t) = -\frac{e\mu_0 c}{4\pi} \frac{v_e^\mu(t')}{c|\vec{r} - \vec{r}_e(t')| - (\vec{r} - \vec{r}_e(t')) \cdot \frac{d\vec{r}_e}{dt'}} = -\frac{e\mu_0 c}{4\pi} \frac{v_e^\mu}{R_\nu v_e^\nu}. \quad (2.40)$$

In the forthcoming sections we derive the details of the radiation emitted by an electron interacting with a counter-propagating wave.

2.4 Thomson backscattering radiation

Far from the electron charge, the field can no longer be a coulomb field, which decays rapidly at large distances, but must be radiation. At a large distance R the time Fourier transform of the Liénard–Wiechert potential is:

$$\tilde{A}^\mu(\vec{r}, \omega) = \mu_0 \int \frac{e^{i\frac{\omega}{c}|\vec{r} - \vec{r}'|}}{4\pi|\vec{r} - \vec{r}'|} \tilde{j}^\mu(\vec{r}', \omega) d^3 r' \simeq \frac{\mu_0 e^{i\frac{\omega R}{c}}}{4\pi R} \int e^{-i\vec{k}\cdot\vec{r}} j^\mu(\vec{r}', \omega) d^3 r'. \quad (2.41)$$

The magnetic part of the radiation field is found from its definition in terms of the potential:

$$\vec{H}(\vec{R}, \omega) = \frac{1}{\mu_0} \vec{\nabla} \times \vec{A}(\vec{R}, \omega) \simeq \frac{i e^{i\frac{\omega R}{c}}}{4\pi R} \vec{k} \times \int e^{-i\vec{k}\cdot\vec{r}} \vec{j}(\vec{r}', \omega) d^3 r' \quad (2.42)$$

where we have used $|\vec{R} - \vec{r}'| \simeq R - \vec{R} \cdot \vec{r}'/R$ and defined $\vec{k} = \omega \vec{R}/cR$. According to the Poynting vector theory, the distribution of the radiated photons is given by:

$$\frac{d^2 E}{d\omega d\Omega} = \frac{\mu_0 c}{\pi} |H(\vec{R}, \omega)|^2 R^2. \quad (2.43)$$

The expression for the space-time Fourier transform of the electron current density is easily found by starting from the domain $(\vec{k}, t)$. The spatial Fourier transform of the current associated with an electron moving along a trajectory $\vec{r}_e$ is, according to equation (2.34):

$$\tilde{\vec{j}}(\vec{k}, t) = \int d^3 r \vec{j}(\vec{r}, t) e^{-i\vec{k}\cdot\vec{r}} = -e\frac{d\vec{r}_e}{dt} \int d^3 r e^{-i\vec{k}\cdot\vec{r}} \delta(\vec{r} - \vec{r}_e) = -e\frac{d\vec{r}_e}{dt} e^{-i\vec{k}\cdot\vec{r}_e}. \quad (2.44)$$

Thus, one obtains:

$$\widetilde{\vec{j}}(\vec{k}, \omega) = \int dt \widetilde{\vec{j}}(\vec{k}, t)e^{i\omega t} = -e \int dt \frac{d\vec{r}_e}{dt} e^{i\omega t - i\vec{k}\cdot\vec{r}_e}. \quad (2.45)$$

By defining the observation direction $\hat{n} = \vec{R}/R = (n_x, n_y, n_z) = (\cos\phi \sin\theta, \sin\phi \sin\theta, \cos\theta)$ and combining equations (2.43) and (2.45), the spectral–angular distribution can be recast into:

$$\frac{d^2E}{d\omega d\Omega} = \frac{\mu_0 \omega^2}{16\pi^3 c} |\hat{n} \times \widetilde{\vec{j}}(\vec{k}, \omega)|^2 = \frac{\mu_0 e^2 \omega^2}{16\pi^3 c} \left| \hat{n} \times \int_{-\infty}^{t} dt' \frac{d\vec{r}_e}{dt'} e^{i\omega t' - i\vec{k}\cdot\vec{r}_e(t')} \right|^2. \quad (2.46)$$

This last equation is sufficient to derive the details of the Thomson radiation's spectral–angular distribution, provided that the trajectory of the electron is the one previously found, namely equation (2.20). The first step we take to perform the calculation is to give a workable expression for the radiation phase in equation (2.46). We make a change of variable in the time integral, replacing t with ζ, observing that the electron trajectory shows a planar symmetry. Let us also define the constant $\eta = \gamma_0 m_0 c + p_z(0)$ (see equation (2.9)). Thus, we obtain:

$$\omega t - \vec{k}\cdot\vec{r}_e = \omega \frac{\zeta - z(\zeta)}{c} - \frac{\omega}{c}\cos\phi\sin\theta x(\zeta) - \frac{\omega}{c}\sin\phi\sin\theta y(\zeta) - \frac{\omega}{c}\cos\theta z(\zeta). \quad (2.47)$$

Regarding the dependence on the phase, we recognize: a) terms independent of ζ, i.e. constant; b) terms linear in ζ; c) terms proportional to the electromagnetic field; d) terms that are proportional to the electromagnetic intensity. In the following, we group these contributions to obtain a smooth approach to the calculation details:

$$\psi_0 = -\frac{\omega}{c}(x_0 \cos\phi \sin\theta + y_0 \sin\phi \sin\theta + z_0(1 + \cos\theta)) \quad (2.48)$$

$$\psi_1 = \frac{\omega}{c}\left(1 - \frac{p_x(0)\cos\phi\sin\theta + p_y(0)\sin\phi\sin\theta + p_z(0)(1+\cos\theta)}{\eta}\right)\zeta \quad (2.49)$$

$$\psi_2 = \frac{\omega}{c}\frac{e}{\eta}\left(-\cos\phi\sin\theta\int_{-\infty}^{\zeta}d\zeta' A_x(\zeta') - \sin\phi\sin\theta\int_{-\infty}^{\zeta}d\zeta' A_y(\zeta') + (1+\cos\theta)\frac{\vec{p}_{\perp 0}}{\eta}\cdot\int_{-\infty}^{\zeta}d\zeta' \vec{A}(\zeta')\right) \quad (2.50)$$

$$\psi_3 = \frac{\omega}{c}(1 + \cos\theta)\frac{e^2}{2\eta^2}\int_{-\infty}^{\zeta}d\zeta' |A(\zeta')|^2. \quad (2.51)$$

The spectral–angular distribution in equation (2.46) can be rewritten as:

$$\frac{d^2E}{d\omega d\Omega} = \frac{\mu_0 e^2 \omega^2}{16\pi^3 c}\left|\hat{n} \times \int_{-\infty}^{\zeta}d\zeta \frac{d\vec{r}_e}{d\zeta}e^{i(\psi_0 + \psi_1 + \psi_2 + \psi_3)}\right|^2. \quad (2.52)$$

The phase term containing ψ_0 does not depend on ζ and can therefore be eliminated from the radiation formula after the application of the squared modulus. Indeed, such a phase term would be important when considering the radiation produced by an electron bunch that is confined in a length comparable to the radiation wavelength. We do not study this case here; therefore, we focus on incoherent radiation. Moreover, by construction, equation (2.52) refers to a single electron, while in the study of coherent emission, one should specify the equation for a current of many electrons all radiating together. In order to proceed, we now have to express the vector potential of the plane wave. For pedagogical reasons, we consider the simplest of the possible forms, i.e. a pulse that is flat from 0 to ζ and null outside this range; thus, we write the potential as:

$$A^\mu(\zeta) = \left(0, A_{x0} \cos\left(\frac{\omega_0 \zeta}{c} + \varphi_x\right), A_{y0} \cos\left(\frac{\omega_0 \zeta}{c} + \varphi_y\right), 0\right) \quad (2.53)$$

where the phases $\varphi_{x,y}$ and the amplitudes A_{x0} and A_{y0} determine the polarization state of the wave. In fact, when the phase difference $\phi_x - \phi_y = 0$, the wave is linearly polarized, while when the phase difference is $\phi_x - \phi_y = \pm\pi/2$, the wave is circularly polarized. In all other cases, the wave is said to be elliptically polarized. Figure 2.2 shows electron trajectories calculated via equations (2.20) for linear (a), circular (b), and elliptic (c) polarizations of the exciting wave. The integral of the field can now be easily worked out:

$$\int_0^\zeta d\zeta' A_{x,y}(\zeta') = \frac{A_{\{x,y\}0}c}{\omega_0}\left[\sin\left(\frac{\omega_0 \zeta}{c} + \varphi_{x,y}\right) - \sin\varphi_{x,y}\right]. \quad (2.54)$$

Similarly, we express the integral of the squared modulus of the field as follows:

$$\int_0^\zeta d\zeta'|A(\zeta')|^2 = \int_0^\zeta d\zeta'\left[A_{x0}^2 \cos^2\left(\frac{\omega_0 \zeta'}{c} + \varphi_x\right) + A_{y0}^2 \cos^2\left(\frac{\omega_0 \zeta'}{c} + \varphi_y\right)\right]$$

$$= \frac{A_{x0}^2 c}{2\omega_0}\left[\frac{\omega_0 \zeta}{c} + \frac{1}{2}\sin\left(2\frac{\omega_0 \zeta}{c} + 2\varphi_x\right) - \frac{1}{2}\sin 2\varphi_x\right] + \quad (2.55)$$

$$+ \frac{A_{y0}^2 c}{2\omega_0}\left[\frac{\omega_0 \zeta}{c} + \frac{1}{2}\sin\left(2\frac{\omega_0 \zeta}{c} + 2\varphi_y\right) - \frac{1}{2}\sin 2\varphi_y\right].$$

We define new phase terms that will be useful in the following calculations:

$$\Psi_0 = -\frac{\omega}{c}(x_0 \cos\phi \sin\theta + y_0 \sin\phi \sin\theta + z_0(1 + \cos\theta)) \quad (2.56)$$

$$\Psi_1 = \frac{\omega}{c}\left(1 - \frac{p_x(0)\cos\phi\sin\theta + p_y(0)\sin\phi\sin\theta + p_z(0)(1+\cos\theta)}{\eta} + (1+\cos\theta)\frac{e^2 A_0^2}{4\eta^2}\right) \quad (2.57)$$

$$\Psi_2 = \Psi_{2x}\left[\sin\left(\frac{\omega_0 \zeta}{c} + \varphi_x\right) - \sin(\varphi_x)\right] + \Psi_{2y}\left[\sin\left(\frac{\omega_0 \zeta}{c} + \varphi_y\right) - \sin(\varphi_y)\right] \quad (2.58)$$

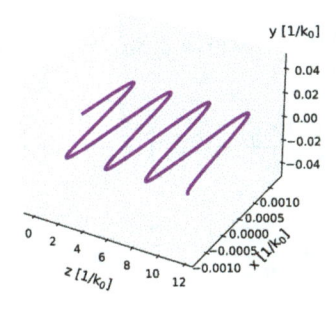

(a) Used parameters: $a_0 = 0.5$, $\gamma_0 = 200$, $\varphi_x = \varphi_y = 0$, $x_0 = y_0 = z_0 = 0$, $p_x(0) = p_y(0) = 0$, $A_{y0} = 0$, $\omega_0 = 2.35 \times 10^{15}$ Hz. Linearly polarized field.

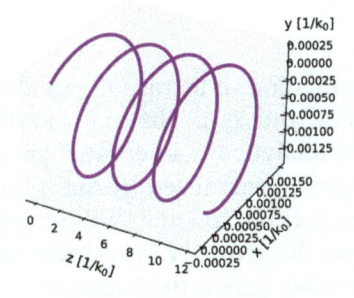

(b) Used parameters: $a_0 = 0.5$, $\gamma_0 = 200$, $\varphi_x = \pi/4 = -\varphi_y = 0$, $x_0 = y_0 = z_0 = 0$, $p_x(0) = p_y(0) = 0$, $A_{x0} = A_{y0}$, $\omega_0 = 2.35 \times 10^{15}$ Hz. Circularly polarized field.

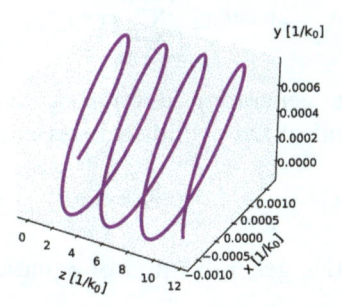

(c) Used parameters: $a_0 = 0.5$, $\gamma_0 = 200$, $\varphi_x = 0$, $\varphi_y = -\pi/4$, $x_0 = y_0 = z_0 = 0$, $p_x(0) = p_y(0) = 0$, $A_{y0} = \tan(\pi/8) A_{y0}$, $\omega_0 = 2.35 \times 10^{15}$ Hz. Elliptically polarized field.

Figure 2.2. Electron trajectories.

$$\Psi_{2x} = \frac{e\omega}{\eta\omega_0}\left(\cos\phi\sin\theta - \frac{p_{x0}}{\eta}(1+\cos\theta)\right)A_{x0}; \quad \Psi_{2y} = \frac{e\omega}{\eta\omega_0}\left(\sin\phi\sin\theta - \frac{p_{y0}}{\eta}(1+\cos\theta)\right)A_{y0} \quad (2.59)$$

$$\Psi_3 = \Psi_{3x}\sin\left(2\frac{\omega_0\zeta}{c} + 2\varphi_x\right) + \Psi_{3y}\sin\left(2\frac{\omega_0\zeta}{c} + 2\varphi_y\right) \quad (2.60)$$

$$\Psi_{3x} = \frac{\omega}{\omega_0}(1+\cos\theta)\frac{e^2 A_{x0}^2}{8\eta^2} \quad ; \quad \Psi_{3y} = \frac{\omega}{\omega_0}(1+\cos\theta)\frac{e^2 A_{y0}^2}{8\eta^2} \quad (2.61)$$

where $A_0^2 = A_{x0}^2 + A_{y0}^2$. The spectral–angular distribution of the Thomson-back-scattered photons is then written as:

$$\frac{d^2 E}{d\omega d\Omega} = \frac{\mu_0 e^2 \omega^2}{16\pi^3 c}$$

$$\left|\hat{n} \times \int_0^\zeta d\zeta' \frac{d\vec{r}_e}{d\zeta'} e^{i\left[\Psi_1 \zeta' - \Psi_{2x}\sin\left(\frac{\omega_0\zeta'}{c}+\varphi_x\right) - \Psi_{2y}\sin\left(\frac{\omega_0\zeta'}{c}+\varphi_y\right) + \Psi_{3x}\sin\left(2\frac{\omega_0\zeta'}{c}+2\varphi_x\right) + \Psi_{3y}\sin\left(2\frac{\omega_0\zeta'}{c}+2\varphi_y\right)\right]}\right|^2 \quad (2.62)$$

The exponent in the radiation integral of equation (2.62) contains oscillating functions at frequencies ω_0 and $2\omega_0$. The mathematical manipulations used to get an expression that can be analytically integrated are greatly simplified by the use of Bessel functions and their generalized forms (for further comments, see the Comments and exercises section and the bibliography at the end of the chapter). To this aim, we recall that the ordinary cylindrical Bessel functions of integer order are specified by the following 'generating function':

$$e^{\frac{x}{2}(t-\frac{1}{t})} = \sum_{n=-\infty}^{n=+\infty} t^n J_n(x) \quad \text{with} \quad J_n(x) = \left(\frac{x}{2}\right)^n \sum_{r=0}^{r=+\infty}(-1)^r \frac{\left(\frac{x}{2}\right)^r}{r!(r+n)!}. \quad (2.63)$$

If we set $t = e^{i\theta}$ in equation (2.63), we find:

$$e^{ix\sin\theta} = \sum_{n=-\infty}^{n=+\infty} e^{in\theta} J_n(x) \quad (2.64)$$

known as a Jacobi–Anger generating function, which can also be exploited to derive the integral representation of the cylindrical Bessel of the first kind:

$$J_n(x) = \frac{1}{2\pi}\int_{-\pi}^{\pi} \cos x\sin(\theta) - n\theta \, d\theta. \quad (2.65)$$

Therefore, the two-variable generalization of equation (2.63) is:

$$e^{\frac{x}{2}(t-\frac{1}{t}) + \frac{y}{2}(t^2 - \frac{1}{t^2})} = \sum_{n=-\infty}^{n=+\infty} t^n J_n(x, y). \quad (2.66)$$

The associated Jacobi–Anger integral representation is written as:

$$J_n(x, y) = \frac{1}{\pi}\int_{-\pi}^{\pi} \cos x\sin(\theta) + y\sin(2\theta) - n\theta \, d\theta. \quad (2.67)$$

An alternative representation in terms of series is also provided by:

$$J_n(x, y) = \sum_{l=-\infty}^{l=+\infty} J_{n-2l}(x)J_l(y) \rightarrow J_n(x, -y) = \sum_{l=-\infty}^{l=+\infty} J_{n+2l}(x)J_l(y). \quad (2.68)$$

However, it is worth noting that it is convenient to arrange the exponential in equation (2.62) as shown below:

$$\mathscr{A} = e^{i[\Psi_1\zeta' - \Psi_{2x}\sin\Omega_x) - \Psi_{2y}\sin(\Omega_y) + \Psi_{3x}\sin(2\Omega_x) + \Psi_{3y}\sin(2\Omega_y)]}$$
$$\Omega_{x,y} = \frac{\omega_0 \zeta'}{c} + \varphi_{x,y}. \quad (2.69)$$

The expansion in terms of a two-variable Bessel yields:

$$\mathscr{A} = e^{i\Psi_1\zeta'} \sum_{n=-\infty}^{n=+\infty} J_n(\Psi_{2x}, -\Psi_{3x})e^{-in\Omega_x} \sum_{l=-\infty}^{l=+\infty} e^{-il\Omega_y} J_l(\Psi_{2y}, -\Psi_{3y}). \quad (2.70)$$

If we set $n + l = h$, rescale the indices, and adjust the series in equation (2.70) as follows:

$$\mathscr{A} = e^{i\Psi_1\zeta'} \sum_{h=-\infty}^{h=+\infty} e^{-ih\Omega_x} \sum_{l=-\infty}^{l=+\infty} J_{h-l}(\Psi_{2x}, -\Psi_{3x}) J_l(\Psi_{2y}, -\Psi_{3y}) e^{il(\phi_x - \phi_y)} \quad (2.71)$$

the function in equation (2.71) can be viewed as a four-variable/one-parameter Bessel written as:

$$\sum_{l=-\infty}^{l=+\infty} J_{h-l}(\Psi_{2x}, -\Psi_{3x}) J_l(\Psi_{2y}, -\Psi_{3y}) e^{il(\phi_x - \phi_y)} \equiv J_h\left(\Psi_{2x}, -\Psi_{3x}; \Psi_{2y}, -\Psi_{3y}|e^{il(\phi_x - \phi_y)}\right). \quad (2.72)$$

A brief summary of the mathematical properties of multivariable Bessel is presented in the comments and exercises (section 2.6) at the end of this chapter, where the spectral–angular distribution 2.62 is written in terms of generalized cylindrical functions. According to the previous manipulation, equation (2.62) can be written as:

$$\frac{d^2E}{d\omega d\Omega} = \frac{\mu_0 e^2 \omega^2}{16\pi^3 c} \left| \hat{n} \times \sum_{h=-\infty}^{h=+\infty} \int_0^\zeta d\zeta' \frac{d\vec{r}_e}{d\zeta'} e^{i(\Psi_1 - h\frac{\omega_0}{c})\zeta'} e^{-ih\phi_x} J_h\left(\Psi_{2x}, -\Psi_{3x}; \Psi_{2y}, -\Psi_{3y}|e^{il(\phi_x - \phi_y)}\right) \right|^2 \quad (2.73)$$

where the index h represents the order of the radiated harmonic. It is worth noting that the integration over the coordinate ζ' is factored out; this is the noticeable simplification introduced by the generalized Jacobi–Anger expansion. It is now important to underline that the structure of the integral of radiation is a consequence of the electron trajectories inside the laser wave. The motion is oscillatory in the transverse directions (see figures 2.2). It exhibits a more complex pattern when the longitudinal components are included (see figures 2.2). The deviation becomes more significant with an increase in the strength parameter a_0 associated with the optical wave power density. We expect, therefore, that the Thomson scattering process we are dealing with is not induced by a purely dipole effect. The presence of an extra component in the z-motion oscillating at double the frequency of the transverse part is responsible for a richer phenomenology in terms of harmonic content, as we detail below. The parameters containing quadratic terms in a_0 are those labeled with $\Psi_{3x,3y}$, and no 'dipole' effects that break scattering play a sizeable role if $a_0 \ll 1$. To better appreciate the previous discussion, we note that the integral in equation (2.73) on the variable ζ' is achieved by expressing $\vec{r}_e$ in terms of $x(\zeta')$, $y(\zeta')$, $z(\zeta')$, using the trajectory components given in equation (2.20), and then proceeding as illustrated below and in more detail in the Comments and exercises section. The strategy is that of explicitly working out the integral in equation (2.73), which can be reduced to:

$$\hat{n} \times \int_0^\zeta d\zeta' \frac{d\vec{r}_e}{d\zeta'} e^{i(\Psi_1 - h\frac{\omega_0}{c})\zeta'} = \int_0^\zeta d\zeta' \begin{pmatrix} \sin\theta \sin\phi \frac{dz}{d\zeta'} - \cos\theta \frac{dy}{d\zeta'} \\ \cos\theta \frac{dx}{d\zeta'} - \sin\theta \cos\phi \frac{dz}{d\zeta'} \\ \sin\theta \cos\phi \frac{dy}{d\zeta'} - \sin\theta \sin\phi \frac{dx}{d\zeta'} \end{pmatrix} e^{i(\Psi_1 - h\frac{\omega_0}{c})\zeta'}. \quad (2.74)$$

The velocity components in equation (2.74) exhibit (see equations (2.2)) an oscillating behavior of the type $d\vec{r}_e/d\zeta' \propto e^{is\omega_0 \zeta'/c}$, and therefore equation (2.74) reduces to a series of straightforward integrals of the type:

$$\int_0^\zeta d\zeta' \frac{d\vec{r}_e}{d\zeta'} e^{is\frac{\omega_0}{c}\zeta'} e^{i(\Psi_1 - h\frac{\omega_0}{c})\zeta'} = \int_0^\zeta d\zeta' \frac{d\vec{r}_e}{d\zeta'} e^{i(\Psi_1 - (h-s)\frac{\omega_0}{c})\zeta'} \quad \text{with} \quad s = 1, 2. \quad (2.75)$$

The harmonic shift produces an analogous shift in the Bessel function index. Let us define $\vec{\mathscr{I}} = (\mathscr{I}_x, \mathscr{I}_y, \mathscr{I}_z)$ as:

$$\mathscr{I}_{x,y,z} = \sum_{h=-\infty}^{h=+\infty} \int_0^\zeta d\zeta' \frac{d\vec{r}_e}{d\zeta'} e^{i(\Psi_1 - h\frac{\omega_0}{c})\zeta'} e^{-ih\phi_x} \mathscr{J}_h\left(\Psi_{2x}, -\Psi_{3x}; \Psi_{2y}, -\Psi_{3y} \middle| e^{i l(\phi_x - \phi_y)}\right). \quad (2.76)$$

The result of the algebraic manipulations eventually leads to (the arguments of the Bessel function have been omitted for conciseness):

$$\begin{aligned}
\mathscr{I}_x &= -i \sum_h \left[u_x \mathscr{J}_h + \frac{\tilde{u}_x}{2}(e^{i\phi_x}\mathscr{J}_{h-1} + e^{-i\phi_x}\mathscr{J}_{h+1}) \right] \frac{e^{i\nu_h \zeta} - 1}{\nu_h} \\
\mathscr{I}_y &= -i \sum_h \left[u_y \mathscr{J}_h + \frac{\tilde{u}_y}{2}(e^{i\phi_y}\mathscr{J}_{h-1} + e^{-i\phi_y}\mathscr{J}_{h+1}) \right] \frac{e^{i\nu_h \zeta} - 1}{\nu_h} \\
\mathscr{I}_z &= -i \sum_h \left[u_z \mathscr{J}_h - \frac{\kappa_x}{2}(\mathscr{J}_{h+1}e^{-i\phi_x} + \mathscr{J}_{h-1}e^{i\phi_x}) - \frac{\kappa_y}{2}(\mathscr{J}_{h+1}e^{-i\phi_y} + \mathscr{J}_{h-1}e^{i\phi_y}) - \frac{\tilde{\kappa}_x}{2}\mathscr{J}_h - \frac{\tilde{\kappa}_y}{2}\mathscr{J}_h + \right. \\
&\qquad \left. - \frac{\tilde{\kappa}_x}{4}(\mathscr{J}_{h-2}e^{2i\phi_x} + \mathscr{J}_{h+2}e^{-2i\phi_x}) - \frac{\tilde{\kappa}_y}{4}(\mathscr{J}_{h-2}e^{2i\phi_y} + \mathscr{J}_{h+2}e^{-2i\phi_y}) \right] \frac{e^{i\nu_h \zeta} - 1}{\nu_h}
\end{aligned} \quad (2.77)$$

where:

$$\nu_h = \Psi_1 - h\frac{\omega_0}{c}$$

$$u_{x,y,z} = \frac{p_{x,y,z}(0)}{\gamma_0 m_0 c + p_z(0)}$$

$$\tilde{u}_{x,y} = \frac{eA_{x,y;0}}{\gamma_0 m_0 c + p_z(0)} \qquad (2.78)$$

$$\kappa_{x,y} = \frac{ep_{x,y}(0)A_{x,y;0}}{\gamma_0 m_0 c + p_z(0)^2}$$

$$\tilde{\kappa}_{x,y} = \frac{e^2 A^2_{x,y;0}}{\gamma_0 m_0 c + p_z(0)^2}.$$

The physical content of equation (2.77) is fairly clear: the contributions containing the 'transitions' $\Delta h = \pm 1$ are associated with dipole effects, while the terms including the $\Delta h = \pm 2$ counterparts are a consequence of the double-frequency, non-dipole oscillations. It is worth noting that the term

$$\frac{e^{i(\Psi_1 - h\frac{\omega_0}{c})\varsigma} - 1}{\Psi_1 - h\frac{\omega_0}{c}} \qquad (2.79)$$

rapidly goes to zero unless $\Psi_1 = h\omega_0/c$. In other words, an efficient mechanism of radiation is only realized for the frequencies that satisfy this condition. Such a condition is also called phase matching, for reasons that we are going to explain soon. This condition is satisfied by frequencies such that:

$$\left(1 - \frac{p_z(0)(1 + \cos\theta)}{\eta} + (1 + \cos\theta)\frac{e^2 A_0^2}{4\eta^2}\right)\omega - h\omega_0 = 0. \qquad (2.80)$$

We then observe that for small angles $\cos\theta \simeq 1 - \theta^2/2$ and for ultrarelativistic electrons, namely $\gamma(0) \equiv \gamma_0 \gg 1$, the phase-matching condition becomes:

$$\left(1 - \frac{\left(1 - \frac{1}{2\gamma_0^2}\right)\left(2 - \frac{\theta^2}{2}\right)}{2 - \frac{1}{2\gamma_0^2}} + \left(2 - \frac{\theta^2}{2}\right)\frac{e^2 A_0^2}{16\gamma_0^2 m_0^2 c^2}\right)\omega - h\omega_0 = 0 \qquad (2.81)$$

where we have used $p_z(0) \simeq \gamma_0(1 - 1/2\gamma_0^2)mc$ because $\beta \simeq 1 - 1/2\gamma_0^2$. Equation (2.81) can be further processed to yield:

$$\left(\frac{1 + \gamma_0^2\theta^2}{4\gamma_0^2} + \frac{e^2 A_0^2}{8\gamma_0^2 m_0^2 c^2}\right)\omega - h\omega_0 = 0 \rightarrow \omega_h = \frac{4\gamma_0^2 h\omega_0}{1 + \gamma_0^2\theta^2 + \frac{e^2 A_0^2}{2m_0^2 c^2}}. \qquad (2.82)$$

Equation (2.82) clearly shows that the radiation mechanism is such as to generate harmonics ω_h of the fundamental frequency ω_0. Moreover, these harmonics are significantly blueshifted due to a double Doppler effect, which results in the $4\gamma_0^2$ factor. In fact, in a typical backscattering configuration, the exciting wave is moving towards the electron, and the source, i.e. the electron, is also moving toward the detector, thus causing a double Doppler effect. Finally, the denominator is purely related to the Doppler effect, where the radiation is redshifted at larger angles. The more violent the oscillation of the electron, the more retarded the forward-emitted radiation ($\theta \simeq 0$), hence the redshifts related to A_0^2. In other words, the solution for z in equation (2.20) predicts that when the exciting field is strong enough, the electron motion along the interaction axis is retarded (see the minus sign) with respect to an observation plane far from the source. It is also important to note that the photon energy at large angles goes to zero; therefore, no radiated energy should be expected off axis: the approximation $\theta \ll 1$ is fully justified for ultrarelativistic electrons. To conclude this section, we would just like to analyze the squared modulus of equation (2.79):

$$\left| \left(\frac{e^{i(\Psi_1 - h\frac{\omega_0}{c})\zeta} - 1}{\Psi_1 - h\frac{\omega_0}{c}} \right) \right|^2 = 4\frac{1 - \cos\left(\Psi_1\zeta - h\frac{\omega_0}{c}\zeta\right)}{2\left(\Psi_1 - h\frac{\omega_0}{c}\right)^2} = \zeta^2 \operatorname{sinc}^2\left[\left(\Psi_1 - h\frac{\omega_0}{c}\right)\frac{\zeta}{2}\right]. \quad (2.83)$$

The sinc function $\operatorname{sinc}(x)$ in equation (2.83) is representative of a line emission. The propagation coordinate ζ increases with the interaction time, where the limit of the sinc function for large arguments is zero. This means that only those frequencies for which the expression in brackets is approximately or exactly zero can be emitted. We have seen already that this condition allows the definition of the emitted harmonics (see equation (2.82)). We have given the name 'phase matching' to the condition in which the expression in brackets is zero. Indeed, the phase of the radiation field appears in the integral of equation (2.62): during propagation, only the frequencies that correspond to the condition expressed by equation (2.82) can grow constructively and be output. Figure 2.3 shows an example of a Thomson-backscattered radiation spectrum for a fixed observation angle. It is possible to observe that the spectrum is composed only of harmonics that satisfy the phase-matching condition during the emission process. Moreover, figure 2.4 shows the typical angular distribution of Thomson radiation scattered by a single electron. The results of figures 2.3 and 2.4 were obtained using the same parameters for the exciting wave and the test electron. It is important to notice that the angular distribution in figure 2.4 contains contributions from all the harmonic lines in the spectrum of figure 2.3. The angular divergence of the radiation spot is comparable to $1/\gamma_0$. This is a rather general feature of Thomson backscattering radiation.

2.4.1 Linear Thomson backscattering

From equations (2.20), we learn that when $a_0 \gg 1$, the electron trajectories are no longer linear in the exciting field: the same applies to the radiation, therefore nonlinear optics is to be expected. Nonlinear optics is responsible for the generation of new frequencies compared to the fundamental, ω_0. Conversely, for $a_0 \ll 1$, a linear response of the system must be expected; in other words, the scattered field is linear in the exciting field. In this case, the electron trajectories are given by

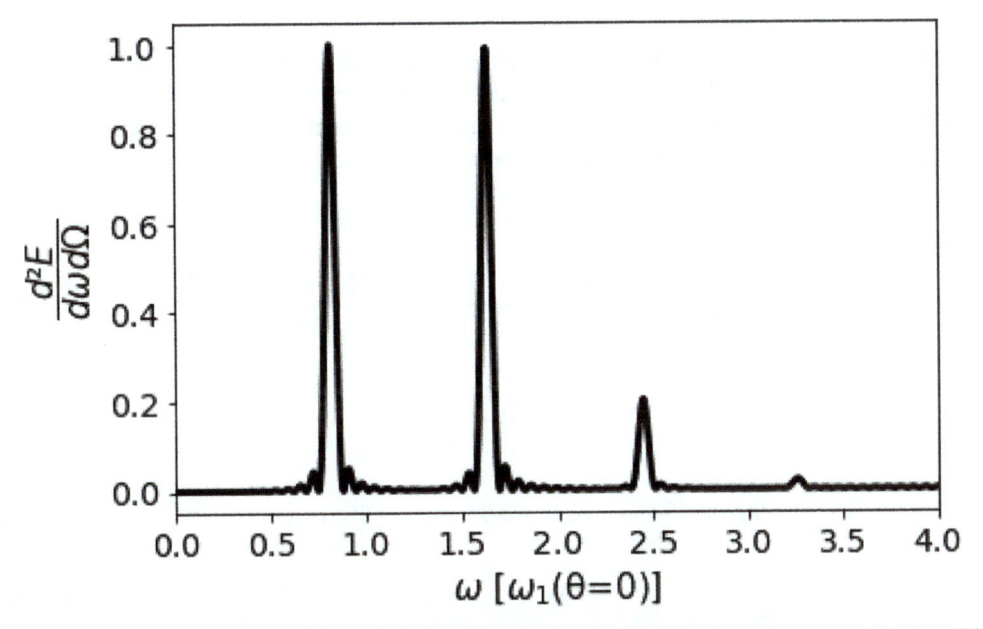

Figure 2.3. The Thomson backscattering radiation spectrum for the following parameters: $a_0 = 0.5$, $\gamma_0 = 200$, $\varphi_x = \varphi_y = 0$, $x_0 = y_0 = z_0 = 0$, $p_x(0) = p_y(0) = 0$, $A_{y0} = 0$, and $\omega_0 = 2.35 \times 10^{15}$ Hz. The observation angle is fixed at: $\theta = 1/2\gamma_0$, $\phi = 0$. The plot is normalized to the peak intensity.

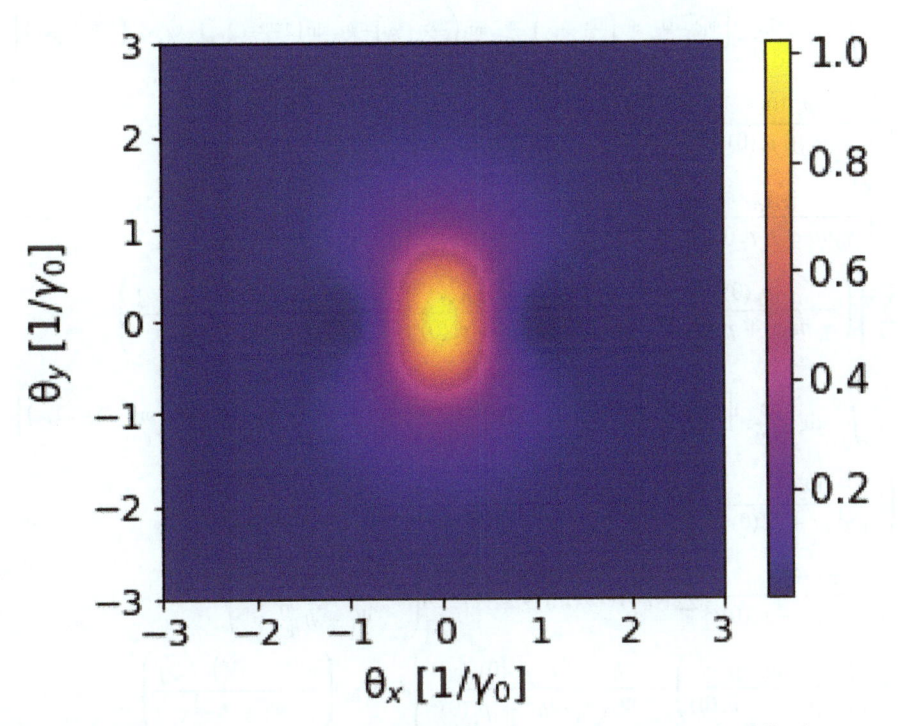

Figure 2.4. The angular distribution of Thomson backscattering radiation for the following parameters: $a_0 = 0.5$, $\gamma_0 = 200$, $\varphi_x = \varphi_y = 0$, $x_0 = y_0 = z_0 = 0$, $p_x(0) = p_y(0) = 0$, $A_{y0} = 0$, $\omega_0 = 2.35 \times 10^{15}$ Hz, and $\zeta = 10$ μm. The plot is normalized to the peak intensity. Moreover, $\theta_x = \theta \cos\phi$, $\theta_x = \theta \sin\phi$.

equations (2.23), assuming that an ultrarelativistic electron is moving only along z before the interaction. For sake of simplicity and to provide a didactic example, let us also assume that the exciting wave is linearly polarized in the x-axis and that $\varphi_x = 0$. Under these conditions, the phase terms are:

$$\Psi_0 \simeq 0 \tag{2.84}$$

$$\Psi_1 \simeq \frac{\omega}{c}\left(1 - \frac{p_z(0)(1+\cos\theta)}{\eta}\right) \simeq \frac{\omega}{c}\left(\frac{1+\gamma_0^2\theta^2}{4\gamma_0^2}\right) \tag{2.85}$$

$$\Psi_2 = \Psi_{2x}\sin\left(\frac{\omega_0\zeta}{c}\right) \tag{2.86}$$

$$\Psi_{2x} = \frac{eA_{x0}\omega}{\eta\omega_0}\cos\phi\sin\theta \quad ; \quad \Psi_{2y} = 0 \tag{2.87}$$

$$\Psi_3 \simeq 0 \tag{2.88}$$

$$\Psi_{3x} \simeq 0 \quad ; \quad \Psi_{3y} = 0. \tag{2.89}$$

Therefore:

$$\begin{aligned}
X &= \int_0^\zeta d\zeta' \frac{dx}{d\zeta'} e^{i\left[\Psi_1\zeta' - \Psi_{2x}\sin\left(\frac{\omega_0\zeta'}{c}+\varphi_x\right) - \Psi_{2y}\sin\left(\frac{\omega_0\zeta'}{c}+\varphi_y\right) + \Psi_{3x}\sin\left(2\frac{\omega_0\zeta'}{c}+2\varphi_x\right) + \Psi_{3y}\sin\left(2\frac{\omega_0\zeta'}{c}+2\varphi_y\right)\right]} \\
&\simeq \left(\frac{-ip_x(0)}{\gamma_0 m_0 c + p_z(0)}\right)\sum_h J_h(\Psi_{2x})\left(\frac{e^{i(\Psi_1-h\frac{\omega_0}{c})\zeta}-1}{\Psi_1 - h\frac{\omega_0}{c}}\right) \\
&+ \frac{1}{2}\left(\frac{-ieA_{x0}}{\gamma_0 m_0 c + p_z(0)}\right)\sum_h [J_{h+1}(\Psi_{2x}) + J_{h-1}(\Psi_{2x})]\left(\frac{e^{i(\Psi_1-h\frac{\omega_0}{c})\zeta}-1}{\Psi_1 - h\frac{\omega_0}{c}}\right) \\
&\simeq \sum_h\left[\left(\frac{-ip_x(0)}{\gamma_0 m_0 c + p_z(0)}\right) + \frac{h}{\Psi_{2x}}\left(\frac{-ieA_{x0}}{\gamma_0 m_0 c + p_z(0)}\right)\right]J_h(\Psi_{2x})\left(\frac{e^{i(\Psi_1-h\frac{\omega_0}{c})\zeta}-1}{\Psi_1 - h\frac{\omega_0}{c}}\right)
\end{aligned} \tag{2.90}$$

$$\begin{aligned}
Z &= \int_0^\zeta d\zeta' \frac{dz}{d\zeta'} e^{i\left[\Psi_1\zeta' - \Psi_{2x}\sin\left(\frac{\omega_0\zeta'}{c}+\varphi_x\right) - \Psi_{2y}\sin\left(\frac{\omega_0\zeta'}{c}+\varphi_y\right) + \Psi_{3x}\sin\left(2\frac{\omega_0\zeta'}{c}+2\varphi_x\right) + \Psi_{3y}\sin\left(2\frac{\omega_0\zeta'}{c}+2\varphi_y\right)\right]} \\
&\simeq \left(\frac{-ip_z(0)}{\gamma_0 m_0 c + p_z(0)}\right)\sum_h J_h(\Psi_{2x})\left(\frac{e^{i(\Psi_1-h\frac{\omega_0}{c})\zeta}-1}{\Psi_1 - h\frac{\omega_0}{c}}\right) \\
&+ \frac{iep_x(0)A_{x0}}{2(\gamma_0 m_0 c + p_z(0))^2}\sum_h [J_{h+1}(\Psi_{2x}) + J_{h-1}(\Psi_{2x})]\left(\frac{e^{i(\Psi_1-h\frac{\omega_0}{c})\zeta}-1}{\Psi_1 - h\frac{\omega_0}{c}}\right) \\
&\simeq \sum_h\left[\left(\frac{-ip_z(0)}{\gamma_0 m_0 c + p_z(0)}\right) + \frac{h}{\Psi_{2x}}\frac{i.\,e.\,p_x(0)A_{x0}}{\gamma_0 m_0 c + p_z(0)^2}\right]J_h(\Psi_{2x})\left(\frac{e^{i(\Psi_1-h\frac{\omega_0}{c})\zeta}-1}{\Psi_1 - h\frac{\omega_0}{c}}\right)
\end{aligned} \tag{2.91}$$

where we have used $J_{h+1}(\Psi_{2x}) + J_{h-1}(\Psi_{2x}) = (2h/\Psi_{2x})J_h(\Psi_{2x})$. The cross product in equation (2.62) can be now given an explicit form:

$$\hat{n} \times (X, 0, Z) = (Z \sin\phi \sin\theta, X \cos\theta - Z \cos\phi \sin\theta, -X \sin\phi \sin\theta). \quad (2.92)$$

The squared modulus for small angles of observation, i.e. for relativistic electrons, yields:

$$|(\cos\phi \sin\theta, \sin\phi \sin\theta, \cos\theta) \times (X, 0, Z)|^2 \simeq X^2(1 - \theta^2 \cos^2\phi) + Z^2\theta^2 - 2Re[X^*Z]\theta \cos\phi. \quad (2.93)$$

Finally, the spectral–angular distribution of the scattered radiation becomes:

$$\frac{d^2E}{d\omega d\Omega} = \frac{\mu_0 e^2 \omega^2}{16\pi^3 c}[X^2(1 - \theta^2 \cos^2\phi) + Z^2\theta^2 - 2Re(X^*Z)\theta \cos\phi]. \quad (2.94)$$

For $a_0 \ll 1$, the Bessel functions of Ψ_{2x} can be considered in the limit of small arguments:

$$J_h\left(\frac{a_0\omega}{2\gamma_0\omega_0}\cos\phi \sin\theta\right) \simeq \delta_{h0} + \frac{a_0\omega}{4\gamma_0\omega_0}\cos\phi \sin\theta(\delta_{h1} - \delta_{h(-1)}). \quad (2.95)$$

Moreover, recalling our assumption $p_x(0) = 0$, neglecting the emission at zero and negative frequencies, and also considering the case of sufficiently long interactions, i.e. $\left(\Psi_1 - \frac{\omega_0}{c}\right)\zeta/2 \gg 1$, it is possible to obtain:

$$\frac{d^2E}{d\omega d\Omega} \simeq \frac{\alpha\hbar\omega^2 a_0^2}{64\pi^2\gamma_0^2 c^2}\left(1 + \frac{\theta^4\omega^2}{4\omega_0^2}\cos^2\phi - \frac{\omega}{\omega_0}\theta^2\cos^2\phi\right)\zeta^2\text{sinc}^2\left[\left(\Psi_1 - \frac{\omega_0}{c}\right)\frac{\zeta}{2}\right]. \quad (2.96)$$

The long interaction case is very likely for an excitation wave length ζ much longer than the electromagnetic period $2\pi/\omega_0$, i.e. for a many-cycle wave. The electron undergoes so many oscillation cycles that the bandwidth of the emitted line becomes extremely narrow. In this limit, the cross terms appearing after squaring the sums in equations (2.90) and (2.91) can be set to zero. In other words, there is no interference between the harmonics. This behavior is very similar to that of magnetic undulators, which are discussed at the end of this chapter. Equation (2.96) gives the spectral–angular distribution of Thomson backscattering radiation in the linear regime, i.e. where the scattered field is linear with respect to the exciting field (hence the radiated energy E is proportional to A_0^2, as in equation (2.96)). In the linear regime, only the first harmonic contributes to the radiation spectrum. Negative frequencies have been neglected because they do not contribute significantly to the emitted energy.

2.4.1.1 Integration over angles

For sufficiently long interactions, i.e. $(\Psi_1 - \frac{\omega_0}{c})\zeta/2 \gg 1$, the sinc function in equation (2.96) can be approximated by:

$$\text{sinc}^2\left[\left(\frac{\omega}{\omega_0}\frac{1+\gamma_0^2\theta^2}{4\gamma_0^2} - 1\right)\frac{\omega_0\zeta}{2c}\right] \simeq \pi\delta\left[\left(\frac{\omega}{\omega_0}\frac{1+\gamma_0^2\theta^2}{4\gamma_0^2} - 1\right)\frac{\omega_0\zeta}{2c}\right] = \frac{4\pi c}{\omega\zeta}\frac{\delta(\theta - \theta^*)}{\theta^*} \quad (2.97)$$

where, from the rule of composition of the Dirac delta with an argument function of θ, θ^* is defined as:

$$\theta^* = \frac{1}{\gamma_0}\sqrt{\frac{4\gamma_0^2\omega_0}{\omega} - 1}. \tag{2.98}$$

Therefore, the spectral–angular distribution of the photon number, obtained from the radiated energy at equation (2.96) after dividing by $\hbar\omega$, can also be approximated by:

$$\frac{d^2N}{d\omega d\Omega} = \frac{\alpha}{16\pi c}\frac{a_0^2}{\gamma_0^2}\left(1 + \frac{\theta^4\omega^2}{4\omega_0^2}\cos^2\phi - \frac{\omega}{\omega_0}\theta^2\cos^2\phi\right)\zeta\frac{\delta(\theta - \theta^*)}{\theta^*} \tag{2.99}$$

where we have defined the fine structure constant $\alpha = e^2/4\pi\varepsilon_0\hbar c$. The integration over the azimuthal angle φ is trivial, yielding:

$$\frac{d^2N}{d\omega d\theta} = \int_0^{2\pi} d\phi \sin\theta \frac{d^2N}{d\omega d\Omega} = \frac{\alpha}{8c}\frac{a_0^2}{\gamma_0^2}\left(1 + \frac{\theta^4\omega^2}{8\omega_0^2} - \frac{\omega}{2\omega_0}\theta^2\right)\zeta\delta(\theta - \theta^*). \tag{2.100}$$

The integration over θ is even more trivial:

$$\frac{dN}{d\omega} = \int_0^\infty d\theta \int_0^{2\pi} d\phi \sin\theta \frac{d^2N}{d\omega d\Omega} = \frac{\alpha\zeta}{64c}\frac{a_0^2}{\gamma_0^6\omega_0^6}\left(8\gamma_0^4\omega_0^2 - 4\gamma_0^2\omega_0\omega + \omega^2\right). \tag{2.101}$$

Equation (2.101) turns to be extremely useful in studying the radiation spectrum integrated over all angles of observation, which may correspond to the experimental case of a spectrometer without spatial/angular resolution. An example of a Thomson backscattering radiation spectrum is provided in figure 2.5 for a maximum observation angle of $\theta_{max} = 3/\gamma_0$. Even though we have integrated the radiation spectrum from zero to infinity over θ, in realistic cases, the detector's aperture imposes a limit on the maximum angle of observation. Given the spectral–angular correlation that defines the harmonic lines ω_1, a limited aperture also corresponds to a minimal detected photon energy. Thus, equation (2.101) should normally be considered valid when starting from a minimum frequency defined as $\omega_{min} = \omega_1(\theta_{max})$. Below this frequency, radiation is not detected and its evaluation is irrelevant; therefore, for $\omega < \omega_{min}$, we also set $dN/d\omega = 0$. With no aperture in the detection system, the integration over the photon energy is performed with the following limits:

$$N = \int_0^{4\gamma_0^2\omega_0} \frac{dN}{d\omega}d\omega = \frac{2\pi}{3}\alpha a_0^2 N_c \tag{2.102}$$

where we have defined the number of cycles of the electron under the action of the wave to be $N_c = \zeta\omega_0/2\pi c$. For the sake of simplicity, in equation (2.102) we have also assumed $0 \simeq \omega_{min} \ll 4\gamma_0^2\omega_0$. In general, the emitted photon number is slightly smaller than $2\pi\alpha a_0^2 N_c$. However, as the radiation is mostly emitted around $4\gamma_0^2\omega_0$,

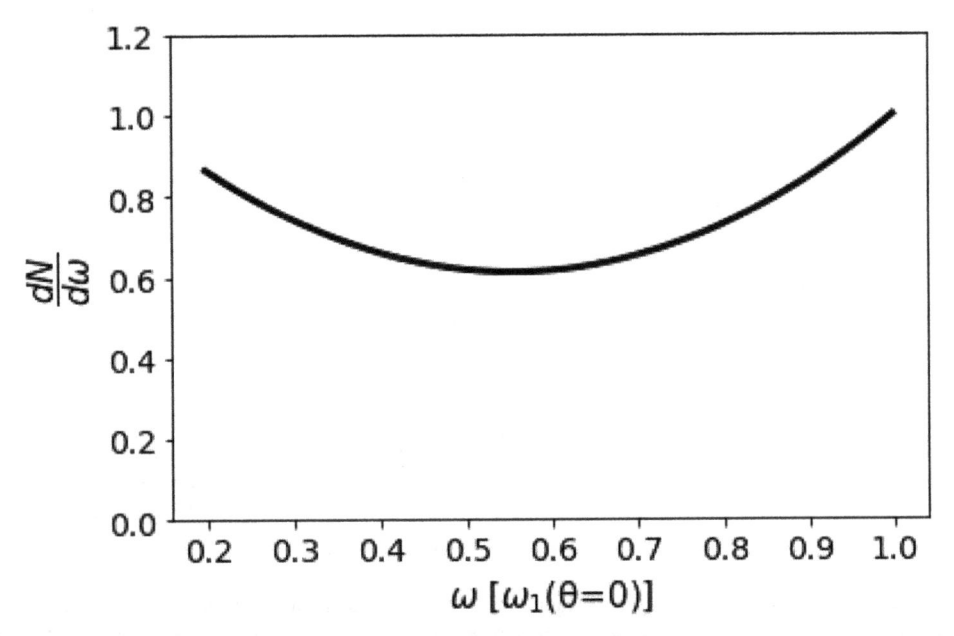

Figure 2.5. The Thomson backscattering radiation spectrum of the first harmonic for the parameters: $a_0 = 0.05$, $\gamma_0 = 200$, $\varphi_x = \varphi_y = 0$, $x_0 = y_0 = z_0 = 0$, $p_x(0) = p_y(0) = 0$, $A_{y0} = 0$, $\omega_0 = 2.35 \times 10^{15}$ Hz, and $\zeta = 10$ μm. The plot is normalized to the peak intensity. The maximum observation angle $\theta_{max} = 3/\gamma_0$.

equation (2.102) can be assumed to be a valid approximation in a rather general sense. Dividing by the initial photon flux $F_0 = \varepsilon_0 c E_0^2 \zeta / 2\hbar\omega_0 c$, it is possible to obtain the cross section for the linear Thomson backscattering process:

$$\sigma_T = \frac{N}{F_0} = \frac{8\pi}{3} r_0^2. \tag{2.103}$$

In other words, the Thomson cross section is retrieved, confirming the validity of our approximations (figure 2.6).

2.4.1.2 Integration over frequencies

For sufficiently long interactions, it is also possible to approximate the sinc function with a Dirac delta in the following way:

$$\text{sinc}^2\left[\left(\frac{\omega}{\omega_0}\frac{1+\gamma_0^2\theta^2}{4\gamma_0^2} - 1\right)\frac{\omega_0\zeta}{2c}\right] \simeq \frac{2\pi c}{\zeta}\frac{4\gamma_0^2}{1+\gamma_0^2\theta^2}\delta\left(\omega - \frac{4\gamma_0^2\omega_0}{1+\gamma_0^2\theta^2}\right). \tag{2.104}$$

Therefore, the spectral–angular distribution of the Thomson scattered radiation in the linear emission regime is simplified to:

$$\frac{d^2N}{d\omega d\Omega} = \frac{\alpha\omega}{2\pi c}\frac{a_0^2}{16\gamma_0^2}\left(1 + \frac{\theta^4\omega^2}{4\omega_0^2}\cos^2\phi - \frac{\omega}{\omega_0}\theta^2\cos^2\phi\right)\zeta\delta\left(\omega\frac{1+\gamma_0^2\theta^2}{4\gamma_0^2} - \omega_0\right). \tag{2.105}$$

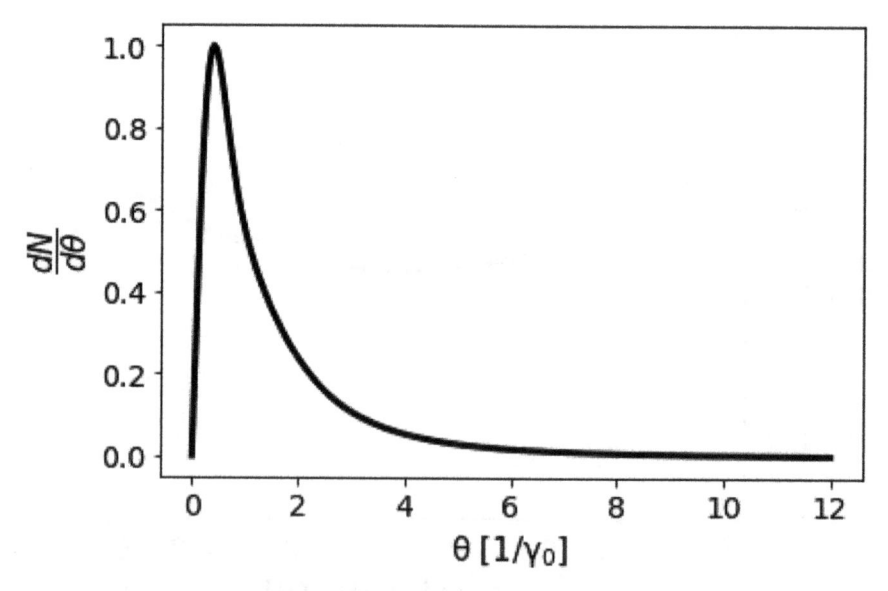

Figure 2.6. The Thomson backscattering radiation φ-integrated angular distribution for the parameters: $a_0 = 0.05$, $\gamma_0 = 200$, $\varphi_x = \varphi_y = 0$, $x_0 = y_0 = z_0 = 0$, $p_x(0) = p_y(0) = 0$, $A_{y0} = 0$, $\omega_0 = 2.35 \times 10^{15}$ Hz, and $\zeta = 10$ μm. The plot is normalized to the peak intensity.

Integrating over the azimuthal angle gives:

$$\frac{d^2N}{d\omega d\theta} = \frac{\alpha\omega}{c}\frac{a_0^2}{16\gamma_0^2}\left(1 + \frac{\theta^4\omega^2}{8\omega_0^2} - \frac{\omega}{2\omega_0}\theta^2\right)\zeta\frac{4\gamma_0^2\theta}{1+\gamma_0^2\theta^2}\delta\left(\omega - \frac{4\gamma_0^2\omega_0}{1+\gamma_0^2\theta^2}\right). \quad (2.106)$$

Integrating over the frequencies, we now obtain:

$$\frac{dN}{d\theta} = \frac{\alpha\omega_0 a_0^2 \zeta\gamma_0^2\theta(1+\gamma_0^4\theta^4)}{c(1+\gamma_0^2\theta^2)^4}. \quad (2.107)$$

Finally, integrating over the polar angle yields $N = 2\pi\alpha a_0^2 N_c/3$, exactly as in equation (2.102). It is possible to guess that the Thomson cross section can again be retrieved by dividing by the incoming flux of radiation F_0. The angular integration, as expected, gives the same result as the frequency integration. This is a valid benchmark for the correctness of our calculations. In conclusion, we can also calculate the width of the angular distribution:

$$<\theta> = \frac{\int d\theta \frac{dN}{d\theta}\theta}{\int d\theta \frac{dN}{d\theta}} = \frac{9\pi}{16\gamma_0}. \quad (2.108)$$

The angle $<\theta>$ corresponds to the measurement of the angular extension of the Thomson backscattering radiation front; it must be matched to the aperture of the detector in any experimental setup if total collection of the radiation is sought. In other words, equation (2.108) can be considered to be the divergence of the

Thomson backscattering radiation beam. However, given the spectral–angular correlation of Thomson backscattering radiation, it is very common to detect the radiation over a narrow angle in such a way as to minimize the spectral bandwidth, thus monochromatizing the source.

2.4.2 Nonlinear Thomson backscattering: harmonic emission

Equation (2.96) is valid for $a_0 \ll 1$. When the relativistic parameter is not too small, higher-order Bessel functions contribute to the radiation spectrum. In fact, the expansion of these functions for small arguments is such that:

$$J_h^2\left(\frac{a_0\omega}{2\gamma_0\omega_0}\cos\phi\sin\theta\right) \propto a_0^{2h}. \tag{2.109}$$

Thus, each harmonic leads to a law of proportionality of the kind $E \propto a_0^{2h}$. We have seen that the linear case is realized for $h = 1$ because the scattered energy is proportional to the energy (and the intensity) of the exciting wave. Conversely, when harmonic emission is important, a nonlinear dependence of the scattered energy upon the incoming energy is obtained, according to equation (2.109). It is clear that 'harmonic emission' and 'the nonlinear regime of Thomson backscattering' are two ways of naming the same phenomenon. For an arbitrarily large a_0, the Bessel functions cannot really be expanded as we did previously for small arguments: this forces us to retain the sum over the harmonics. In the most general case, the radiation spectrum associated with the Thomson backscattering process is composed of harmonic lines. In this section we work out an expression for the spectral–angular distribution of radiation emitted in the nonlinear/harmonic regime. We do this under the same conditions as those used in the previous section, i.e. for a linear polarization of the exciting wave and for an electron moving, just before the interaction, only along the main axis z and with an initial phase of oscillation equal to zero. In this case, we need to retain the terms that are of second order in the relativistic parameter. Under these conditions, the phase terms are:

$$\Psi_0 \simeq 0 \tag{2.110}$$

$$\Psi_1 \simeq \frac{\omega}{c}\left(1 - \frac{p_z(0)(1+\cos\theta)}{\eta} + \frac{e^2 A_0^2(1+\cos\theta)}{4\eta^2}\right) \simeq \frac{\omega}{c}\left(\frac{1 + \gamma_0^2\theta^2 + \frac{a_0^2}{2}}{4\gamma_0^2}\right) \tag{2.111}$$

$$\Psi_2 = \Psi_{2x}\sin\left(\frac{\omega_0\zeta}{c}\right) \tag{2.112}$$

$$\Psi_{2x} = \frac{a_0\omega}{2\gamma_0\omega_0}\cos\phi\sin\theta \quad;\quad \Psi_{2y} = 0 \tag{2.113}$$

$$\Psi_3 = \Psi_{3x} \sin\left(2\frac{\omega_0 \zeta}{c}\right) \tag{2.114}$$

$$\Psi_{3x} \simeq \frac{\omega}{\omega_0}(1 + \cos\theta)\frac{a_0^2}{32\gamma_0^2} \quad ; \quad \Psi_{3y} = 0. \tag{2.115}$$

Using a method similar to that of the previous section, equations (2.90) and (2.91) are now written without approximations:

$$\begin{aligned}
X &= \left(\frac{-ip_x(0)}{\gamma_0 m_0 c + p_z(0)}\right) \sum_{h,l} J_{h+2l}(\Psi_{2x}) J_l(\Psi_{3x}) \left(\frac{e^{i(\Psi_1 - h\frac{\omega_0}{c})\zeta} - 1}{\Psi_1 - h\frac{\omega_0}{c}}\right) \\
&+ \frac{1}{2}\left(\frac{-ieA_{x0}}{\gamma_0 m_0 c + p_z(0)}\right) \sum_{h,l} [J_{h+2l+1}(\Psi_{2x}) + J_{h+2l-1}(\Psi_{2x})] J_l(\Psi_{3x}) \left(\frac{e^{i(\Psi_1 - h\frac{\omega_0}{c})\zeta} - 1}{\Psi_1 - h\frac{\omega_0}{c}}\right) \\
&= \sum_{h,l} \left[\left(\frac{-ip_x(0)}{\gamma_0 m_0 c + p_z(0)}\right) + \frac{h+2l}{\Psi_{2x}}\left(\frac{-ieA_{x0}}{\gamma_0 m_0 c + p_z(0)}\right)\right] J_{h+2l}(\Psi_{2x}) J_l(\Psi_{3x}) \left(\frac{e^{i(\Psi_1 - h\frac{\omega_0}{c})\zeta} - 1}{\Psi_1 - h\frac{\omega_0}{c}}\right)
\end{aligned} \tag{2.116}$$

$$\begin{aligned}
Z &= \sum_{h,l} \left(\frac{-ip_z(0)}{\gamma_0 m_0 c + p_z(0)} + \frac{h+2l}{\Psi_{2x}} \frac{iep_x(0)A_{x0}}{(\gamma_0 m_0 c + p_z(0))^2}\right) J_{h+2l}(\Psi_{2x}) J_l(\Psi_{3x}) \left(\frac{e^{i(\Psi_1 - h\frac{\omega_0}{c})\zeta} - 1}{\Psi_1 - h\frac{\omega_0}{c}}\right) \\
&+ \frac{ie^2 A_{x0}^2}{2(\gamma_0 m_0 c + p_z(0))^2} \sum_{h,l} J_{h+2l}(\Psi_{2x}) J_l(\Psi_{3x}) \left(\frac{e^{i(\Psi_1 - h\frac{\omega_0}{c})\zeta} - 1}{\Psi_1 - h\frac{\omega_0}{c}}\right) \\
&+ \frac{ie^2 A_{x0}^2}{4(\gamma_0 m_0 c + p_z(0))^2} \sum_{h,l} [J_{h+2l+2}(\Psi_{2x}) + J_{h+2l-2}(\Psi_{2x})] J_l \Psi_{3x} \left(\frac{e^{i(\Psi_1 - h\frac{\omega_0}{c})\zeta} - 1}{\Psi_1 - h\frac{\omega_0}{c}}\right).
\end{aligned} \tag{2.117}$$

Thus, the spectral–angular distribution of the scattered radiation in the nonlinear regime becomes equation (2.94) through the use of equations (2.116) and (2.117). It is worth noting that for $a_0 \ll 1$, equations (2.116) and (2.117) reduce to equations (2.90) and (2.91), since in that case $J_m^2(\Psi_{3x}) \simeq J_m^2(0) = \delta_{m0}$. Figure 2.7 shows the spectral–angular pattern of the nonlinear Thomson scattered radiation after integration over the angular variable φ. It is possible to notice that the Doppler effect redshifts each harmonic line on a separate path of the spectral–angular domain. In figure 2.8, we display the result shown in figure 2.7 using a 3D plot: this is a customary way to represent the spectral–angular distribution of the radiation in the research literature.

2.4.2.1 Second and third harmonics

We now focus our attention on the emission of the second and third harmonics, specifically on the spectral–angular properties of the latter. The topics and tools provided here are intended to be introductory for the study of any higher harmonics of interest. In the slightly nonlinear regime ($a_0 \lesssim 1$) of Thomson backscattering and for interactions that are long enough, we will assume that the harmonic lines are well

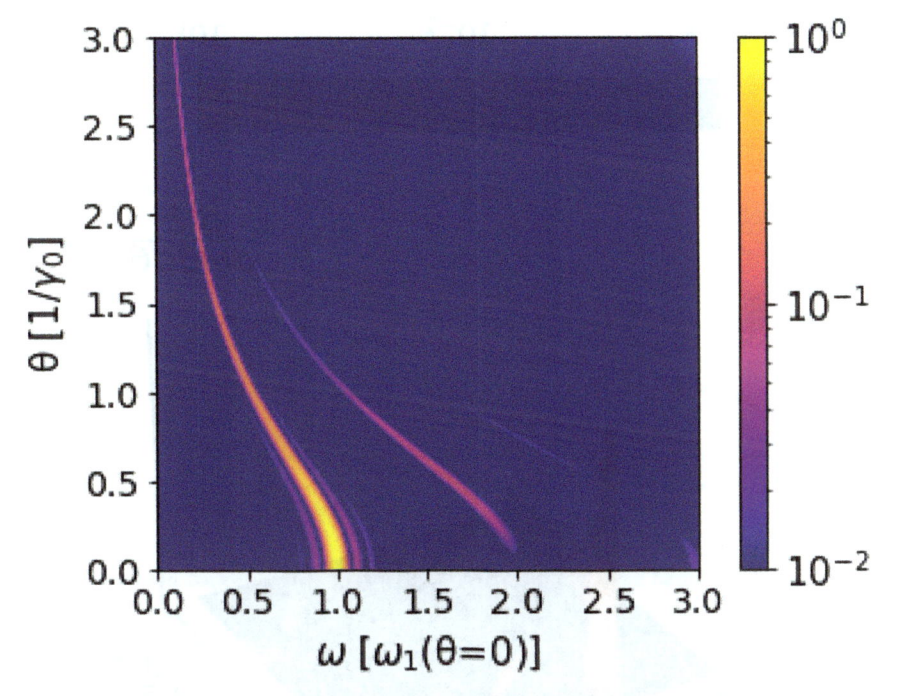

Figure 2.7. The spectral–angular distribution of nonlinear Thomson backscattering radiation after integration over φ for the parameters: $a_0 = 0.5$, $\gamma_0 = 200$, $\varphi_x = \varphi_y = 0$, $x_0 = y_0 = z_0 = 0$, $p_x(0) = p_y(0) = 0$, $A_{y0} = 0$, $\omega_0 = 2.35 \times 10^{15}$ Hz, and $\varsigma = 10$ μm. The plot is normalized to the peak intensity. The calculation is truncated at the third harmonic, which is already two orders of magnitude lower in energy compared to the first.

separated. Within the latter approximation, we write the spectral–angular distribution of the second harmonic according to equation (2.94) as:

$$\frac{d^2N}{d\omega d\Omega}\bigg|_2 \simeq \frac{\alpha \omega \varsigma^2}{64\pi^2 c^2} \frac{a_0^2}{\gamma_0^2}\left[(J_3 + J_1)^2 + \frac{4J_2^2}{a_0^2}\gamma_0^2\theta^2 - \frac{4\gamma_0 J_2(J_3 + J_1)}{a_0}\theta \cos\phi\right]\mathrm{sinc}^2\left[\left(\frac{\omega}{\omega_1} - 2\right)\frac{\omega_0\varsigma}{2c}\right] \quad (2.118)$$

where $J_h = J_h(\Psi_{2x}, -\Psi_{3x})$. Using the definition of the generalized Bessel function, we can expand to the first orders in Ψ_{2x} and Ψ_{3x}:

$$J_{h+1} + J_{h-1} \simeq \frac{\Psi_{2x}^{-3+h}}{2^{5+h}}\left(-\frac{128\Psi_{3x}}{\Gamma(2-h)} + \frac{96h(1+h)(2+h)(3+h)(2+(-2+h)\Psi_{3x})\Psi_{2x}^2}{3\Gamma(4+h)}\right.$$
$$\left. + \frac{-12(2+h)(3+h)(4h+(-4+(-1+h)h)\Psi_{3x})\Psi_{2x}^4 + (6h(3+h)+(-3+h)(2+h)(4+h)\Psi_{3x})\Psi_{2x}^6}{3\Gamma(4+h)}\right) \quad (2.119)$$

$$J_h \simeq \frac{\Psi_{2x}^h}{2^{4+h}}\left(\frac{8\Psi_{3x}(4-4h+\Psi_{2x}^2)}{\Psi_{2x}^2\Gamma(h)} + \frac{16}{\Gamma(1+h)} - \frac{(8+h(4+(3+h)\Psi_{3x}))\Psi_{2x}^2}{\Gamma(3+h)}\right).$$

The above expansions work pretty well up to the third harmonic. To study higher-order harmonics, it is necessary to expand to higher orders in Ψ_{2x} and Ψ_{3x}. Then, by integrating equation (2.118) by means of the same techniques used for the first

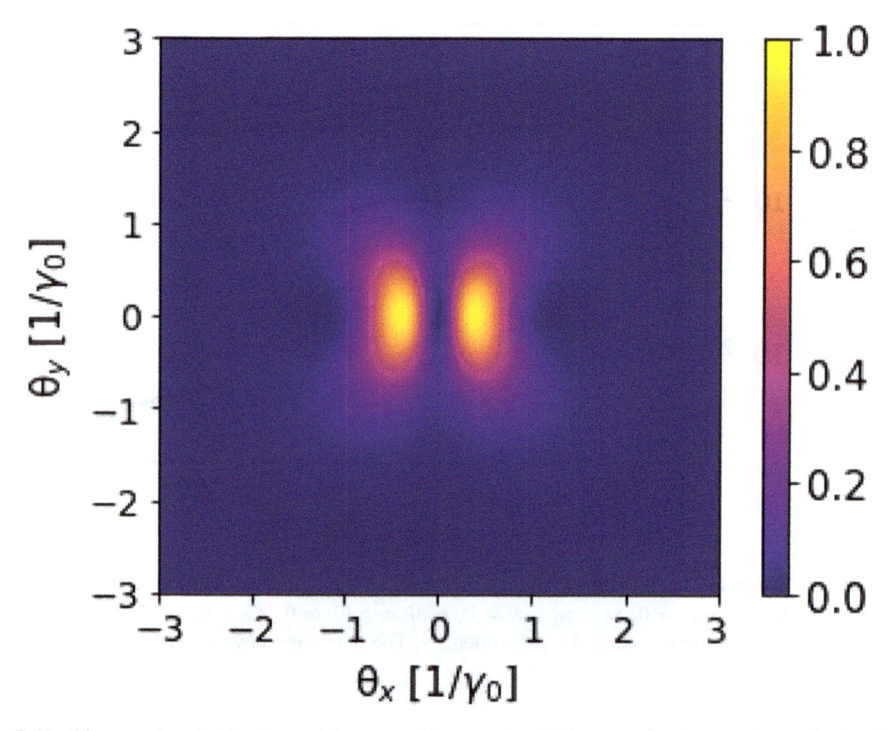

Figure 2.10. The angular distribution of the second harmonic of Thomson backscattering radiation for the parameters: $a_0 = 0.5$, $\gamma_0 = 200$, $\varphi_x = \varphi_y = 0$, $x_0 = y_0 = z_0 = 0$, $p_x(0) = p_y(0) = 0$, $A_{y0} = 0$, $\omega_0 = 2.35 \times 10^{15}$ Hz, and $\zeta = 10$ μm. The plot is normalized to the peak intensity. Moreover, $\theta_x = \theta \cos \phi$ and $\theta_x = \theta \sin \phi$.

At fixed observation angles, the ratio between the harmonics may be different (see figure 2.3). A plot of the angular distribution relative to the spectrum in figure 2.9 is provided in figure 2.4, where the contribution from the first harmonic is rather dominant. Indeed, we can consider that the pattern in figure 2.4 is essentially equal to the pattern of the first harmonic of Thomson backscattering radiation.

2.4.3 The extremely nonlinear regime: emission of a continuum

We have already found that given a linearly polarized exciting wave (along x), the Thomson backscattering radiation is described by equation (2.62). In this section the aim is to find an approximate expression for the Thomson backscattering radiation distribution in the extremely nonlinear regime of interaction realized when $a_0 \gg 1$. We neglect, as usual, the initial phase of oscillation φ_x. As a result, equation (2.62) can be expressed in terms of the generalized Bessel functions:

$$\frac{d^2E}{d\omega d\Omega} = \frac{\mu_0 e^2 \omega^2}{16\pi^3 c} \left| \hat{n} \times \sum_h J_h(\Psi_{2x}, -\Psi_{3x}) \int_0^{\zeta} d\zeta' \frac{d\vec{r}_e}{d\zeta'} e^{i(\Psi_1 - h\frac{\omega_0}{c})\zeta'} \right|^2. \quad (2.124)$$

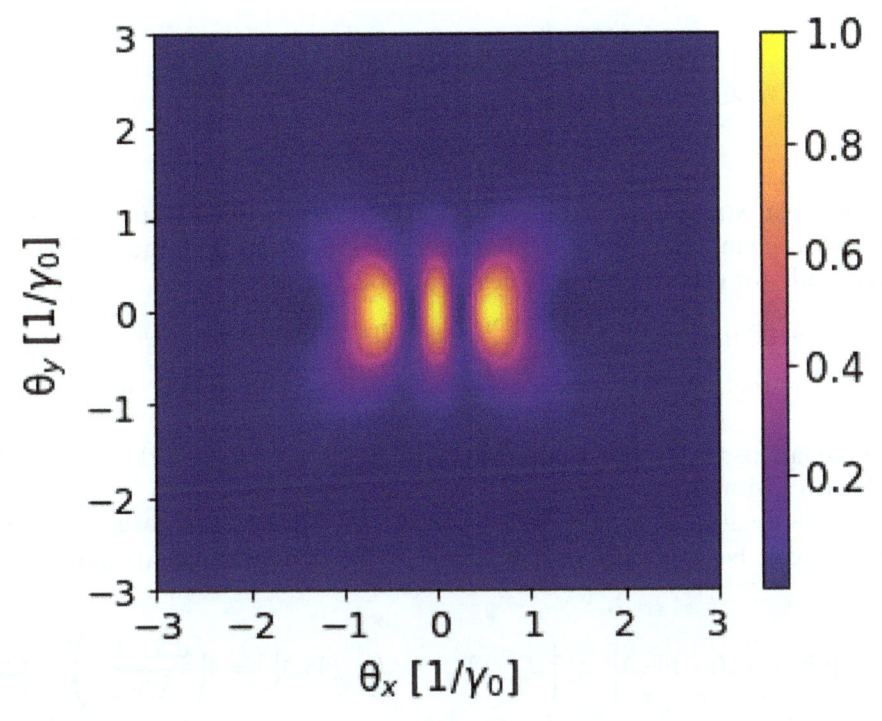

Figure 2.11. The angular distribution of the third harmonic of Thomson backscattering radiation for the parameters: $a_0 = 0.5$, $\gamma_0 = 200$, $\varphi_x = \varphi_y = 0$, $x_0 = y_0 = z_0 = 0$, $p_x(0) = p_y(0) = 0$, $A_{y0} = 0$, $\omega_0 = 2.35 \times 10^{15}$ Hz, and $\zeta = 10$ μm. The plot is normalized to the peak intensity. Moreover, $\theta_x = \theta \cos \phi$ and $\theta_x = \theta \sin \phi$.

For the case under study, the integrals appearing in equation (2.124) can be solved as follows:

$$\int_0^\zeta d\zeta' \frac{dx}{d\zeta'} e^{i(\Psi_1 - h\frac{\omega_0}{c})\zeta'} = \frac{1}{2}\left(\frac{-ieA_{x0}}{\gamma_0 m_0 c + p_z(0)}\right)\left[\left(\frac{e^{i(\Psi_1 - (h+1)\frac{\omega_0}{c})\zeta} - 1}{\Psi_1 - (h+1)\frac{\omega_0}{c}}\right) + \left(\frac{e^{i(\Psi_1 - (h-1)\frac{\omega_0}{c})\zeta} - 1}{\Psi_1 - (h-1)\frac{\omega_0}{c}}\right)\right] \quad (2.125)$$

$$\int_0^\zeta d\zeta' \frac{dz}{d\zeta'} e^{i(\Psi_1 - h\frac{\omega_0}{c})\zeta'} \simeq \frac{-ip_z(0)}{\gamma_0 m_0 c + p_z(0)}\left(\frac{e^{i(\Psi_1 - h\frac{\omega_0}{c})\zeta} - 1}{\Psi_1 - h\frac{\omega_0}{c}}\right). \quad (2.126)$$

where we have assumed $a_0 \ll \gamma_0$. Equation (2.125) can be further simplified via:

$$J_h \Psi_{2x}, -\Psi_{3x} \int_0^\zeta d\zeta' \frac{dx}{d\zeta'} e^{i(\Psi_1 - h\frac{\omega_0}{c})\zeta'} = \frac{-ia_0}{4\gamma_0}[J_{h+1}\Psi_{2x}, -\Psi_{3x} + J_{h-1}\Psi_{2x}, -\Psi_{3x}]\left(\frac{e^{i(\Psi_1 - h\frac{\omega_0}{c})\zeta} - 1}{\Psi_1 - h\frac{\omega_0}{c}}\right) \quad (2.127)$$

where, in the case of long interactions, the Dirac delta approximation can be applied for the phase-matching term in brackets, both for equation (2.127) and equation (2.126). In order to work out an analytical result for the asymptotic behavior of the Thomson backscattering radiation spectrum, we consider the on-axis radiation. For $\theta = 0$, the generalized Bessel function simply becomes:

$$J_h(\Psi_{2x}, -\Psi_{3x}) = J_h\left(0, -\frac{\omega}{\omega_0}\frac{a_0^2}{16\gamma_0^2}\right) \simeq J_h\left(0, -\frac{ha_0^2}{4\left(1+\frac{a_0^2}{2}\right)}\right) \quad (2.128)$$

where the last part is obtained after approximating the phase-matching term as a Dirac delta, giving $\omega = \omega_h$. It is possible to demonstrate that equation (2.128) is nonzero only for odd h. This is achieved by noticing that:

$$J_{h+1}(0, -\Psi_{3x}) + J_{h-1}(0, -\Psi_{3x}) = \sum_m [J_{h+2m+1}(0)J_m(\Psi_{3x}) + J_{h+2m-1}(0)J_m(\Psi_{3x})]$$
$$= \sum_m [\delta_{h+2m+1,0}J_m(\Psi_{3x}) + \delta_{h+2m-1,0}J_m(\Psi_{3x})] = J_{\frac{-h-1}{2}}(\Psi_{3x}) + J_{\frac{-h+1}{2}}(\Psi_{3x}) \quad (2.129)$$

where the last passage implies odd h in order to get an integer index for the Bessel functions. Moreover, for $h > 0$, it is easy to verify that $|J_{\frac{h-1}{2}}(\Psi_{3x}) + J_{\frac{-h+1}{2}}(\Psi_{3x})|^2 = |J_{\frac{h-1}{2}}(\Psi_{3x}) - J_{\frac{h+1}{2}}(\Psi_{3x})|^2$, where, for large h, the last expression can be approximated in terms of the derivative of a standard Bessel function:

$$\left|J_{\frac{h-1}{2}}(\Psi_{3x}) - J_{\frac{h+1}{2}}(\Psi_{3x})\right|^2 \simeq \left|J_{\frac{h}{2}-1}(\Psi_{3x}) - J_{\frac{h}{2}+1}(\Psi_{3x})\right|^2 = 4\left(\frac{dJ_{h/2}\Psi_{3x}}{d\Psi_{3x}}\right)^2. \quad (2.130)$$

Therefore, using the asymptotic expression for the derivative of the Bessel function in terms of the modified Bessel function K, it is possible to obtain:

$$\left(\frac{dJ_{h/2}\Psi_{3x}}{d\Psi_{3x}}\right)^2 \simeq \frac{h^2\left(1-\frac{4\Psi_{3x}^2}{h^2}\right)^2}{3\pi^2\Psi_{3x}^2}K_{2/3}^2\left[\frac{h}{6}\left(1-\frac{4\Psi_{3x}^2}{h^2}\right)^{3/2}\right]. \quad (2.131)$$

It is possible to find an expression for the square of equation (2.116) in the limit $a_0 \gg 1$ and $\theta = 0$:

$$X^2 = \frac{\pi^2 a_0^2 N_c c^2}{4\gamma_0^2 \omega_0^2}\frac{h^2\left(1-\frac{4\Psi_{3x}^2}{h^2}\right)^2}{3\pi^2\Psi_{3x}^2}K_{2/3}^2\left[\frac{h}{6}\left(1-\frac{4\Psi_{3x}^2}{h^2}\right)^{3/2}\right] \quad (2.132)$$

where we have used the approximation:

$$\left|\left(\frac{e^{i(\Psi_1-h\frac{\omega_0}{c})\zeta}-1}{\Psi_1-h\frac{\omega_0}{c}}\right)\right|^2 \simeq \frac{2\pi\zeta c}{\omega_0}\delta\left(\frac{\omega}{\omega_0}\frac{1+\gamma_0^2\theta^2}{4\gamma_0^2}-h\right) = \frac{4\pi^2 N_c c^2}{\omega_0^2}\delta\left(\frac{\omega}{\omega_0}\frac{1+\gamma_0^2\theta^2}{4\gamma_0^2}-h\right). \quad (2.133)$$

We can now work out the Ψ_{3x}^2/h^2 term:

$$1 - \frac{4\Psi_{3x}^2}{h^2} = 1 - \left(\frac{2a_0^2}{4+2a_0^2}\right)^2 \simeq \frac{4}{a_0^2}. \quad (2.134)$$

After recognizing that $h = a_0^2\omega/8\gamma_0^2\omega_0$ (this comes from the condition $\omega = \omega_h$ by virtue of equation (2.133) and the sum in equation (2.124)), it is eventually possible to recast equation (2.132) into:

$$X^2 = \frac{16N_c c^2}{3a_0^2\gamma_0^2\omega_0^2} K_{2/3}^2\left[\frac{\omega}{6\gamma_0^2 a_0\omega_0}\right]. \tag{2.135}$$

The radiation spectrum is easily found via equation (2.94):

$$\frac{d^2E}{d\omega d\Omega} = \frac{\hbar\alpha\omega^2 X^2}{4\pi^2 c^2} = \frac{4\hbar\alpha N_c}{3\pi^2} \frac{\omega^2}{a_0^2\gamma_0^2\omega_0^2} K_{2/3}^2\left[\frac{\omega}{6\gamma_0^2 a_0\omega_0}\right]. \tag{2.136}$$

A plot of a typical continuum spectrum is provided in figure 2.12 (right). Unlike the harmonic case, the anharmonic case yields a broadband radiation spectrum that can extend to much higher frequencies than the first harmonics. For the provided example, the γ-ray region of the electromagnetic spectrum is reached, where quantum corrections start to be important. We will come back to the latter topic later in this work. Using an analogy with synchrotron radiation, we can interpret the Bessel K function as $K_{2/3}(\omega/2\omega_c)$, where ω_c denotes the critical frequency:

$$\omega_c = 3\gamma_0^2 a_0\omega_0. \tag{2.137}$$

The critical frequency divides the spectrum in two regions of radiation, each carrying the same amount of energy. Figure 2.12 shows the transition from a nonlinear regime to an extremely nonlinear regime. In the former case, for $a_0 = 2$, the harmonics are still well separated and confined to a spectral region extending to a maximum harmonic number below 30. In the latter case, for $a_0 = 10$, the number of harmonics can reach a few thousand, and the spacing approaches zero, giving a continuum. The Bessel $K_{2/3}$ approximation yields a good envelope approximation for the on-axis radiation. Figure 2.13 shows a close-up of the right-hand plot in figure 2.12, focusing on the first radiated harmonics. The envelope approximation

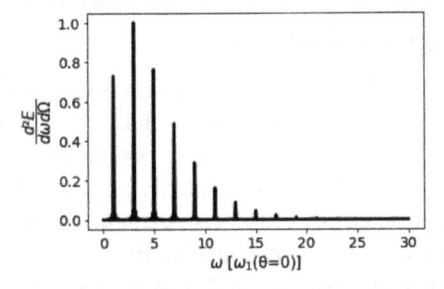

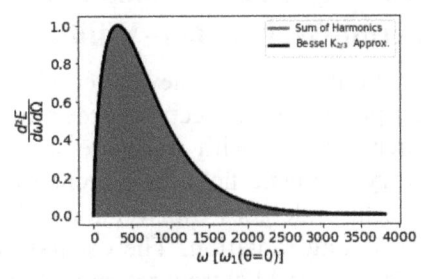

Figure 2.12. Left: A Thomson backscattering radiation spectrum in the nonlinear regime of interaction for $a_0 = 2$. **Right:** A Thomson backscattering radiation spectrum in the extremely nonlinear regime of interaction for $a_0 = 10$. Critical frequency $\omega_c \simeq 1.7$ MeV. The parameters used for both plots are: $\gamma_0 = 200$, $\varphi_x = \varphi_y = 0$, $x_0 = y_0 = z_0 = 0$, $p_x(0) = p_y(0) = 0$, $A_{y0} = 0$, $\omega_0 = 2.35 \times 10^{15}$ Hz, and $\zeta = 10$ μm. The observation angle is fixed at: $\theta = 0$, $\phi = 0$. The plots are normalized to the peak intensity.

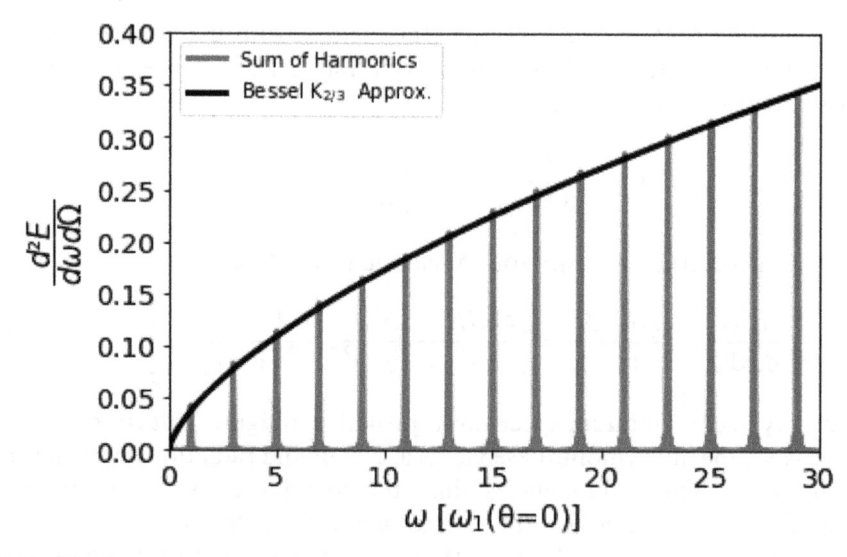

Figure 2.13. First harmonics of a Thomson backscattering radiation spectrum in the extremely nonlinear regime of interaction for the following parameters: $a_0 = 10$, $\gamma_0 = 200$, $\varphi_x = \varphi_y = 0$, $x_0 = y_0 = z_0 = 0$, $p_x(0) = p_y(0) = 0$, $A_{y0} = 0$, $\omega_0 = 2.35 \times 10^{15}$ Hz, and $\zeta = 10$ μm. The observation angle is fixed at: $\theta = 0$, $\phi = 0$. The plot is normalized to the peak intensity. The critical frequency $\omega_c \simeq 1.7$ MeV.

fails somewhat when h is not too large, in agreement with the degree of accuracy associated with the asymptotic expansion in equation (2.131). In particular, for $h < 30$, the envelope underestimates the level of the peaks but overestimates the amount of radiated energy, since the intervals between harmonics should actually be vacuum. However, since the most of the radiated energy is contained by the high-frequency spectrum, the integration of equation (2.136) can give a good approximation of the energy radiated on-axis. A deeper insight into the angular distribution of Thomson backscattering radiation in the highly nonlinear regime of interaction is provided in the Comments and exercises section of this chapter.

2.5 Analogy with the emission in magnetic undulators and the Fermi–Weizsäcker–Williams approximation

Magnetic undulators are devices employed in the operation of x-ray sources driven by high-performance electron accelerators. They consist of a chain of magnetic permanent dipoles with alternate polarity. This arrangement produces a static oscillatory magnetic field, as sketched in figure 2.14. An electron passing through such a field undergoes transverse oscillations, causing the consequent emission of bremsstrahlung radiation. The oscillations of an electron under the action of this static magnetic field closely resemble the trajectories reported in equations (2.20). If we assume that the field is directed in the vertical (y) direction, the electron oscillates in the x–z plane with a spatial period specified by the periodicity fixed by the dipole array. The associated static field in the y-direction can indeed be written as $B_0 \cos(k_0 z)$, which is an approximation valid only in the proximity of the magnet

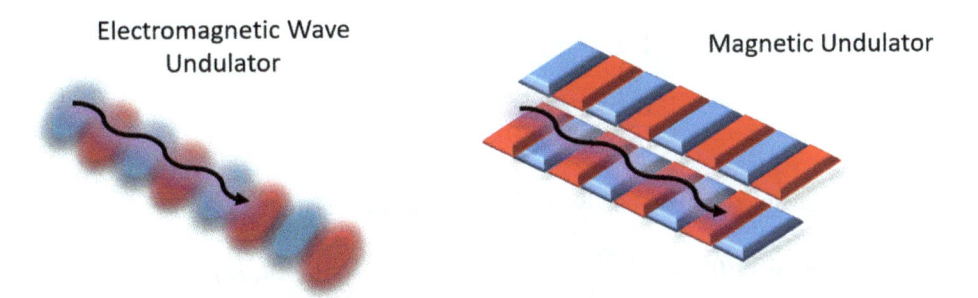

Figure 2.14. **Left:** charge oscillations under the action of an intense electromagnetic wave. Red and blue colors correspond to different signs of the phase of the field. **Right:** charge oscillations between two series of permanent magnetic dipoles with alternate polarity.

axis. To understand this point, we note that, as the field is static, Maxwell's equations impose the $\vec{\nabla} \times \vec{B} = 0$ condition fulfilled by the previously guessed field distribution in the vicinity of x=0, y=0 (for further details, see section 2.6 at the end of the chapter). The static oscillating approximation is fairly effective in describing the electron wiggling around the central part of the undulator. The electron trajectory can be derived by expressing the field in terms of the corresponding vector potential and then proceeding as already done in this chapter. We note indeed that $\vec{B} = \vec{\nabla} \times \vec{A}$; thus, $B_0 \hat{y} = \partial_z A_x$, assuming $A_z = 0$. Similarly to equations (2.6), the electron motion in the three directions of space is given by the differential equations:

$$\frac{dp_x}{dt} = ev_z \partial_z A_x$$
$$\frac{dp_y}{dt} = 0$$
$$\frac{dp_z}{dt} = -ev_x \partial_z A_x$$
$$m_0 c^2 \frac{d\gamma}{dt} = 0. \tag{2.138}$$

Equations (2.138) state that the motion occurs in the x–z plane and that kinetic energy is conserved (the modulus of the velocity is indeed unaffected by the magnetic field). Since $\gamma = \gamma_0$, the equations of motion can be recast into:

$$\frac{dv_x}{dt} = \frac{eB_0 \cos(k_0 z)}{\gamma_0 m_0} v_z$$
$$\frac{dv_z}{dt} = -\frac{eB_0 \cos(k_0 z)}{\gamma_0 m_0} v_x.$$

For ultrarelativistic particles entering straight into an undulator placed along z, one normally has $v_x \ll v_z$ and $v_z \simeq c$, thus:

$$\frac{dv_z}{dt} = 0 \rightarrow v_z = v_0 \simeq c$$
$$\frac{dv_x}{dt} = \frac{eB_0 c}{\gamma_0 m_0} \cos(k_0 ct). \tag{2.139}$$

The motion along x can be easily found after integration:

$$v_x(t) = v_x(0) + \frac{eB_0}{\gamma_0 m_0 k_0} \sin(k_0 ct) \simeq \frac{eB_0}{\gamma_0 m_0 k_0} \sin(k_0 ct)$$
$$x(t) = x_0 - \frac{eB_0}{\gamma_0 m_0 k_0^2 c} \cos(k_0 ct) = x_0 - \frac{K}{\gamma_0 k_0} \cos(k_0 ct) \tag{2.140}$$

where we have assumed that the oscillatory motion imposed by the undulator is much larger than the initial velocity along x (or, similarly, that the particle is moving only along z before entering the undulator) and also defined the so-called undulator *strength parameter* $K = eB_0/m_0 c k_0$. At this point, we may be not satisfied with the rough approximation $v_z \simeq c$; therefore, we obtain a better approximation for the axial velocity of the electron:

$$\frac{dv_z}{dt} = -\frac{eB_0 \cos(k_0 ct)}{\gamma_0 m_0} v_x \simeq -\frac{K^2 c^2 k_0}{\gamma_0^2} \cos(k_0 ct)\sin(k_0 ct) \rightarrow v_z(t) \simeq c - \frac{K^2 c}{2\gamma_0^2} \sin^2(k_0 ct)$$
$$z(t) \simeq ct - \frac{K^2 c}{2\gamma_0^2} \int_0^t \sin^2(k_0 ct')dt'. \tag{2.141}$$

With the replacement $K \rightarrow a_0/2$, the electron dynamics has significant analogies with the discussion already developed for the Thomson backscattering problem. Therefore, we have established an analogy between magnetic and electromagnetic undulators in terms of the electron dynamics. Concerning the radiation, it should be noted that the independent variable here is not ζ but t: this leads to somewhat relevant differences in the emitted spectrum. The coefficient Ψ_1 is now:

$$\Psi_1 \simeq \frac{\omega}{c}\left(1 - \frac{v_0}{c}\cos\theta + \frac{K^2}{4\gamma_0^2}\cos\theta\right) \simeq \frac{\omega}{c}\left(\frac{1 + \gamma_0^2\theta^2 + \frac{K^2}{2}}{2\gamma_0^2}\right). \tag{2.142}$$

Thus, following an analogous calculation to that of the Thomson scattering case, the emitted harmonics are defined by:

$$\omega_h = \frac{2\gamma_0^2 h \omega_0}{1 + \gamma_0^2 \theta^2 + \frac{K^2}{2}}. \tag{2.143}$$

We cannot conclude this brief description of the photon emission in magnetic undulators without noting the similarity between equations (2.142) and the expression for the frequency of the backscattered photons (and the associated harmonics). The dependence of the radiated frequency on the square of the electron energy and a shift linked to the field intensity suggest that the processes can be traced back to similar mechanisms. However, a snap conclusion in this direction may be misleading if not properly framed within the proper environment, which goes beyond classical electromagnetism; it is touched upon in the following subsection and in the exercises at the end of the chapter.

2.5.1 The Fermi–Weizsäcker–Williams approximation

The emission process due to ultrarelativistic electrons in magnetic undulators was originally treated by Madey using a CBS point of view, which assumed that the static magnetic field could be treated as an ensemble of (pseudo) photons carrying a (near) transverse field. The idea was pursued, and it was pointed out that the photons associated with this field are massive, which is also the case for photons in a waveguide. For further comments, see the exercises in section 2.6 and the bibliography at the end of the chapter. The Madey treatment was later recognized as a kind of virtual quanta method, even though, worded in these terms, this statement may generate some confusion. The Thomson method described in chapter 1 yields a clear picture of the mechanisms underlying the emission of radiation by accelerated charges within the context of a classical picture. The relevant quantum extension was achieved, in almost contemporary investigations, by Fermi, Weizsäcker, and Williams, who realized that the field of a relativistic electron is predominantly transverse (and thus amenable to canonical quantization), and for this reason it behaves similarly to an electromagnetic plane wave in scattering problems. A Lorentz-boosted transverse magnetostatic field is also transverse and similar to an electromagnetic plane wave. In this section, we demonstrate that an undulator field witnessed by the electron in its own reference frame behaves like an electromagnetic wave. Therefore, it can be scattered as such, similarly to the process of Thomson scattering. We consider the magnetic field $B_0 \cos(k_0 z)\hat{y}$ in the reference frame of an electron moving toward the field at velocity $v_0 \hat{z}$. Via the Lorentz transformation, it is possible to obtain:

$$z' = \gamma_0(z - v_0 t) \simeq \gamma_0(z - ct)$$
$$\vec{E}' = \gamma_0 v_0 \hat{z} \times B_0 \cos(k_0 z)\hat{y} \simeq -\gamma_0 c B_0 \cos(k_0 z)\hat{x} = -\gamma_0 c B_0 \cos(\gamma_0 k_0 z' + \gamma_0 \omega_0 t')\hat{x} \quad (2.144)$$
$$\vec{B}' = \gamma_0 B_0 \cos(\gamma_0 k_0 z' + \gamma_0 \omega_0 t')\hat{y}$$

where $\omega_0 = k_0 c$. The electromagnetic field in the reference frame of the electron can be rewritten in terms of a vector potential:

$$\vec{A}' = \frac{B_0}{k_0} \sin(\gamma_0 k_0 z' + \gamma_0 \omega_0 t')\hat{x}. \quad (2.145)$$

In the reference frame of the electron, in the linear regime of interaction, the scattered field is just proportional to the incident field $\vec{A}'$ and has the same frequency, $\gamma_0 \omega_0$. However, the frequency emitted toward the observer in the laboratory frame, where the electron is moving along the positive axis z toward the detector, must be found with a Lorentz boost:

$$\omega_1 = \gamma_0(\gamma_0\omega_0 + \gamma_0 k_0 v_0 \cos\theta) \simeq \gamma_0^2 \omega_0 (1 + \beta_0 \cos\theta) \simeq \frac{\omega_0}{1 - \beta_0 \cos\theta} \simeq \frac{2\gamma_0^2 \omega_0}{1 + \gamma_0^2 \theta^2}. \quad (2.146)$$

In conclusion, with a Lorentz boost, we have treated the magnetostatic field of the undulator interacting with the traveling electron as an electromagnetic wave impinging on an electron at rest in its own frame. We have called this procedure the Fermi–Weizsäcker–Williams approximation, by analogy with the work of these eminent scientists who found a smart method to calculate bremsstrahlung spectra by treating the Coulomb field of a relativistic electron as an electromagnetic plane wave impinging on an atomic field. The result confirms the findings of previous sections, in which the Lorentz contraction and the relativistic Doppler effect determine the emitted radiation spectrum in magnetic undulators.

2.6 Comments and exercises

2.6.1 Exercises

Exercise 2.1. Derive equation (2.73).
  Hint: Use the Jacobi–Anger expansion (equation (2.64)) and suitable reindexing.

Exercise 2.2. Derive equation (2.98) from the properties of the Dirac delta of a function.
  Hint:
  The Dirac delta of a function is defined as:

$$\delta(f(x)) = \sum_n \frac{\delta(x - x_n)}{|f'(x_n)|} \quad (2.147)$$

where x_n are the zeros of the function $f(x)$, i.e. $f(x_n) = 0$.

Exercise 2.3. Evaluate the critical value of the relativistic parameter a_0 for which classical arguments are no longer valid.
  Solution:
  A naïve way to make the requested estimation is to impose the condition that the critical photon energy is lower than the electron energy, i.e. $\hbar\omega_c < \gamma_0 m_0 c^2$. Thus, after inversion of the expression, one obtains:

$$a_0 < \frac{2 m_0 c^2}{3 \gamma_0 \hbar \omega_0}. \quad (2.148)$$

For $\gamma_0 \simeq 200$ (i.e. for ∼100 MeV electrons) and a Ti:Sa laser wavelength centered at 800 nm, the relativistic parameter must be smaller than ∼1100 for the classical description to be applicable. In the opposite case, i.e. $a_0 > 1100$, a quantum description is surely needed (actually, a quantum description is needed as soon as the emitted photon energy starts being comparable with the electron energy, so it is needed even before the threshold value of 1100 is reached). However, a value as large as 1000 is unreachable with current technology based on Ti:Sa, since it would require an energy per pulse of the order of a fraction of a megajoule.

Exercise 2.4. Calculate the Thomson backscattering spectral–angular distribution of a generic harmonic h for the slightly nonlinear regime of interaction.
Solution:
We assume that the harmonics are well separated, which we know to be the case for sufficiently long interaction times. The spectral–angular distribution for the generic harmonic is then, according to equation (2.94):

$$\left.\frac{d^2N}{d\omega d\Omega}\right|_h \simeq \frac{\alpha\omega\zeta^2}{64\pi^2c^2}\frac{a_0^2}{\gamma_0^2}\left[(J_{h+1}+J_{h-1})^2 + \frac{4J_h^2}{a_0^2}\gamma_0^2\theta^2 - \frac{4\gamma_0 J_h(J_{h+1}+J_{h-1})}{a_0}\theta\cos\phi\right]\text{sinc}^2\left[\left(\frac{\omega}{\omega_s}-h\right)\frac{\omega_0\zeta}{2c}\right] \quad (2.149)$$

where $J_h = J_h(\Psi_{2x}, -\Psi_{3x})$. Indeed, we have left an explicit dependence upon Ψ_{3x}, since, in the study of harmonics where $h > 1$, the approximation $\Psi_{3x} = 0$ leads to divergent results.

Exercise 2.5. Verify that equation (2.149) is $\propto a_0^{2h}$, as discussed for equation (2.109).
Hint:
$\Psi_{2x}^h \propto a_0^h$ according to equation (2.112).

Exercise 2.6. Verify that, for $h = 1$ and $a_0 \ll 1$, $\hbar\omega$ times equation (2.149) reduces to equation (2.96).
Hint:
The relevant value of the gamma function for $h = 1$ is $\Gamma(1 + h) = \Gamma(2) = 1$.

Exercise 2.7. Calculate the Thomson backscattering spectral distribution of a generic harmonic h for the slightly nonlinear regime of interaction.
Hint:
Use the following equation to approximate the sinc function in equation (2.149):

$$\text{sinc}^2\left[\left(\frac{\omega}{\omega_0}\frac{1+\gamma_0^2\theta^2}{4\gamma_0^2}-h\right)\frac{\omega_0\zeta}{2c}\right] \simeq \pi\delta\left[\left(\frac{\omega}{\omega_0}\frac{1+\gamma_0^2\theta^2}{4\gamma_0^2}-h\right)\frac{\omega_0\zeta}{2c}\right] = \frac{4\pi c}{\omega\zeta}\frac{\delta(\theta-\theta^*)}{\theta^*} \quad (2.150)$$

where $\theta_h^* = \sqrt{4h\gamma_0^2\omega_0/\omega - 1}/\gamma_0$. The integration over θ is trivial, while the integration over φ can only be performed using an asymptotic expansion of the generalized

Bessel functions, which is possibly more accurate than the expansion provided in equation (2.119).

Exercise 2.8. Verify that, for $h = 1$, the angular integration of equation (2.149) reduces to equation (2.101).
Hint:
Consider the limits $\Psi_{3x} = 0$ and $\Psi_{2x} \ll 1$. This allows us to write: $J_1(\Psi_{2x}, -\Psi_{3x}) \simeq J_1(\Psi_{2x}) \simeq \Psi_{2x}/2$.

Exercise 2.9. Calculate the number of Thomson backscattering photons emitted in the second harmonic for the slightly nonlinear regime of interaction.
Solution:
We integrate equation (2.120) or equation (2.121), obtaining:

$$N_2 = \frac{21 a_0^2}{40} \frac{2\pi \alpha a_0^2 N_c}{3} = \frac{21 a_0^2}{40} N_1. \tag{2.151}$$

The above result has been obtained by integrating over the whole radiation spectrum, i.e. including the Doppler redshifted radiation which actually overlaps with the first harmonic. Therefore, the presence of the second harmonic also implies more radiation in the region of the first harmonic, but this is a correction of order a_0^2. However, if one considers the integral upon frequencies in the region from the first to the second harmonic, one can end up with:

$$N_2 = \frac{57 a_0^2}{160} N_1. \tag{2.152}$$

Exercise 2.10. Calculate the ratio between the numbers of Thomson backscattering photons emitted in the first two harmonics.
Solution:
Using equations (2.102) and (2.151), we get:

$$\frac{N_2}{N_1} = \frac{21 a_0^2}{40} = 0.525 a_0^2. \tag{2.153}$$

Therefore, for the slightly nonlinear regime of interaction, the number of photons emitted in the second harmonic is lower than that emitted in the first harmonic (for $a_0 \sim 1$, it is about half). If we instead only consider the radiation above the first-harmonic threshold, then:

$$\frac{N_2}{N_1} = \frac{57 a_0^2}{160} \simeq 0.356 a_0^2. \tag{2.154}$$

Exercise 2.11. Calculate the nonlinear Thomson cross section for the second-harmonic radiation.

Solution:
We use the definition of the cross section given in equation (2.103) together with equation (2.153):

$$\sigma_2 = \frac{N_2}{F_0} = \frac{N_2}{N_1}\frac{N_1}{F_0} = \frac{N_2}{N_1}\sigma_T = \frac{21a_0^2}{40}\frac{8\pi}{3}r_0^2. \qquad (2.155)$$

If we instead only consider the radiation above the first-harmonic threshold, then:

$$\sigma_2 = \frac{57a_0^2}{160}\sigma_T. \qquad (2.156)$$

Exercise 2.12. Calculate the Thomson backscattering angular distribution of a generic harmonic h for the slightly nonlinear regime of interaction.
  Hint:
  Use the following equation to approximate the sinc function in equation (2.149):

$$\text{sinc}^2\left[\left(\frac{\omega}{\omega_0}\frac{1+\gamma_0^2\theta^2}{4\gamma_0^2} - h\right)\frac{\omega_0\zeta}{2c}\right] \simeq \frac{2\pi c}{\zeta}\frac{4\gamma_0^2}{1+\gamma_0^2\theta^2}\delta\left(\omega - \frac{4h\gamma_0^2\omega_0}{1+\gamma_0^2\theta^2}\right). \qquad (2.157)$$

The integration over ω is trivial, while the integration over φ can only be performed using an asymptotic expansion of the generalized Bessel functions, which is possibly more accurate than the one provided in equation (2.119).

Exercise 2.13. Verify that for $h = 1$, the spectral integration of equation (2.149) reduces to equation (2.107).
  Hint:
  Consider the limits $\Psi_{3x} = 0$ and $\Psi_{2x} \ll 1$. This allows us to write: $J_1(\Psi_{2x}, -\Psi_{3x}) \simeq J_1(\Psi_{2x}) \simeq \Psi_{2x}/2$.

Exercise 2.14. Calculate the angular width of the second-harmonic radiation.
  Solution:
  Using equation (2.120), we calculate the width from its definition:

$$<\theta>_2 = \frac{\int d\Omega \frac{dN}{d\Omega}\big|_2 \theta}{\int d\Omega \frac{dN}{d\Omega}\big|_2} = \frac{5\pi}{16\gamma_0}. \qquad (2.158)$$

Exercise 2.15. Find the spectral–angular distribution of Thomson backscattering radiation in the highly nonlinear regime of interaction.

Solution:
We express the currents $\mathscr{J}$ in a general form, obtaining:

$$\mathscr{J}_z = \sum_h \mathscr{J}_h(\Psi_{2x}, -\Psi_{3x}) \int_0^\zeta d\zeta' \frac{dz}{d\zeta'} e^{i(\Psi_1 - h\frac{\omega_0}{c})\zeta'} \simeq \sum_l \mathscr{J}_{h(\omega)+2l}(\Psi_{2x}) \mathscr{J}_l(\Psi_{3x}) \frac{\pi\sqrt{N_c}\,c}{\omega_0} \quad (2.159)$$

$$\mathscr{J}_x \simeq \sum_l [\mathscr{J}_{h(\omega)+2l+1}(\Psi_{2x}) + \mathscr{J}_{h(\omega)+2l-1}(\Psi_{2x})]\mathscr{J}_l(\Psi_{3x}) \frac{\pi a_0 \sqrt{N_c}\,c}{2\gamma_0 \omega_0} \quad (2.160)$$

where we have summed over h using the Dirac delta approximation in equation (2.133) and defined $h(\omega)$ via the condition:

$$\omega = \omega_h \rightarrow h(\omega) = \frac{a_0^2 \omega}{8\gamma_0^2 \omega_0}. \quad (2.161)$$

The spectral–angular distribution is, finally:

$$\frac{d^2 E}{d\omega d\Omega} = \frac{e^2 \omega^2}{16\pi^3 \varepsilon_0 c^3} |\hat{n} \times \vec{\mathscr{J}}|^2. \quad (2.162)$$

Exercise 2.16. Find the spectral–angular distribution of Thomson backscattering radiation in the highly nonlinear regime of interaction for $\phi = \pi/2$, when the oscillation is along x.
Solution:
For $\phi = \pi/2$, i.e. for an observation angle which is orthogonal to the plane of charge oscillation, $\Psi_{2x} = 0$, since $\Psi_{2x} \propto \cos\phi$. Let us now work out the Ψ_{3x}^2/h^2 term:

$$1 - \frac{4\Psi_{3x}^2}{h^2} = 1 - 4\left(\frac{a_0^2}{4\left(1 + \gamma_0^2 \theta^2 + \frac{a_0^2}{2}\right)}\right)^2 \simeq \frac{4}{a_0^2}(1 + \gamma_0^2 \theta^2). \quad (2.163)$$

The horizontal current $\mathscr{J}_x$, in analogy with the approach of section 2.4.3, is simply:

$$\mathscr{J}_x \simeq -i \frac{4\sqrt{N_c}\,c}{\gamma_0 \omega_0} \frac{(1 + \gamma_0^2 \theta^2)}{\sqrt{3}\, a_0} K_{2/3}\left[\frac{\omega}{2\omega_c}(1 + \gamma_0^2 \theta^2)^{3/2}\right] \quad (2.164)$$

where it is now possible to define a θ-dependent critical frequency, i.e.:

$$\omega_c(\theta) = \frac{3 a_0 \gamma_0^2 \omega_0}{(1 + \gamma_0^2 \theta^2)^{3/2}}. \quad (2.165)$$

For $\mathscr{J}_z$, starting from equation (2.159), it is possible to obtain:

$$\mathscr{J}_z \simeq \sum_{h,l} \mathscr{I}_{h(\omega)+2l(0)} \mathscr{I}_l(\Psi_{3x}) \frac{\pi\sqrt{N_c}c}{\omega_0} = \sum_{h,l} \delta_{h(\omega)+2l,0} \mathscr{I}_l(\Psi_{3x}) \frac{\pi\sqrt{N_c}c}{\omega_0} = \mathscr{I}_{\frac{h(\omega)}{2}}(\Psi_{3x}) \frac{2\pi\sqrt{N_c}c}{\omega_0} \quad (2.166)$$

where we have used the fact that only even harmonics are associated with $\mathscr{J}_z$, therefore $\mathscr{I}_{-h/2} = \mathscr{I}_{h/2}$. Equation (2.166) can be further simplified via an asymptotic expansion similar to the one used for $\mathscr{J}_x$:

$$\mathscr{J}_z \simeq \frac{2\pi\sqrt{N_c}c}{\omega_0} \frac{\sqrt{1-\frac{4\Psi_{3x}^2}{h^2}}}{\sqrt{3}\pi} K_{1/3}\left[\frac{h}{6}\left(1-\frac{4\Psi_{3x}^2}{h^2}\right)^{3/2}\right] = \frac{4\sqrt{N_c}c}{\sqrt{3}\omega_0 a_0}\sqrt{1+\gamma_0^2\theta^2}\, K_{1/3}\left[\frac{\omega}{2\omega_c(\theta)}\right]. \quad (2.167)$$

The spectral–angular distribution for $\phi = \pi/2$ is, finally:

$$\frac{d^2E}{d\omega d\theta} = \frac{4\hbar\alpha N_c\omega^2\theta}{3\pi^2 a_0^2 \gamma_0^2 \omega_0^2}(1+\gamma_0^2\theta^2)^2\left(K_{2/3}^2\left[\frac{\omega}{2\omega_c(\theta)}\right] + \frac{\gamma_0^2\theta^2}{1+\gamma_0^2\theta^2}K_{1/3}^2\left[\frac{\omega}{2\omega_c(\theta)}\right]\right) \quad (2.168)$$

where $d\Omega$ has been replaced by $d\theta$, since ϕ is fixed and θ just represents the vertical angle. The angular width of the backscattered radiation for $\phi = \pi/2$, by analogy with synchrotron radiation, is of the order of $1/\gamma_0$.

2.6.2 Undulator radiation and Compton backscattering

We have underscored that the emission in magnetic undulators is a mechanism that generates short-wavelength photon beams. An ultrarelativistic electron going through an undulator emits, in the forward direction, radiation at a wavelength $\lambda_s = \frac{\lambda_u}{2\gamma_0^2}\left(1+\frac{K_u^2}{2}\right)$. We have already underscored the perfect analogy with the CBS wavelength. In particular, it has been noted that, upon replacing the undulator period λ_u with $\lambda_0/2$ and the undulator parameter strength K_u with a_0, the CBS wavelength is recovered from the undulator counterpart.

Exercise 2.17. Explain why the undulator period and the laser wavelength are not equivalent in the definition of the scattered wavelength.
  Solution:
  The following argument is usually exploited to get the wavelength of the undulator radiation. Electrons and light move at different velocities. At the end of an undulator period, the emitted radiation is ahead by a factor of $(1-\beta_0)\lambda_u$. This is a phase shift, and to avoid negative interference effect we must have

$$(1-\beta_0)\lambda_u = \lambda_s. \quad (2.169)$$

Hence, for ultrarelativistic electrons, we find $\frac{1}{2\gamma^2}\lambda_u = \lambda_s$. In the case of CBS, the laser field is not static and is counter-propagating with respect to the electrons. Reiterating the same argument as before, we find:

$$\frac{(1-\beta_0)}{1+\beta_0}\lambda_0 = \lambda_s \quad (2.170)$$

which, for ultrarelativistic electrons, eventually yields

$$\frac{1}{4\gamma_0^2}\lambda_0 = \lambda_s. \tag{2.171}$$

Exercise 2.18. Explain why the definition of the strength parameter for CBS is the same as for an undulator, in spite of the fact that the laser equivalent u-period is $\lambda_0/2$.
Solution:
Recall that

$$K_u = \frac{eB_u\lambda_u}{2\pi\, m_e c}, \qquad a_0 = \frac{eE_0\lambda_0}{2\pi\, m_e c^2}. \tag{2.172}$$

Accordingly, we convert from one parameter to the other via the substitutions

$$B_u \to \frac{E_0}{c} \qquad \lambda_u \to \lambda_0, \tag{2.173}$$

which are in contradiction with the results of the previous discussion.

It should again be stressed that the electron and laser are counter-propagating and therefore the electric and magnetic forces enhance each other. The equivalent undulator field (namely the magnetic field associated with the laser wave) can accordingly be written as

$$B_0 = B + \frac{E}{\beta_0 c} = \frac{1+\beta_0}{\beta_0}B \tag{2.174}$$

for ultrarelativistic electrons $B_0 \cong 2B$, and therefore the contradiction is resolved.

Exercise 2.19. What is the equivalent Poynting vector of the undulator.
Hint:
Use the notion acquired in chapter 1 and prove that

$$I_u = \frac{c^2 B_u^2}{Z_0}. \tag{2.175}$$

Exercise 2.20. Find the equivalent magnetic field for a laser with intensity $I_0 = 10^{19} \text{Wm}^{-2}$.
Hint:
Note that in **exercise 2.6.19**, we found:

$$B_0 = \frac{\sqrt{Z_0 I_0}}{c} \cong 204.5\text{T}. \tag{2.176}$$

Exercise 2.21. The on-axis field of an undulator realized with permanent magnets is no larger than 10 T. According to the previous exercise, the laser magnetic field can

easily reach tens of teslas. With such a magnetic field, the undulator would be dominated by nonlinear effects. Why does this not happen in the case of CBS?
Hint:
Recall the definition of the strength parameter that includes the laser wavelength.

Exercise 2.22. Use the notions learned in this chapter to derive the number of photons emitted by a single electron in an undulator magnet of length $L_u = N_u \lambda_u$.
Hint:
It is easy to confirm that:

$$N_{\mathrm{ph}} = \frac{4}{3}\pi\, \alpha\, K_u^2 N_u. \tag{2.177}$$

2.6.3 The properties of generalized Bessel functions

The proposal to generalize the ordinary Bessel functions to forms with more than one variable can be traced back to the beginning of the twentieth century. They were forgotten for a while and then rediscovered, almost five decades later, within the framework of physical problems involving non-dipolar scattering problems. The dipole interaction applies, for example, to problems related to atom–radiation interaction where the wavelength is significantly longer than the atomic structure. The validity of the approximation has been shown to break for large radiation intensities, which are nowadays easily seen during investigations of the process of atomic ionization in strong laser fields. In this process, the electric laser field induces large electron velocities, with the consequence that the electron trajectory is modified by the magnetic component of the laser field, which is responsible for a distortion of the electron trajectory itself (figure 2.15). This is a different flavor of the discussion we had in the main body of the chapter. We have indeed underscored that strong field effects manifest themselves through so-called figure-of-eight motion, which is due to the combination of transverse and horizontal components of the motion. The importance of the latter is that it is the direct signature of high-intensity effects. We have also seen that, from the

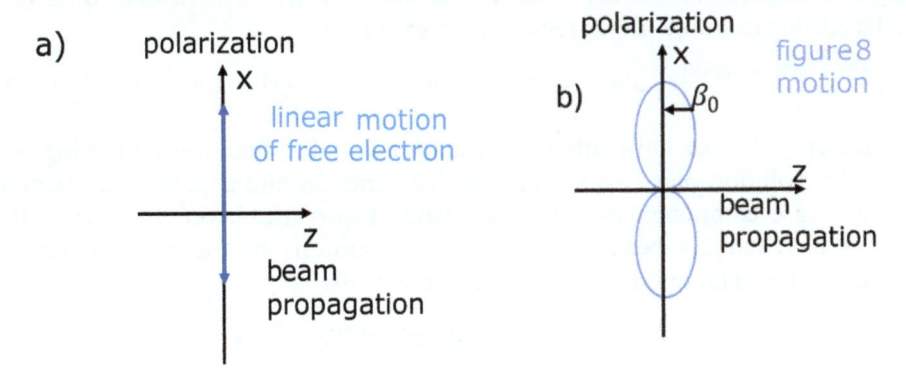

Figure 2.15. A summary of (a) the dipole approximation and (b) figure-of-eight motion.

mathematical point of view, this motion distortion is responsible for the emergence of new terms in the radiation integral, which can be treated using an extension of the Bessel function given in equations (2.68). Here, we summarize various properties used during the previous discussion (even though they were not explicitly mentioned), listing recurrences, integral representation, etc. For a more accurate treatment, the reader is addressed to the specialized literature on this topic.

Reflection properties

$$J_n(-x, y) = (-1)^n J_n(x, y)$$
$$J_n(x, -y) = \sum_l J_{n-2l}(x) J_l(-y) = \sum_l (-1)^l J_{n+2l}(x) J_l(-y) \qquad (2.178)$$
$$J_{-n}(x, y) = (-1)^n J_n(x, -y)$$

Limit cases

$$\lim_{y \to 0} J_n(x, y) = J_n(x)$$

$$\lim_{x \to 0} J_n(x, y) = \begin{pmatrix} J_{\frac{n}{2}}(y) & n - \text{even} \\ 0 & n - \text{odd} \end{pmatrix}$$

$$J_n(x, y)\Big|_{x \ll 1} \simeq \begin{pmatrix} J_{\frac{n}{2}}(y) + \frac{x^2}{8}\left[J_{\frac{n+2}{2}}(y) + J_{\frac{n-2}{2}}(y) - 2J_{\frac{n}{2}}(y)\right] & n - \text{even} \\ J_1(x)\left[J_{\frac{n-1}{2}}(y) - J_{\frac{n+1}{2}}(y)\right] + J_3(x)\left[J_{\frac{n-3}{2}}(y) - J_{\frac{n+3}{2}}(y)\right] & n - \text{odd} \end{pmatrix} \qquad (2.179)$$

$$J_n(x, y)\Big|_{y \ll 1} \simeq J_n(x)\left(1 - \frac{y^2}{4}\right) + \frac{y}{2}\left(1 - \frac{y^2}{8}\right)[J_{n-2}(x) - J_{n+2}(x)] + \frac{y^2}{8}[J_{n-4}(x) + J_{n+4}(x)]$$

Recurrence identities

$$\frac{\partial}{\partial x} J_n(x, y) = \frac{1}{2}[J_{n-1}(x, y) - J_{n+1}(x, y)]$$
$$\frac{\partial}{\partial y} J_n(x, y) = \frac{1}{2}[J_{n-2}(x, y) - J_{n+2}(x, y)] \qquad (2.180)$$
$$2n J_n(x, y) = x[J_{n-1}(x, y) + J_{n+1}(x, y)] + 2y[J_{n-2}(x, y) + J_{n+2}(x, y)]$$

Along with the previous family of Bessel functions, further generalized forms can be introduced; for example, the three-variable extension:

$$e^{ix \sin \theta + iy \sin 2\theta + iz \sin 3\theta} = \sum_n J_n(x, y, z) e^{in\theta} \quad \text{where} \quad J_n(x, y, z) = \sum_l J_{n-3l}(x, y) J_l(z). \qquad (2.181)$$

In equation (2.72) we have introduced a four-variable function consisting of the discrete convolution of a two-variable Bessel function and a parameter associated with the phase components of the laser field. Two-variable/one-parameter Bessel functions have largely been employed in applications of the physical problems discussed so far. The relevant series definition reads:

$$J_n(x, y | \tau) = \sum_l \tau^l J_{n-2l}(x) J_l(y). \qquad (2.182)$$

The associated generating function can be written as:

$$\sum_{n=-\infty}^{n=+\infty} t^n J_n(x, y|\tau) = e^{\frac{x}{2}(t-\frac{1}{t})+\frac{y}{2}(\tau t^2 - \frac{1}{\tau t^2})}$$

$$\sum_{n=-\infty}^{n=+\infty} e^{in\theta} J_n(x, y|e^{i\varphi}) = e^{ix\sin(\theta)+\frac{y}{2}\sin(2\theta+\varphi)}$$

(2.183)

and its integral representation is:

$$J_n(x, y|e^{i\varphi}) = \frac{1}{2\pi}\int_{-\pi}^{+\pi} \cos[x\sin(\theta) + y\sin(2\theta + \varphi) - n\theta]d\theta.$$

(2.184)

Regarding the example we have introduced to study Thomson scattering in fairly general terms, we note that:

$$J_n(x, y; z, w|\tau) = \sum_{l=-\infty}^{l=+\infty} J_{n-l}(x, y) J_l(z, w|\tau).$$

(2.185)

2.6.4 The Hamilton–Jacobi approach to the dynamics of electrons interacting with a plane wave

In this section, we use the first principles of relativistic mechanics to calculate the trajectories of electrons when they interact with a plane electromagnetic wave (which is not necessarily monochromatic). For this purpose, the Hamilton–Jacobi formalism will be adopted. The four-dimensional form of the Hamilton–Jacobi equation for the electron–field interaction is:

$$g^{\mu\nu}\left(\frac{\partial S}{\partial x^\mu} - \frac{e}{c}A_\mu\right)\left(\frac{\partial S}{\partial x^\nu} - \frac{e}{c}A_\nu\right) = m_0^2 c^2.$$

(2.186)

We make planar symmetry explicit in the four-vector potential argument by setting $A_\mu \equiv A_\mu(\xi)$, with $\xi = k_\mu x^\mu = \omega_0(t + z/c)$, where the positive sign in the phase expression indicates a head-on interaction in the z direction; we then impose the Lorentz gauge condition:

$$\frac{\partial A^\mu}{\partial x^\mu} = \frac{\partial A^\mu}{\partial \xi} k_\mu = 0$$

(2.187)

which, due to the planarity of the field, is equivalent to $A^\mu k_\mu = 0$, i.e. to a transversality condition. To find Hamilton's principal function, we look for a solution of the type:

$$S = -f_\mu x^\mu + F(\xi)$$

(2.188)

where f_μ is the four-vector that satisfies the condition $f_\mu f^\mu = m_0^2 c^2$, i.e. the generalized four-momentum of the electron under the action of an electromagnetic field. Inserting equation (2.188) into equation (2.186), we obtain:

$$2\eta\frac{\partial F}{\partial \xi} - 2\frac{e}{c}f_\mu A^\mu - \frac{e^2}{c^2}A_\mu A^\mu = 0$$

(2.189)

where $\eta = f_\mu k^\mu$. From the above, we can deduce the expression for F and therefore for S.

$$S = -f_\mu x^\mu + \frac{ef_\mu}{\eta} \int A^\mu d\xi + \frac{e^2}{2\eta} \int A^\mu A_\mu d\xi. \qquad (2.190)$$

Since $k^\mu = (\omega_0/c, 0, 0, -\omega_0/c)$, we obtain the following for η:

$$\eta = \frac{\omega_0}{c} f_0 + f_3. \qquad (2.191)$$

Expanding the square of f_μ, we have:

$$f_0^2 - f_3^2 - \kappa^2 = m_0^2 c^2 \qquad (2.192)$$

where κ denotes the modulus of the generalized transverse four-momentum of the particle: the planarity of the field, in fact, entails the conservation of $\vec{\kappa}$, which is naturally interpreted as the initial transverse momentum of the electron just before it interacts with the wave. At this point we introduce the following identity:

$$f^3 x^3 - f^0 x^0 = \frac{(f^3 + f^0)(x^3 - x^0)}{2} + \frac{(f^3 - f^0)(x^3 + x^0)}{2}$$

and by means of the previous equations, we obtain the following expression for S:

$$S = \vec{p}_\perp(0) \cdot \vec{r}_\perp - \frac{c\eta}{2\omega_0}(ct - z) - \frac{p_\perp^2(0) + m^2 c^2}{2\eta}\xi + \frac{ef_\mu}{\eta}\int A^\mu d\xi + \frac{e^2}{2\eta}\int A_\mu A^\mu d\xi. \qquad (2.193)$$

By equating the derivatives of S with respect to the $\vec{p}_\perp(0)$ components and with respect to η to some constants (initial positions), we obtain the following for the electrons' trajectories:

$$\vec{r}_\perp = \vec{r}_\perp(0) + \frac{\vec{p}_\perp(0)}{\eta}\xi + \frac{e}{\eta}\int \vec{A} d\xi$$

$$z = z_0 + \left[\frac{c}{2\omega_0} - \frac{p_\perp^2(0) + m^2 c^2}{2\eta^2}\left(\frac{\omega_0}{c}\right)\right]\xi - \frac{\omega_0 e}{c\eta^2}\vec{p}_\perp(0) \cdot \int \vec{A} d\xi - \frac{\omega_0 e^2}{2c\eta^2}\int A_\mu A^\mu d\xi. \qquad (2.194)$$

We can work out the time derivative of S (which corresponds to the opposite of the electron's energy $E = \gamma(t)mc^2$) to arrive at an expression for the Lorentz invariant η, evaluated at the time's origin:

$$\eta = \left(\frac{\omega_0}{c}\right)\gamma_0 m_0 c + p_z(0). \qquad (2.195)$$

By performing the integrations in equations (2.194), the electrons' trajectories can be established for any plane temporally shaped wave packet. It is eventually possible to verify that equations (2.194) are equivalent to equations (2.20).

2.6.5 Beyond the plane wave approximation

Equations (2.20) express the dynamics of electrons under the action of an intense plane electromagnetic wave. However, in most real cases of the Thomson backscattering interaction, the incident electromagnetic pulse is not perceived as a plane wave by the electrons. In fact, to maximize the yield of photons, it is customary to focus the electromagnetic wave at the point of interaction, which is achieved by bending the incident wave front, which, consequently, is far from flat. In this section, we will adopt a perturbative approximation, since the problem of electron dynamics under the action of an intense non-plane wave cannot be solved analytically otherwise. The aim is to understand the differences that a non-plane wave makes to the dynamics of electrons and scattered radiation. To account for a linearly polarized wave moving along x, equations (2.6) can be recast into:

$$\frac{dp_x}{dt} = -e\left(-\frac{\partial A_x}{\partial t} - v_z \partial_z A_x - v_y \partial_y A_x\right)$$

$$\frac{dp_y}{dt} = -ev_x \partial_y A_x$$

$$\frac{dp_z}{dt} = -ev_x \partial_z A_x$$

$$m_0 c^2 \frac{d\gamma}{dt} = e\frac{\partial A_x}{\partial t} v_x.$$

(2.196)

We can observe that the last two equations of (2.196) are identical to the last two equations of (2.6); therefore, they give the same result for the electron dynamics along z and the same constant of motion η. The electron energy is essentially conserved (see equation (2.24) and the discussion below equation (2.27)). Therefore, it is possible to write the second of equations (2.196) as:

$$\frac{dv_y}{dt} = -\frac{ev_x}{\gamma_0 m_0}\partial_y A_x.$$

(2.197)

We recognize that the x-velocity can be approximated by:

$$v_x \simeq v_x(0) + \frac{a_x c}{2\gamma_0}$$

(2.198)

where $a_x = eA_x/m_0 c$. Therefore, plugging equation (2.198) into equation (2.197) gives:

$$\frac{dv_y}{dt} = -\frac{c}{\gamma_0}\left(v_x(0)\partial_y a_x + \frac{c}{4\gamma_0}\partial_y a_x^2\right).$$

(2.199)

The first of equations (2.196) can be recast by expressing the total time derivative:

$$\frac{\partial v_x}{\partial t} + \frac{1}{2}\partial_x v_x^2 = -\frac{e}{\gamma_0 m_0}\left(-\frac{\partial A_x}{\partial t} - v_z \partial_z A_x - v_y \partial_y A_x\right).$$

(2.200)

Assuming that $v_x(x, y, \zeta) = v_x(\zeta) + \delta v_x$, where $v_x(\zeta)$ is the solution for the velocity when $a_x = a_x(\zeta)$ (as in equation (2.198)) and δv_x is the correction for a non-plane wave, we obtain an equation for δv_x:

$$\frac{\partial \delta v_x}{\partial t} + \frac{1}{2}\partial_x v_x^2 = \frac{\partial \delta v_x}{\partial t} + \frac{c}{2\gamma_0}\left(v_x(0)\partial_x a_x + \frac{c}{4\gamma_0}\partial_x a_x^2\right) \simeq \frac{c}{\gamma_0}v_y\partial_y a_x. \qquad (2.201)$$

For the sake of simplicity, let us hereafter consider that the electron enters the interaction region with the axial velocity ($v_x(0)$, $v_y(0) \ll v_z(0)$). We also neglect terms that are of order higher than a_0^2. We thus obtain:

$$\frac{\partial \delta v_x}{\partial \zeta} = -\frac{c}{8\gamma_0^2}\partial_x a_x^2$$
$$\frac{dv_y}{d\zeta} = -\frac{c}{8\gamma_0^2}\partial_y a_x^2. \qquad (2.202)$$

Normally, equations (2.202) cannot be solved analytically. However, a simplistic approach would be that of considering the average of a_x^2 over an oscillation period, thus replacing $a_x^2 - > a_0^2(x, y)/2$. Moreover, a perturbative approach would be that of considering an a_0 small enough that the maximum displacement during the interaction time is smaller than the size of the incident wave. In this case, for example, if the wave is Gaussian, then $\partial_{x,y} a_0^2(x, y) = \partial_{x,y} a_0^2 e^{-2\{x,y\}^2/w_0^2} \simeq -4a_0^2\{x, y\}/w_0^2$. The electron dynamics is then expressed by:

$$\frac{\partial^2 \delta x}{\partial \zeta^2} = \frac{a_0^2}{4\gamma_0^2 w_0^2}x$$
$$\frac{d^2 y}{d\zeta^2} = \frac{a_0^2}{4\gamma_0^2 w_0^2}y \qquad (2.203)$$

where $\delta x = \int dt \delta v_x$. The solution for the electron dynamics in the case of a non-plane incident wave, as calculated by a perturbative approach, is:

$$x \simeq x_0 \cosh\left(\frac{a_0 \zeta}{2\gamma_0 w_0}\right) + \frac{1}{2\gamma_0}\int_0^\zeta d\zeta' a_x(\zeta')$$
$$y = y_0 \cosh\left(\frac{a_0 \zeta}{2\gamma_0 w_0}\right). \qquad (2.204)$$

For the vector potential in equation (2.53), equations (2.204) become:

$$x \simeq x_0 \cosh\left(\frac{a_0 \zeta}{2\gamma_0 w_0}\right) + \frac{a_0 c}{2\gamma_0 \omega_0}\sin\left(\frac{\omega_0 \zeta}{c}\right)$$
$$y = y_0 \cosh\left(\frac{a_0 \zeta}{2\gamma_0 w_0}\right). \qquad (2.205)$$

We observe that for a relatively high intensity of the incident pulse, a new force appears in the electron dynamics, which is $\vec{F}_p \propto \vec{\nabla} a_x^2$ (see equation (2.202)). This force is known as the *ponderomotive force* and tends to expel particles from the the region of maximum electromagnetic intensity.

Further reading

For CBS in the nonlinear Thomson regime, see the following:

Esarey E, Ride S K and Sprangle P 1993 Nonlinear Thomson scattering of intense laser pulses from beams and plasmas *Phys. Rev. E* **48** 3003

Dattoli G, Gallardo J C and Ottaviani P L 1994 Free-electron laser intracavity light as a source of hard x-ray production by Compton backscattering *J. Appl. Phys.* **76** 1399–404

Chen S Y, Maksimchuk A and Umstadter D 1998 Experimental observation of relativistic nonlinear Thomson scattering *Nature* **396** 653–5

Ta Phuoc K *et al* 2003 X-ray radiation from nonlinear Thomson scattering of an intense femtosecond laser on relativistic electrons in a helium plasma *Phys. Rev. Lett.* **91** 195001

Tomassini P *et al* 2005 Thomson backscattering X-rays from ultra-relativistic electron bunches and temporally shaped laser pulses *Appl. Phys. B* **80** 419–36

Phuoc K T *et al* 2005 Nonlinear Thomson scattering from relativistic laser plasma interaction *Eur. Phys. J. D* **33** 301–6

Iinuma M *et al* 2005 Observation of second harmonics in laser-electron scattering using low energy electron beam *Phys. Lett. A* **346** 255–60

Babzien M *et al* 2006 Observation of the second harmonic in Thomson scattering from relativistic electrons *Phys. Rev. Lett.* **96** 054802

Kumita T *et al* 2006 Observation of the nonlinear effect in relativistic Thomson scattering of electron and laser beams *Laser Phys.* **16** 267–71

Boca M and Florescu V 2009 Nonlinear Compton scattering with a laser pulse *Phys. Rev. A* **80** 053403

Zhao S *et al* 2012 Surpassing one x-ray photon per electron in nonlinear Thomson scattering in 180° geometry *Phys. Plasmas* **19** 013111

Khrennikov K *et al* 2015 Tunable all-optical quasimonochromatic Thomson X-ray source in the nonlinear regime *Phys. Rev. Lett.* **114** 195003

Sakai Y *et al* 2015 Observation of redshifting and harmonic radiation in inverse Compton scattering *Phys. Rev. ST Accel. Beams* **18** 060702

Yan W *et al* 2017 High-order multiphoton Thomson scattering *Nat. Photon.* **11** 514–20

Harvey C *et al* 2012 Intensity-dependent electron mass shift in a laser field: Existence, universality, and detection *Phys. Rev. Lett.* **109** 100402

Palma E D and Dattoli G and Sabchevski S 2022 Comments on the physics of microwave-undulators *Appl. Sci.* **12** 10297

Dattoli G, Palma E D and Petrillo V 2023 A collection of formulae for the design of Compton back-scattering X-ray sources *Appl. Sci.* **13** URL: https://www.mdpi.com/2076-3417/13/4/2645

Abramowitz M and Stegun I A 1968 *Handbook of mathematical functions with formulas, graphs, and mathematical tables* **vol 55** (Washington, DC: US National Bureau of Standards)

Lau Y Y *et al* Nonlinear Thomson scattering: a tutorial *Phys. Plasmas* **10** 2155–62

He F et al 2003 Backscattering of an intense laser beam by an electron *Phys. Rev. Lett.* **90** 055002

Williams E J 1935 *Correlation of Certain Collision Problems with Radiation Theory* (Copenhagen: Levin & Munksgaard)

For CBS, undulator, and synchrotron radiation, see the following:

Madey J M J 2003 Stimulated emission of Bremsstrahlung in a periodic magnetic field *J. Appl. Phys.* **42** 1906–13

Dattoli G and Renieri A 1985 1—Experimental and theoretical aspects of the free-electron laser *Laser Handbook* **4** ed Stitch M L and Bass M (Amsterdam: Elsevier). pp 1–141 https://dx.doi.org/10.1016/B978-0-444-86927-2.50005-X

Lieu R and Axford W I 1993 Synchrotron radiation: an inverse Compton effect *Astrophys. J.* **416** 700

de Jager O C and Mastichiadis A 1997 A relativistic Bremsstrahlung/inverse Compton origin for 2EG J1857+0118 associated with supernova remnant W44 *Astrophys. J.* **482** 874

Ng K Y 2009 *The equivalence of inverse Compton scattering and the undulator concept* FERMILAB-FN-0840-AD; TRN: US0904247 Fermi National Accelerator Lab 10.2172/966795

For the Fermi–Weiszacker–Williams approximation, see the following:

Fermi E 1924 Über die Theorie des Stoßes zwischen Atomen und elektrisch geladenen Teilchen *Z. Phys.* **29** 315–27

Weizsäcker C F 1934 Ausstrahlung bei Stößen sehr schneller Elektronen *Z. Phys.* **88** 612–25

Williams E J 1934 Nature of the high energy particles of penetrating radiation and status of ionization and radiation formulae *Phys. Rev.* **45** 729–30

Jackson J D 1998 *Classical Electrodynamics* 3rd edn (New York: Wiley)

McDonald K and Zolotorev M S 2003 Hertzian dipole radiation via the Weizsacker–Williams model doi: https://api.semanticscholar.org/CorpusID:119438135

Dattoli G and Nguyen F 2018 *Prog. Part. Nucl. Phys.* **99** 1–28

For the retarded potentials and the Liénard–Wiechert method for synchrotron/undulator radiation and CBS, see the following:

Alferov D, Bashmakov Y A and Bessonov E 1975 Undulator radiation *Tr. Fiz. Inst., Akad. Nauk. SSSR* **80** 100–24

Krinsky S, Perlman M L and Watson R E 1979 *Characteristics of Synchrotron radiation and of its sources* BNL-27678; TRN: 80-009587 Brookhaven National Lab. doi: https://dx.doi.org/10.2172/5395804

Kim K J 1989 Characteristics of synchrotron radiation *AIP Conf. Proc.* **184** (New York, NY: AIP Publishing). pp 565–632. doi: https://dx.doi.org/10.1063/1.38046

Dattoli G, Renieri A and Torre A 1993 *Lectures on the Free Electron Laser Theory and Related Topics* G—Reference, Information and Interdisciplinary Subjects Series (Singapore: World Scientific). doi: https://dx.doi.org/10.1142/1334

Ciocci F et al 2000 *Insertion Devices for Synchrotron Radiation and Free Electron Laser* (Singapore: World Scientific). doi: https://dx.doi.org/10.1142/4066

Kim K J, Huang Z and Lindberg R 2017 *Synchrotron Radiation and Free-Electron Lasers: Principles of Coherent X-Ray Generation* (Cambridge: Cambridge University Press). doi: https://dx.doi.org/10.1017/9781316677377

Dattoli G et al 2017 *Charged Beam Dynamics, Particle Accelerators and FreeElectron Lasers* (Bristol: IOP Publishing). pp 2053–563. doi: https://dx.doi.org/10.1088/978-0-7503-1239-4

For the generalized Bessel functions, see the following:

Appell P and de Feriet J K 1926 *Fonctions hypergéométriques et hypersphériques: polynomes d'Hermite* (Paris: Gauthier-Villars). doi: https://worldcat.org/title/1345284

More recent articles:

Reiss H R 2004 Absorption of light by light *J. Math. Phys.* **3** 59–67

Reiss H R 1992 Theoretical methods in quantum optics: S-matrix and Keldysh techniques for strong-field processes *Prog. Quant. Electron.* **16** 1–71

Nikishov A I and Ritus V I 1964 Quantum processes in the field of a plane electromagnetic wave and in a constant field. 2 *Sov. Phys. JETP* **19** 1191–9 http://jetp.ras.ru/cgi-bin/e/index/e/19/2/p529?a=list

Dattoli G et al 1990 Theory of generalized Bessel functions *Nuov. Cim.* B **105** 327–48

Dattoli G and Torre A 1996 *Theory and Applications of Generalized Bessel Functions* (Rome: Aracne)

Chiccoli C et al 1992 Theory of one-parameter generalized Bessel functions *Monograph, Gruppo Nazionale Informatica Mathematica* (Rome: CNR)

Reiss H R and Krainov V P 2003 Generalized Bessel functions in tunnelling ionization *J. Phys.* A **36** 5575

Dattoli G et al 2002 Two harmonic undulators and harmonic generation in high gain free electron lasers *Nucl. Instrum. Methods Phys. Res.* **495** 48–57

Chapter 3

Charged beam transport

3.1 Introduction

The present and following chapters deal with charged and optical beam transport. These transports have significant similarities and can be treated using analogous mathematical formalisms. We have chosen to present the relevant theoretical elements in two separate chapters and use sections 3.5 and 4.6 to underscore the contiguity between the relevant physics.

This introduction is intended as a gentle transition from the subjects already treated to the forthcoming matter. Regarding the physics of electron transport, we have so far provided the essential notions to handle particles moving at large velocities, namely at speeds v that are non-negligible with respect to c. We have seen that their velocity depends on the total energy and that high-energy particles should therefore be considered relativistic. This last statement is only 'relatively' true and needs clarification.

In figure 1.1, we proposed a geometrical example. Exercises of this type are sometimes useful, albeit academic. The example proposed here is useful as a mnemonic tool. Figure 3.1 shows the geometric translation of the following relativistic identity, called the relativistic triangle:

$$E^2 = (m_0 c^2)^2 + (pc)^2. \tag{3.1}$$

The use of elementary geometrical means allows the following correspondence between kinematics and geometrical observables:

$$\cos(\theta) = \frac{e_0}{E} = \frac{1}{\gamma}, \quad \cos(\varphi) = \frac{pc}{E} = \beta$$

$$\overline{HB} = pc\cos(\varphi) = \frac{(pc)^2}{E} = e_0\gamma\beta^2 \tag{3.2}$$

$$\overline{CH} = pc\sin(\varphi) = e_0\beta$$

$$E = e_0 + K, \quad e_0 = m_0 c^2.$$

$$E = m_0c^2 + K$$

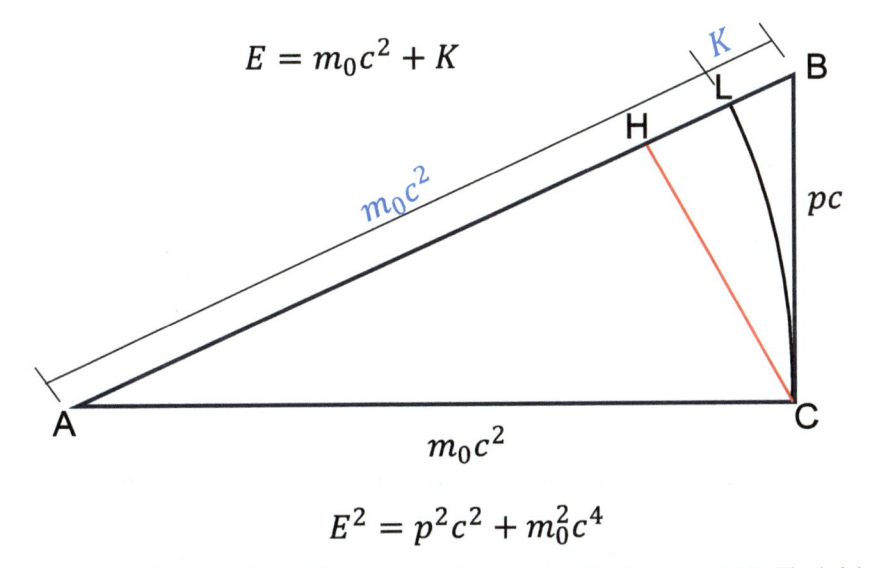

$$E^2 = p^2c^2 + m_0^2c^4$$

Figure 3.1. The relativistic triangle. The kinetic energy is represented by the segment LB. The height CH is linked to the classical momentum (see section 3.5).

It is evident that, within this picture, high-energy particles are characterized by θ and φ angles close to $\frac{\pi}{2}$ and 0, respectively. The reverse holds true for the low-energy case. Moreover, the kinetic energy K is geometrically visualized by subtracting the rest energy m_0c^2 from the total energy E.

If we normalize all sides to e_0, the relativistic triangle can be specified in terms of the kinematic quantities (γ, $\gamma\beta$) only. We have left the mass term to characterize the side of the triangle as representative of quantities with the dimensions of energy. For convenience's sake, we recall that:

$$\gamma = \frac{E}{e_0} \rightarrow v = c\sqrt{1 - \left(\frac{m_0c^2}{E}\right)} \qquad (3.3)$$

$$E \geqslant m_0c^2.$$

According to the above relations, particles with high energy are not necessarily relativistic. We note indeed that an electron with mass 0.511 MeV is ultrarelativistic at a (total) energy of 1 GeV, while a proton (mass 938 MeV) at the same energy is ruled by classical mechanics. The relativistic triangles drawn in figures 3.2 express the geometrical content of the previous statement.

Considerations based on geometrical correspondences, such as those shown in figures 3.1 and 3.2, do not hide any new physical information. They may be useful to visualize the interplay between various kinematic quantities, and we will further comment on their use in the exercises at the end of the chapter. In the following, we

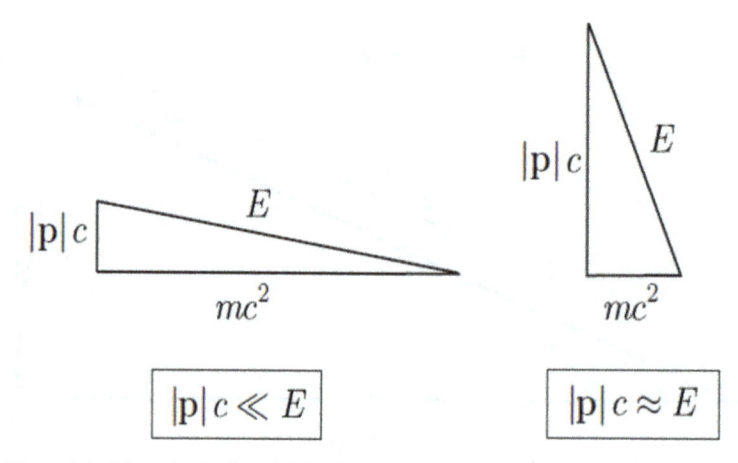

Figure 3.2. The 'classical' and 'ultra'-relativistic limits of the relativistic triangle.

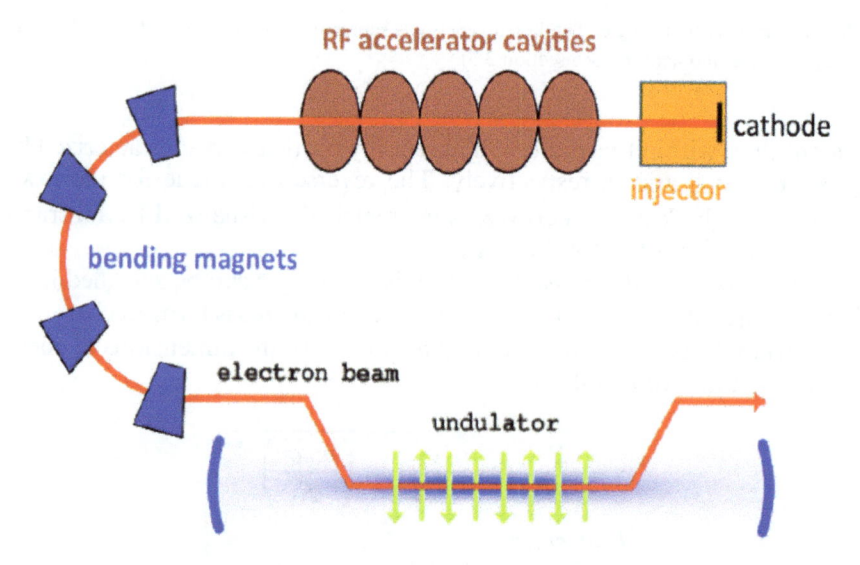

Figure 3.3. A schematic layout of an accelerator producing an electron beam 'transported' along a reference path (realized using bending magnets) to an undulator.

deal with relativistic electrons. For the range of energies we are interested in, the regime can safely be considered ultrarelativistic.

The main topic of this chapter is (electron) beam transport, which is realized through a system of devices constituting a (magnetic) transport channel in accelerator jargon.

An example of such transport is given in figure 3.3, which is a sketch of an accelerating device which delivers a high-energy electron (e-) beam. The beam is then extracted and injected into a region free of accelerating cavities, where it is

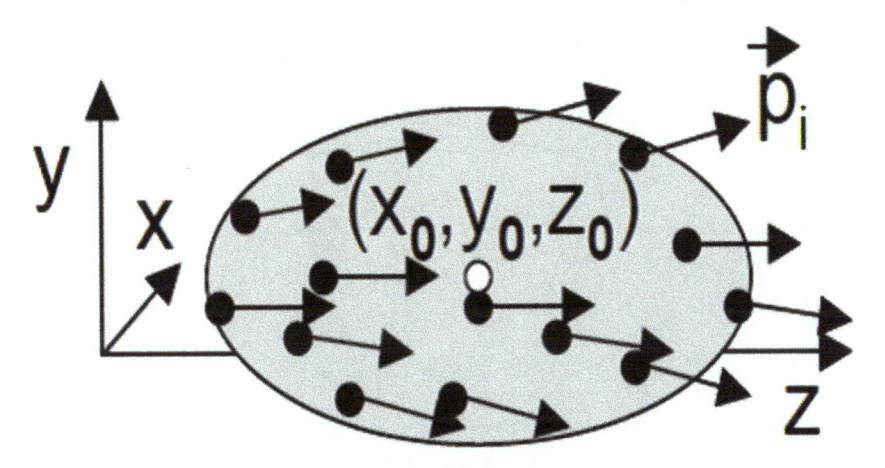

Figure 3.4. An electron beam, showing the relevant distribution in terms of longitudinal and transverse momenta and identification of the reference particle.

manipulated for eventual injection into a magnetic undulator. We have shown a *reference-path* trajectory followed by an e-beam[1] deflected (subjected to bending) by items which deviate the particle direction in different regions. By 'reference path,' we mean the trajectory followed by an ideal electron inside the beam that has the average characteristics of the whole ensemble (average energy, average transverse momentum, etc.; see figure 3.4) We assume that the magnetic field is constant, at least within the region encountered by the propagating electrons; we accordingly conclude that their trajectory inside the magnet is a circle. We can therefore compute the bending radius by equating the Lorentz and centrifugal forces. In the presence of the magnetic field only, the modulus of the electron velocity (hence the relativistic factor) is a conserved quantity and therefore we find:

$$evB = m_0\gamma\frac{v^2}{\rho} \to \rho = \frac{pc}{evB}\beta. \quad (3.4)$$

It is useful to express ρ in practical units and write the bending radius (for ultrarelativistic electrons) as:

$$\rho[\text{m}] \simeq 3.3356 \times 10^{-3} \frac{p\left[\frac{\text{MeV}}{c}\right]}{B[\text{T}]} \quad (3.5)$$

(see section 3.5 for further comments). The previous relationship yields an idea of the magnetic field intensity that should be superimposed on a beam with a given momentum to produce the desired bending radius. For example, a field intensity of 1 T acting on a 100 MeV e-beam produces a bending radius slightly larger than 30 cm. Since the associated bending angle (see figure 3.5 and the section 3.5) is:

---

[1] For the moment, the notion of the beam is fairly abstract; accordingly, we mean an ensemble of electrons with almost identical energies and momenta.

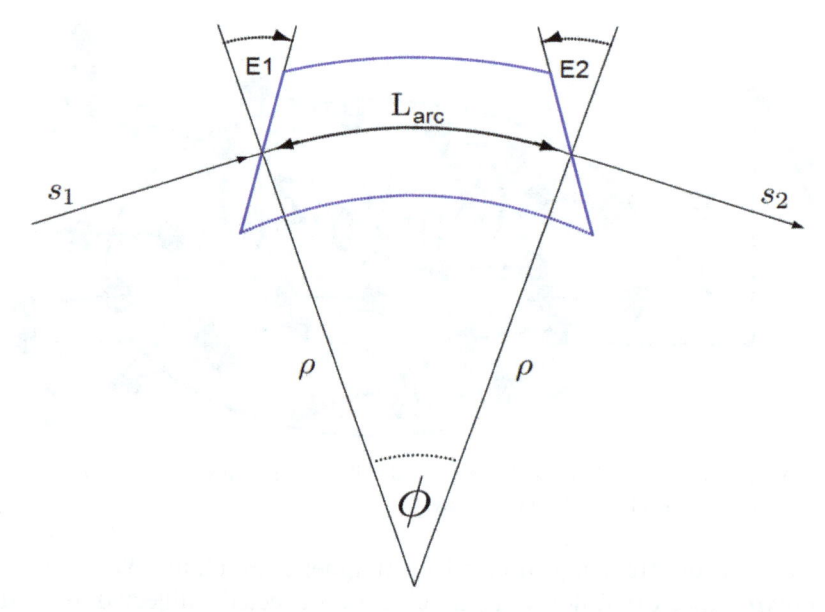

Figure 3.5. The bending radius and bending angle induced by a magnetic field.

$$\vartheta[\text{rad}] \simeq 2.9979 \times 10^2 \frac{B[\text{T}] L_{\text{ar}}[\text{m}]}{p\left[\frac{\text{MeV}}{c}\right]} \tag{3.6}$$

we obtain $\vartheta = 0.3$ rad for $L_{\text{ar}} = 0.1$ m. The role of the transport is not only to modify the trajectory but also to constrain the transverse dimensions of the beam itself.

As shown in figure 3.4, a beam is characterized by transverse distributions along the vertical (y) and horizontal (x) axes. It is therefore important to envision a device that prevents the angular and spatial spread of the electron ensemble above a certain limit.

Suppose that we realize this kind of device by arranging the north and south poles of a magnet as shown in figure 3.6. The resulting field lines in the transverse x and y directions, orthogonal to the electrons' trajectory, are those shown in figure 3.6(b). The effect of the field on the electrons (recall that the beam moves along the z-axis, directed inward relative to the figure plane) produces the force diagram shown in figure 3.6(b). Under these conditions, the net effect on the e-beam is to focus it in the vertical direction and to defocus it in the horizontal direction (see the next section, 3.5, for the necessary details). If, furthermore, the field affects the electron's x and y motions independently, the device has separated functions and no coupling occurs between the vertical components of the motion. Under these conditions, the focusing/defocusing effects can be accounted for by linear differential equations of the elastic type, and the beam dynamics can be treated using straightforward mathematical tools.

It is evident that if the device is rotated (in such a way that the north and south poles are interchanged) as shown in figure 3.7, the focusing/defocusing effects are reversed. A magnetic system that matches the described characteristics does exist. It is the so-called **quadrupole lens** and will be carefully discussed in the forthcoming sections.

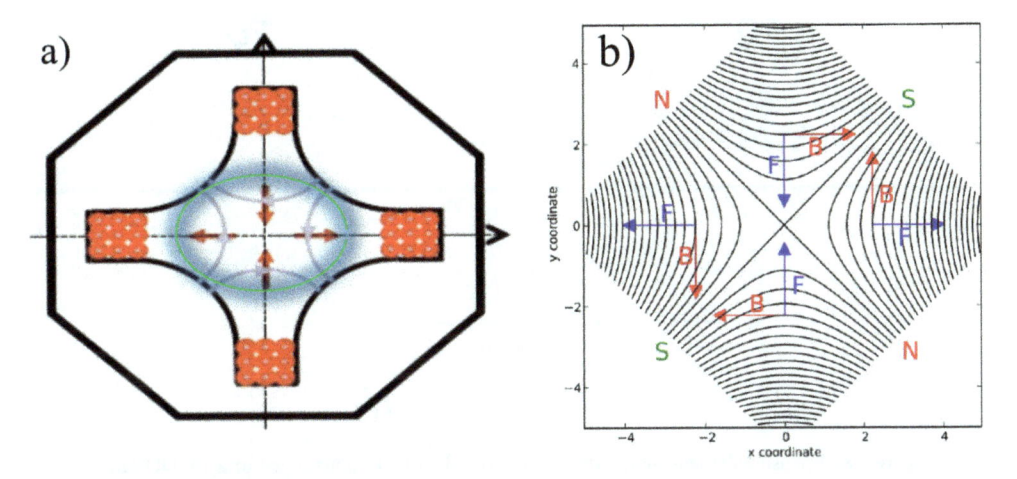

Figure 3.6. (a) The field line distribution of a magnetic device controlling the transverse dimensions of an e-beam (the physical region occupied by the electrons is inside the circle denoted by a dashed border). Adapted from Wolski A 2019 Maxwell's equations for magnets arXiv:1103.0713 (b) A diagram of the focusing/defocusing force in a magnetic quadrupole. Reproduced from Dattoli *et al* 2017 *Charged Beam Dynamics, Particle Accelerators and Free Electron Lasers* (Bristol: IOP Publishing). Copyright IOP Publishing Ltd. All rights reserved.

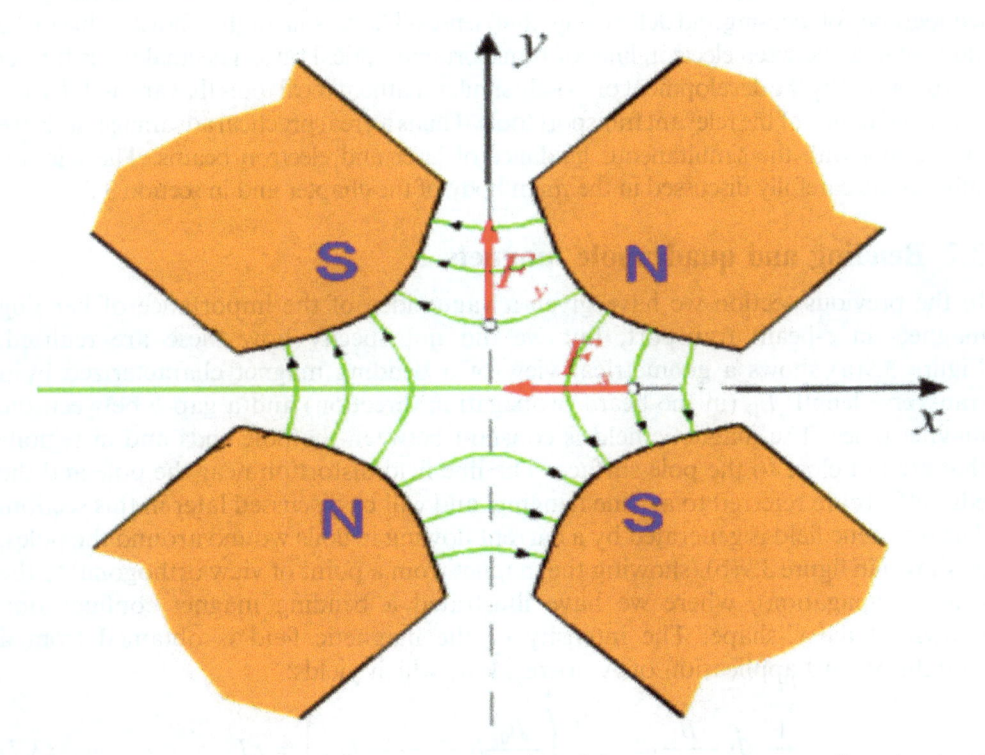

Figure 3.7. The same as figure 3.6 but with a $\pi/2$ rotation of the magnet arrangement. Reproduced with permission from Hillert W 2021 Transverse linear beam dynamics arXiv:2107.02614.

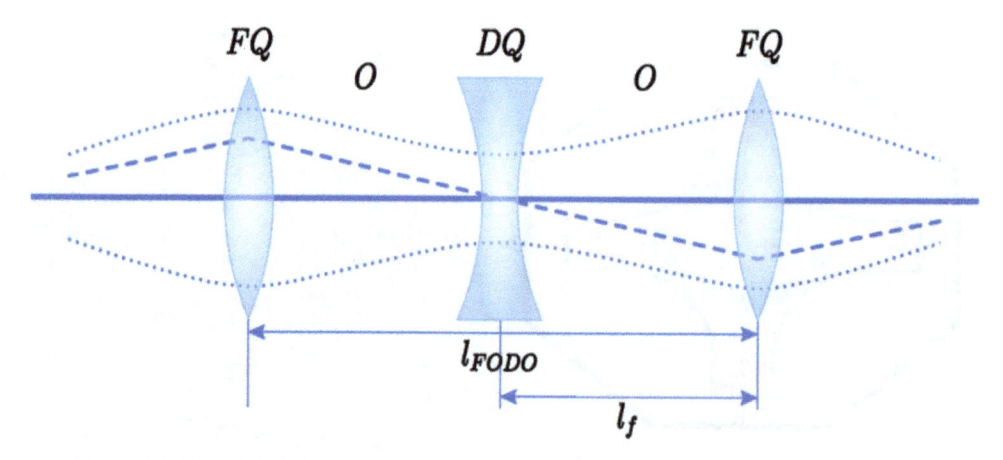

Figure 3.8. Focusing/defocusing quadrupoles and a drift section arranged in a FODO lattice.

Due to its separate focusing/defocusing properties, a single device of the outlined type cannot be used to transport a beam. At least two successive quadrupoles should be exploited to avoid the handling of largely asymmetric beams. Figure 3.8 shows a quadrupole arrangement known as the FODO lattice. It consists of a focusing (F) quadrupole (Q) a free section (O, meaning no magnetic field), a defocusing (D) quadrupole, and another drift section. The magnet lattice shown in figure 3.8 is sketched using the iconography of focusing and defocusing (thin) lenses. The reason for this choice is due to the close analogy between electron lines of transport and optical rays. The analogy is further corroborated by the development of closely similar mathematical tools that are useful in the study and design of the relevant transport tools. This is a great practical advantage, since we are dealing with the simultaneous guidance of laser and electron beams. The relevant subjects are carefully discussed in the main body of the chapter and in section 3.5.

3.2 Bending and quadrupole magnets

In the previous section we have given a vague idea of the importance of bending magnets in e-beam transport, but we did not specify how these are realized. Figure 3.9(a) shows a geometrical view of a bending magnet characterized by a transverse length L_B (in the beam propagation direction) and a gap h between the magnet poles. The magnetic field is constant between the pole ends and in regions that are not close to the pole surface. The line field distortion near the pole and the edge effects are referred to as 'line fringing' and will be discussed later in this section. The magnetic field is generated by a current flowing in coils wound around the poles, as shown in figure 3.9(b) (showing the magnet from a point of view orthogonal to the beam propagation), where we have illustrated a bending magnet configuration known as the C-shape. The intensity of the magnetic field is obtained from a straightforward application of Ampère's law, which yields:

$$\frac{1}{\mu_0} \oint \frac{B}{\mu_r(l)} dl = \frac{1}{\mu_0} \left(\frac{B_0}{\mu_{r,\text{air}}} h + \frac{B_\text{iron}}{\mu_{r,\text{iron}}} l_\text{iron} \right) = nI \qquad (3.7)$$

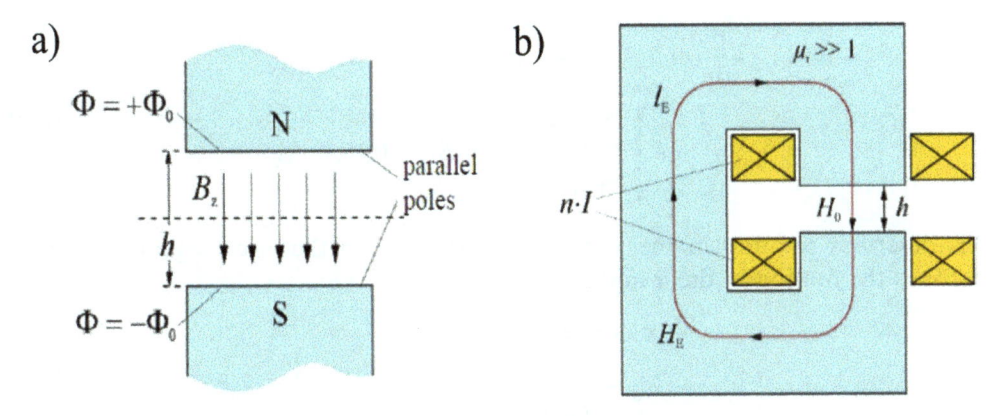

Figure 3.9. (a) Field lines inside a bending magnet. (b) A C-shaped magnet. Reproduced with permission from Hillert W 2021 Transverse linear beam dynamics arXiv:2107.02614.

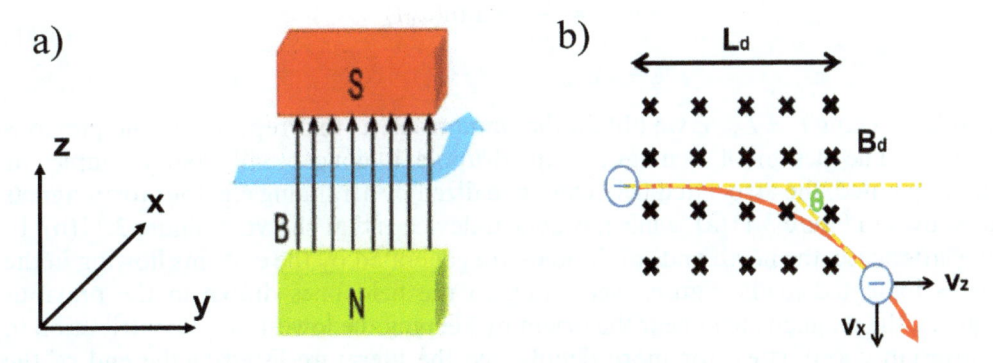

Figure 3.10. (a) A magnetic field and the magnetic field lines inside a bending magnet. (b) A top view of the trajectory in a bending magnet.

where n is the number of coil turns and I is the current flowing through them, $\mu_r(l)$ is the relative field permeability, and the integration path is that provided in figure 3.9(b). Neglecting the magnetic contribution inside the iron ($\mu_{r,\text{iron}} \gg \mu_{r,\text{air}} \simeq 1$), we end up with:

$$\frac{B_0}{\mu_0} h = nI. \tag{3.8}$$

After this short introduction to the bending magnet design, we provide a few elements that can be used to evaluate the electrons' motion inside the magnetic field. In figure 3.10 we have shown the magnetic lines and the axis convention used in the evaluation of the equations of motion.

The electron trajectory is determined by the Lorentz force and written (see the exercises at the end of the chapter) as the following differential equation:

$$\frac{\mathrm{d}}{\mathrm{d}t}(m_0\gamma\,\vec{v}) = -e\,\vec{v}\times\vec{B} =$$

$$= -e\begin{vmatrix} \hat{x} & \hat{y} & \hat{z} \\ v_x & v_y & v_z \\ 0 & B & 0 \end{vmatrix} = eB(v_z, 0, v_x). \quad (3.9)$$

This describes rotation in the x–z plane and leads to the circle equation, which describes the motion in the x and z directions:

$$x + iz = \rho\, e^{i\omega_c t}$$
$$\omega_c = \frac{eB}{m_0\gamma}, \quad \rho = \frac{\beta c}{\omega_c}. \quad (3.10)$$

In equations (3.10), ω_c and ρ are the cyclotron frequency and the previously introduced curvature radius. The bending angle is evidently given by:

$$\tan(\vartheta) = \frac{v_x}{v_z} = \tan(\omega_c t)$$
$$\vartheta = \omega_c t \quad (3.11)$$

and by keeping $t = L_{ar}/c$ we obtain the same result as that reported in the previous section. The design of a magnetic quadrupole is conceptually fairly simple: as already noted, the associated field can be realized by arranging e.g. four toy magnets as shown in figure 3.11(a), while a practical device is that shown in figure 3.11(b). In the latter case, the north and south poles are generated by the currents flowing in the wires indicated in the figure. According to the field lines drawn in the previous figures, the magnetic field near the origin by keeping the lowest order contribution in Cartesian coordinates (for more details, see the literature listed at the end of the chapter and the exercises section 3.5) can be written as:

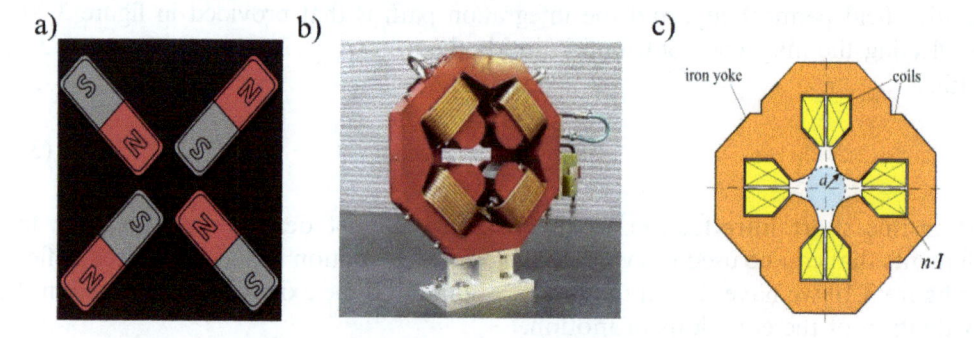

$$\vec{B} \simeq g(y\hat{x} + x\hat{y}) \quad (3.12)$$

Figure 3.11. (a) A conceptual sketch of a quadrupole magnet, realized with ordinary permanent magnets. (b) A realization of a quadrupole. (c) A section of a quadrupole displaying the minimum distance from the magnet's pole to the origin. Reproduced with permission from Hillert W 2021 Transverse linear beam dynamics arXiv:2107.02614.

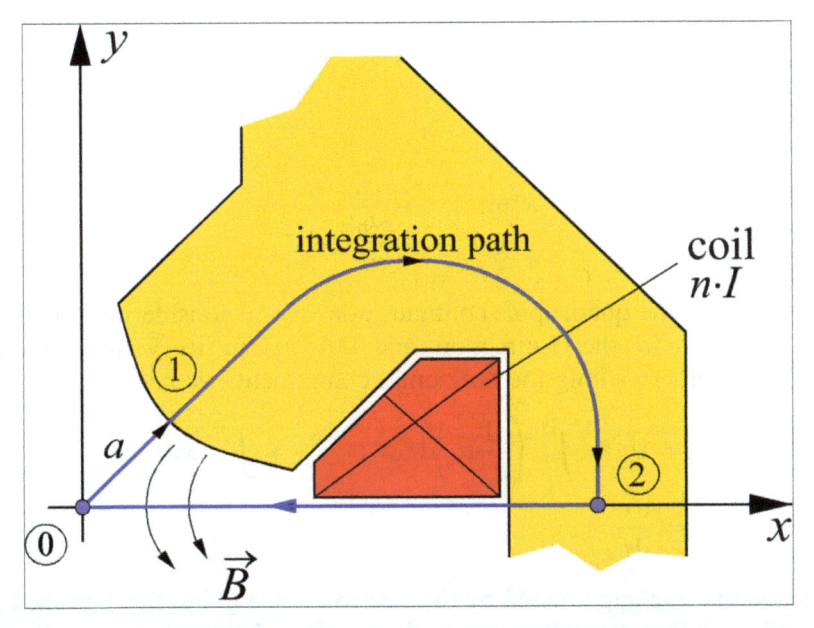

Figure 3.12. The integration path used for the evaluation of the quadrupole constant g. Reproduced with permission from Hillert W 2021 Transverse linear beam dynamics arXiv:2107.02614.

which ensures that $\vec{\nabla}\cdot\vec{B} = 0$. The coefficient g, called the quadrupole gradient, is determined using Ampère's law. For this purpose, we proceed as indicated below:

1. The most convenient integration path is indicated in figure 3.12 and accordingly we find:

$$\frac{1}{\mu_0}\oint \frac{B}{\mu_r(l)}\mathrm{d}l = \frac{1}{\mu_0}\int_0^1 B\mathrm{d}l + \frac{1}{\mu_0}\int_1^2 \frac{B}{\mu_{r,\mathrm{iron}}}\mathrm{d}l + \frac{1}{\mu_0}\int_2^0 \frac{B}{\mu_r}\mathrm{d}l. \quad (3.13)$$

2. For the reasons already discussed, we are allowed to neglect the last two integrals appearing on the right-hand side; therefore, using polar coordinates for the integration, we obtain:

$$\int_0^a B\mathrm{d}l = g\int_0^a r\mathrm{d}r = \mu_0 nI \rightarrow$$
$$\rightarrow g = \frac{2\mu_0 nI}{a^2}. \quad (3.14)$$

Given these clarifications, we can provide an explicit evaluation of the electron motion equations inside the magnet. The Lorentz equations of motion for an electron passing through a magnet with a field specified by equation (3.12) yield the following differential equations for the transverse components (for details, see the exercise section 3.5):

$$\frac{d^2}{d\tau^2}x = -\kappa x$$
$$\frac{d^2}{d\tau^2}y = \kappa y,$$
$$\kappa\,[\mathrm{m}^{-2}] = \frac{e\mu_0 nI}{a^2 p},$$
$$\tau \simeq ct.$$
(3.15)

The prediction for the quadrupole configuration we are considering was that it was expected to focus in the *x*-direction and defocus in its vertical counterpart. Regarding the velocity along the horizontal component, we have:

$$v_{x_2} = v_{x_1} + \int_{\tau_1}^{\tau_2}\left(\frac{d}{d\tau}v_x\right)d\tau \simeq v_{0,x} - \kappa\int_{\tau_1}^{\tau_2} x\,d\tau$$
$$v_x = \frac{dx}{d\tau}.$$
(3.16)

If we assume that the gap crossed by the electron while going through the magnetic lens is small, such that its position inside the quadrupole is constant (see figure 3.13), we find:

$$v_{x_2} \simeq v_{x_1} - \kappa(\tau_2 - \tau_1)x$$
$$\tau_2 - \tau_1 = L_q$$
$$L_q \equiv \text{quadrupole} - \text{length}.$$
(3.17)

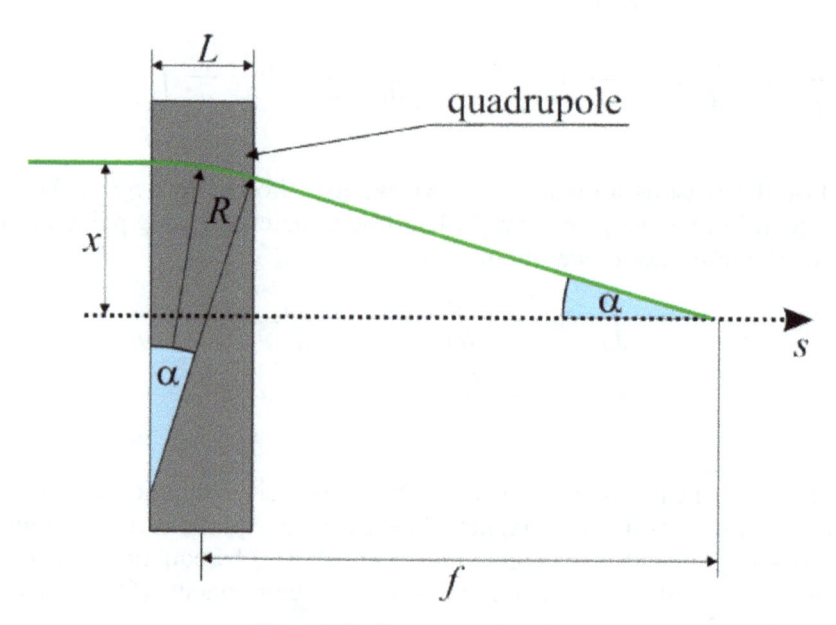

Figure 3.13. Electron-ray focusing.

The approximation implicit in 3.17 will be referred to as the **thin lens** approximation (see below). If we set $\kappa L_q = 1/f$, the solution of the electron equations of motion in the horizontal direction can be written as:

$$\begin{aligned} x_2 &\simeq x_1 \\ v_{x_2} &\simeq v_{x_1} - \frac{x_1}{f} \end{aligned} \qquad (3.18)$$

which can also be cast in matrix form as:

$$\begin{pmatrix} x_2 \\ v_{x_2} \end{pmatrix} \simeq \begin{pmatrix} 1 & 0 \\ -\frac{1}{f} & 1 \end{pmatrix} \begin{pmatrix} x_1 \\ v_{x_1} \end{pmatrix}. \qquad (3.19)$$

This is reminiscent of the equation describing ray deflection in a lens with focal length f, as more carefully detailed in the forthcoming chapter. The effect of a quadrupole magnet is therefore that of deflecting 'electron rays,' which happens in geometrical optics in the thin lens approximation. Unlike optical lenses, the quadrupole defocuses in the y-direction, according to:

$$\begin{pmatrix} y_2 \\ v_{y_2} \end{pmatrix} \simeq \begin{pmatrix} 1 & 0 \\ \frac{1}{f} & 1 \end{pmatrix} \begin{pmatrix} y_1 \\ v_{y_1} \end{pmatrix}. \qquad (3.20)$$

Pushing the geometrical optics analogy further (see the forthcoming section), we can assume that the electron propagation in a drift section (namely a region free of any magnetic influence/electrons) can be written as:

$$\begin{pmatrix} x_2 \\ v_{x_2} \end{pmatrix} = \begin{pmatrix} 1 & l \\ 0 & 1 \end{pmatrix} \begin{pmatrix} x_1 \\ v_{x_1} \end{pmatrix}. \qquad (3.21)$$

The usefulness of the matrix formalism is easily understood from the example reported below. Figure 3.14 shows the abovementioned FODO cell, consisting of a 'half-focusing' quadrupole, a drift section, a defocusing element, a further drift, and another half-focusing quad. The e-beam crosses the magnet device as indicated by the red arrow. The matrix describing the global effect of the FODO cell on the horizontal part of the electron motion can be written as follows (exercise for the reader: explain the factor of two):

$$\hat{M}_x = \begin{pmatrix} 1 & 0 \\ -\frac{1}{2f} & 1 \end{pmatrix} \begin{pmatrix} 1 & L \\ 0 & 1 \end{pmatrix} \begin{pmatrix} 1 & 0 \\ \frac{1}{f} & 1 \end{pmatrix} \begin{pmatrix} 1 & L \\ 0 & 1 \end{pmatrix} \begin{pmatrix} 1 & 0 \\ -\frac{1}{2f} & 1 \end{pmatrix} = \begin{pmatrix} 1 - \frac{L^2}{2f^2} & 2L\left(1 + \frac{L}{2f}\right) \\ -\frac{L}{2f^2} + \frac{L^2}{4f^3} & 1 - \frac{L^2}{2f^2} \end{pmatrix}. \quad (3.22)$$

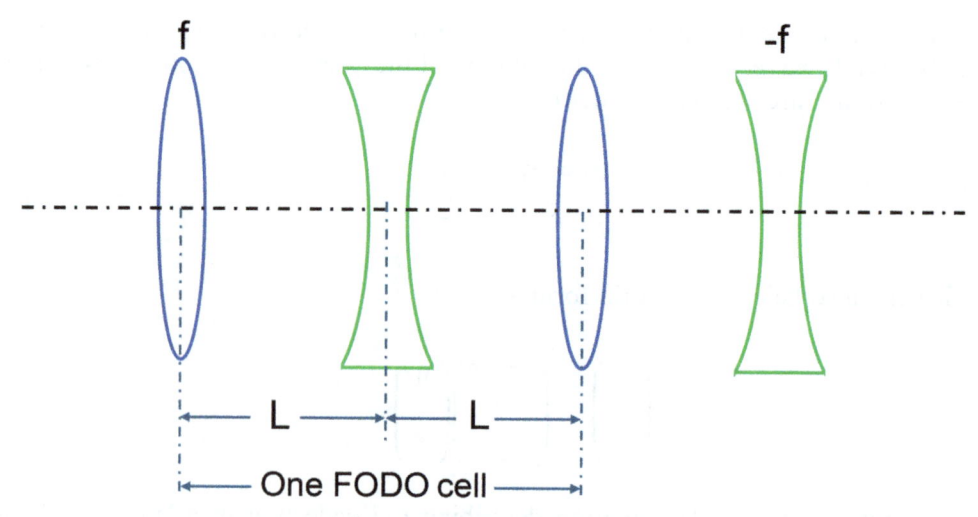

Figure 3.14. A FODO cell.

Write the corresponding matrix for the y-direction and show that:

$$\hat{M}_y = \begin{pmatrix} 1 - \dfrac{L^2}{2f^2} & 2L\left(1 - \dfrac{L}{2f}\right) \\ -\dfrac{L}{2f^2} - \dfrac{L^2}{4f^3} & 1 - \dfrac{L^2}{2f^2} \end{pmatrix} \quad (3.23)$$

and deduce that the net effect is focusing in both planes.

3.3 Beam envelope evolution

In the previous sections we have considered the evolution of a single electron, which represents a kind of beam reference particle. As we have already stressed, an actual beam is a collection of electrons, all having different positions, energies, and transverse momenta.

If each particle is characterized by the variables η, η', $\eta' = \frac{d\eta}{d\tau}$ $\eta = x, y$ (referred to as the phase-space variables from now on), a beam can be described using the statistical properties of an ensemble of particles in (η, η'). In figure 3.15(a) we have denoted a region of these spaces that contains (almost all) the particles constituting a beam. It is evident that the discussion is limited to uncoupled transverse dynamics, and, within this context, we can make the assumption that the distribution inside this space is represented by the product of Gaussian distributions, written as:

$$f(\eta, \eta') = \frac{1}{2\pi\varepsilon} \exp\left(-\frac{\gamma\eta^2 + 2\alpha\eta\eta' + \beta\eta'^2}{2\varepsilon}\right). \quad (3.24)$$

The quantities γ, α, β, called Twiss parameters, satisfy the condition (see the exercises in section 3.5)

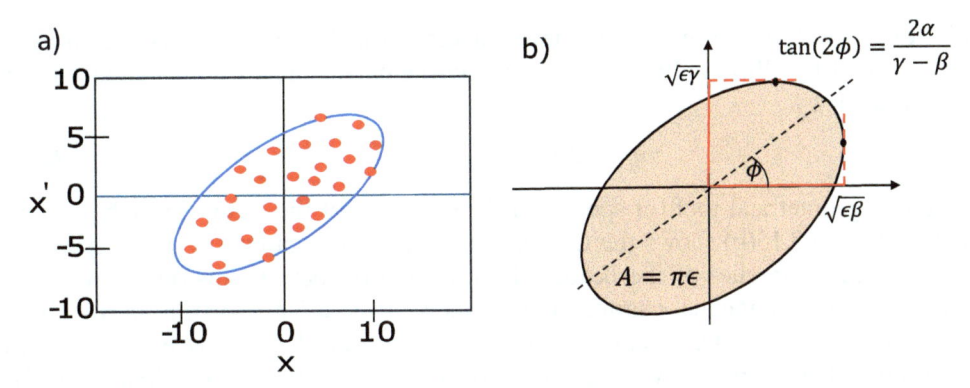

Figure 3.15. (a) Particles inside a Courant–Snyder ellipse. (b) An ellipse with relevant meaningful coordinates.

$$\beta\gamma - \alpha^2 = 1 \tag{3.25}$$

which is necessary to ensure the normalization ($\int f(\eta, \eta')\mathrm{d}\eta \mathrm{d}\eta' = 1$) of the beam's phase-space distribution. Furthermore, the relevant statistical meaning is specified by the definition of the beam's root mean square (rms) quantities specified below (for the explicit calculations, see the exercise section 3.5):

$$\begin{aligned}(\sigma'_\eta)^2 &= \langle \eta'^2 \rangle - \langle \eta' \rangle^2 = \varepsilon\gamma, \\ (\sigma_\eta)^2 &= \langle \eta^2 \rangle - \langle \eta \rangle^2 = \varepsilon\beta \\ \sigma_{\eta,\eta'} &= \langle \eta\eta' \rangle - \langle \eta \rangle\langle \eta' \rangle = -\varepsilon\alpha \\ \langle \rangle &\equiv \text{average on the distribution}\end{aligned} \tag{3.26}$$

Specifically, these provide the rms beam divergence, beam section, and transverse angular correlation, and it is furthermore evident that:

$$\sigma'_\eta \sigma_\eta = \sqrt{1 + \alpha^2}\,\varepsilon \tag{3.27}$$

which exhibits a minimum for an uncorrelated ($\alpha = 0$) distribution.

It is also important to emphasize that:

$$\begin{aligned}\int_{-\infty}^{+\infty} f(\eta, \eta')\mathrm{d}\eta &= \frac{1}{\sqrt{2\pi}\,\sigma'_\eta} \exp\left(-\frac{1}{2}\frac{\eta'^2}{(\sigma'_\eta)^2}\right) \\ \int_{-\infty}^{+\infty} f(\eta, \eta')\mathrm{d}\eta' &= \frac{1}{\sqrt{2\pi}\,\sigma_\eta} \exp\left(-\frac{1}{2}\frac{\eta^2}{(\sigma_\eta)^2}\right).\end{aligned} \tag{3.28}$$

It is now important to note that the level curves of the distribution in equation (3.24) are provided by ellipses. In particular, the curve called the Courant–Snyder (C–S) ellipse, given by

$$\gamma\eta^2 + 2\alpha\eta\eta' + \beta\eta'^2 = \varepsilon \tag{3.29}$$

is a useful geometrical tool for specifying the evolution of the beam through magnetic devices. Figure 3.15(b) shows the C–S ellipse along with meaningful points specifying the statistical properties of the beam. The variation of these points during the beam transit into the magnetic channel represents a geometrical portrait of the beam evolution. The parameter ε represents the beam emittance/phase-space area, which is a constant of motion that will be more adequately discussed in the following. The following exercise is useful to understand how the C–S ellipse can be exploited to specify the beam evolution. Consider a beam traversing a drift section, namely:

$$\begin{pmatrix}\eta\\ \eta'\end{pmatrix} \simeq \begin{pmatrix}1 & l\\ 0 & 1\end{pmatrix}\begin{pmatrix}\eta_0\\ \eta'_0\end{pmatrix} \rightarrow$$
$$\eta = \eta_0 + l\eta'_0$$
$$\eta'_0 = \eta'_0. \tag{3.30}$$

The rms beam section at the end of the drift part is simply given by:

$$\langle \eta^2 \rangle = \left\langle \eta_0^2 + l^2(\eta'_0)^2 + 2l\eta_0\eta'_0 \right\rangle. \tag{3.31}$$

Keeping the average on the distribution equation (3.24), we find:

$$\sigma_\eta^2 = {}_0\sigma_{\eta,0}^2 + l^2\left({}_0\sigma'_{\eta,0}\right)^2 + 2l\, {}_0\sigma_{\eta,\eta'} \tag{3.32}$$

which, written in terms of Twiss parameters, reads:

$$\beta_l = \beta_0 + \gamma_0 l^2 - 2\alpha_0 l. \tag{3.33}$$

The reader can easily show, using the same procedure, that the evolution of the remaining two parameters is fixed by:

$$\gamma_l = \gamma_0,$$
$$\alpha_l = \alpha_0 - \gamma l. \tag{3.34}$$

Regarding, for example, the case of the thin lens:

$$\begin{pmatrix}\eta\\ \eta'\end{pmatrix} \simeq \begin{pmatrix}1 & 0\\ \pm\dfrac{1}{f} & 1\end{pmatrix}\begin{pmatrix}\eta_0\\ \eta'_0\end{pmatrix} \rightarrow$$
$$\eta = \eta_0$$
$$\eta' = \eta'_0 \pm \frac{1}{f}\eta_0. \tag{3.35}$$

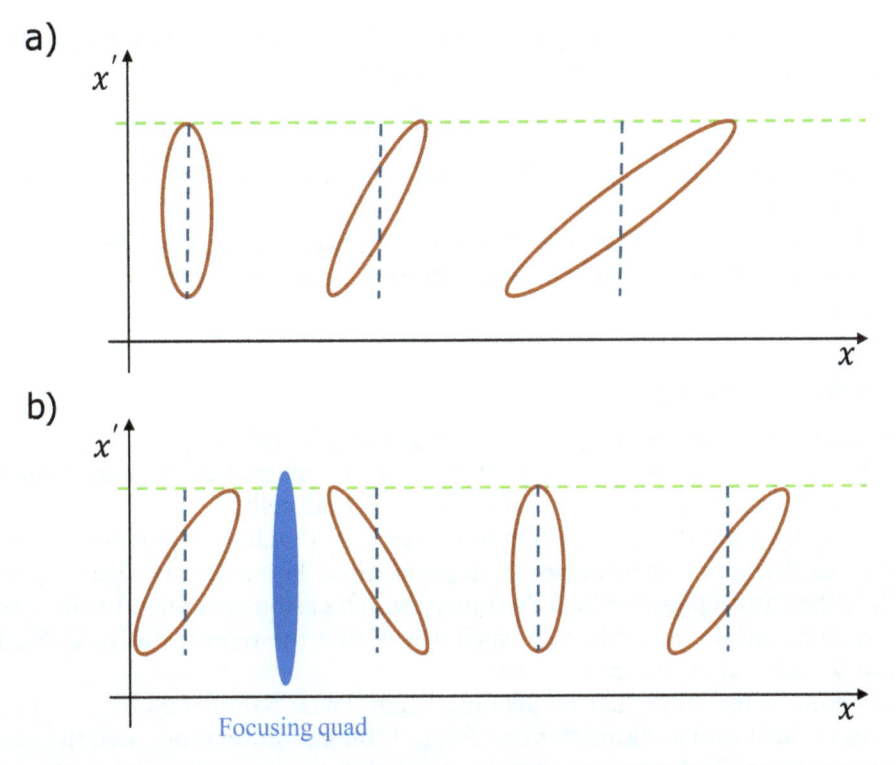

Figure 3.16. C–S ellipse modifications induced by: (a) drift space and (b) a focusing quadrupole.

The effect on the Twiss parameters at the quadrupole output is:

$$\beta_q = \beta_0$$
$$\gamma_q = \gamma_0 + \frac{1}{f^2}\beta_0 \mp \frac{2}{f}\alpha_0 \qquad (3.36)$$
$$\alpha_q = \alpha_0 \pm \frac{1}{f}\beta_0$$

as the reader can easily check. In figure 3.16 we illustrate the behavior of the C–S ellipse along different transport lines.

A less naïve procedure for evaluating the evolution of the Twiss parameters in terms of the transport element is outlined below and further detailed in the exercise section 3.5. We define the following **beta** matrix:

$$\hat{B} = \begin{pmatrix} \beta & -\alpha \\ -\alpha & \gamma \end{pmatrix} \qquad (3.37)$$

linked to the beam covariance matrix by

$$\hat{\Sigma} = \begin{pmatrix} \sigma_{\eta,\eta} & \sigma_{\eta,\eta'} \\ \sigma_{\eta,\eta'} & \sigma_{\eta',\eta'} \end{pmatrix} = \varepsilon \hat{B}. \qquad (3.38)$$

If $\hat{M}$ and $\hat{M}^T$ denote the transport matrix and its transpose, respectively, we can specify the evolution of the beta matrix as follows:

$$\hat{B}' = \hat{M}\hat{B}\hat{M}^T \qquad (3.39)$$

which is a faster computational method and easily amenable to numerical implementation.

In the next chapter, we discuss the complete analogy and show the quantitative correspondence between optical and electron beam propagation computation techniques.

3.4 Beam matching

The concept of beam matching can be worded in different ways.

In the following we refer to transport realized with magnetic devices arranged in a straight line, such as a number of consecutive FODO cells.

This configuration is less demanding then the FODO arrangements used in circular accelerators (synchrotrons or storage rings) but shares the same strategy, which is that of keeping the beam's transverse dimensions confined to the transfer line, while its envelope oscillates in such a way that the representative C–S ellipse reproduces itself after any cell.

The game is therefore that of defining beam Twiss parameters (α, β, γ) at the entrance of the line that realize this condition. If the condition is not met, the beam is said to be mismatched, meaning that its oscillations are irregular; its envelope (and also its emittance) may grow in an uncontrolled way, causing, in the worst case, loss of the beam.

Figure 3.17 illustrates the evolution of the phase-space profile along a DOFO channel. The beam is unmatched; it enters the line with a C–S ellipse characterized by $\alpha = 0$ and leaves the line with the same α but a larger β. Before proceeding further, let us note that given a transport line realized in terms of consecutive FODO cells, we can define the corresponding periodic transfer matrix as:

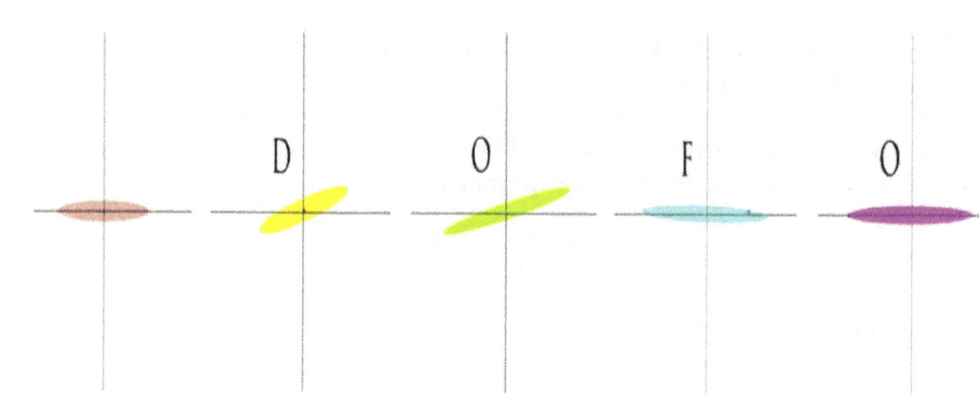

Figure 3.17. An example of the evolution of the phase-space profile of a mismatched beam.

$$\hat{M} = \begin{pmatrix} \cos(\mu) + \alpha_c \sin(\mu) & \beta_c \sin(\mu) \\ -\gamma_c \sin(\mu) & \cos(\mu) - \alpha_c \sin(\mu) \end{pmatrix} \qquad (3.40)$$

where the Twiss coefficients α_c, β_c, and γ_c are characteristic quantities defining the transport itself. It is easily shown (see the exercise section 3.5) that equation (3.40) satisfies the condition $\det(\hat{M}) = 1$. Using the previous relationship, it can be verified that, at the end of the transfer line, the beam covariance matrix is left unchanged by the transformation (3.40) if (see exercise section 3.5)

$$\hat{B}_1 = \begin{pmatrix} \beta_c & -\alpha_c \\ -\alpha_c & \gamma_c \end{pmatrix}. \qquad (3.41)$$

To clarify what we mean, we consider the transfer matrix FODO cell parameterized according to:

$$\hat{M}_x = \begin{pmatrix} 1 - \frac{L^2}{2f^2} & 2L\left(1 + \frac{L}{2f}\right) \\ -\frac{L}{2f^2}\left(1 - \frac{L}{2f}\right) & 1 - \frac{L^2}{2f^2} \end{pmatrix} = \begin{pmatrix} \cos(\mu) & \beta_{x,c} \sin(\mu) \\ -\gamma_{x,c} \sin(\mu) & \cos(\mu) \end{pmatrix} \qquad (3.42)$$

$$\cos(\mu) = 1 - \frac{L^2}{2f^2}, \ \beta_{x,c} = 2L\frac{1 + \sin\left(\frac{\mu}{2}\right)}{\sin(\mu)}, \ \gamma_{x,c} = -\frac{1}{2L}\frac{\sin(\mu)}{1 + \sin\left(\frac{\mu}{2}\right)}, \ \alpha_{x,c} = 0.$$

The output beta coefficient of a beam transported through a transfer line characterized by the matrix (3.42) can be computed using the methods we have just outlined. We find indeed that (o = output, i = input):

$$\begin{aligned} x_0 &= \cos(\mu) x_i + \beta_{x,c} \sin(\mu) x'_i \rightarrow \\ &\rightarrow \beta_0 = \cos(\mu)^2 \beta_i + \beta_{x,c}{}^2 \gamma_i \sin(\mu)^2. \end{aligned} \qquad (3.43)$$

If the input Twiss coefficients coincide with those of the transfer line, we obtain:

$$\begin{aligned} \beta_0 &= \cos(\mu)^2 \beta_{x,c} + (\beta_{x,c})^2 \gamma_{x,c} \sin(\mu)^2 = \beta_{x,c}[\cos(\mu)^2 + (\beta_{x,c})\gamma_{x,c} \sin(\mu)^2] = \\ &= \beta_{x,c}. \end{aligned} \qquad (3.44)$$

Regarding the vertical motion, the transfer matrix can be parameterized using an analogous procedure. We invite the reader to prove that:

$$\beta_{y,0} = 2L\frac{1 + \sin\left(\frac{\mu}{2}\right)}{\sin(\mu)}, \ \gamma_{y,0} = -\frac{1}{2L}\frac{\sin(\mu)}{1 + \sin\left(\frac{\mu}{2}\right)}, \ \alpha_y = 0 \qquad (3.45)$$

($\cos(\mu)$ is defined as in equation (3.42)). In this chapter we have presented a pragmatic view of electron beam propagation in a transport channel. Although simplified, the notions we have provided are sufficient for the design of Compton backscattering sources. Where necessary, further notions will be introduced in the following chapters. Some of the concepts we have developed here will be discussed further in the next chapter, albeit from a different point of view. In the exercise section 3.5 at the end of this chapter, we give a more a more definite idea of how electron and ray transport systems should be conceived in the design of a CBS source.

3.5 Comments and exercises

3.5.1 Hamiltonian mechanics and beam transport

In the previous parts of this chapter, we discussed electron beam transport-related issues. We presented the relevant matter using a kind of ray geometry method. This point of view is sufficiently elaborate to include concepts such as phase-space dynamics and the associated conserved quantities such as emittance, a figure of merit of paramount importance in specifying beam brightness, for example.

It is important to emphasize that we have offered a simplified picture instead of framing the problem within a safer mathematical context.

In the following we summarize a few steps that allow the previous discussion to be translated into Hamiltonian terms.

Exercise 3.1. Prove that the Hamiltonian describing the interaction between a classical relativistic charged particle and an electromagnetic field characterized by the four-vector potential $A^\mu \equiv (\Phi, \vec{A})$ can be written as:

$$H = c\sqrt{(\vec{p} - q\vec{A})^2 + (m_0 c)^2} + q\Phi$$
$$\vec{A} \equiv (A_x, A_y, 0).$$
(3.46)

Use equation (3.46) to derive the particle equations of motion.
Solution:
We assume that an optical ray moves in a straight line and that its strategy when moving from point A to another point B while crossing two media with different refractive indices (see figure 4.11) is that of choosing the most convenient path to minimize the transit time. Bearing in mind that the canonical variables in this case are

$$(x, p_x), (y, p_y), (z, p_z)$$
(3.47)

we find:

$$\dot{x}_\alpha = \frac{\partial H}{\partial p_\alpha} = c\frac{p_\alpha - qA_\alpha}{\sqrt{(\vec{p} - q\vec{A})^2 + (m_0 c)^2}}$$

$$\dot{p}_\alpha = -\frac{\partial H}{\partial x_\alpha} = \frac{cq}{\sqrt{(\vec{p} - q\vec{A})^2 + (m_0 c)^2}}\left[\sum_{\beta=1}^{3}(p_\beta - qA_\beta)\frac{\partial A_\beta}{\partial x_\alpha}\right] - q\frac{\partial \Phi}{\partial x_\alpha}$$
(3.48)

$$\alpha = 1, 2, 3; \quad p_{1,2,3} \equiv p_{x,y,z}; \quad x_{1,2,3} \equiv x, y, z.$$

Using the first of equations (3.48), we can write the derivative of the momentum as:

$$\dot{p}_\alpha = q\left[\sum_{\beta=1}^{3} \dot{x}_\beta \frac{\partial A_\beta}{\partial x_\alpha}\right] - q\frac{\partial \Phi}{\partial x_\alpha}. \tag{3.49}$$

The Hamilton equation defining the velocity is important enough to deserve further comment. It is evident that $\dot{x}_\alpha$ is not simply proportional to the canonical momentum p_α, and indeed the quantity $p_\alpha - qA_\alpha$ defines the kinetic momentum related to the particle velocity by:

$$\begin{aligned} p_\alpha - qA_\alpha &= m\gamma c \beta_\alpha \\ \beta_\alpha &= \frac{\dot{x}_\alpha}{c}. \end{aligned} \tag{3.50}$$

Even though trivial, we underscore that the canonical momentum is obtained from the first of equations (3.50) as follows:

$$p_\alpha = qA_\alpha + m\gamma c \beta_\alpha. \tag{3.51}$$

Exercise 3.2. Use the equations obtained in the previous exercise to derive the Lorentz equations of motion.
Solution:
The problem is easily solved using equation (3.49), which can be written as:

$$q\dot{A}_\alpha + mc\frac{\mathrm{d}}{\mathrm{d}t}(\beta_\alpha \gamma) = q\left[\sum_{\beta=1}^{3} \dot{x}_\beta \frac{\partial A_\beta}{\partial x_\alpha}\right] - q\frac{\partial \Phi}{\partial x_\alpha}. \tag{3.52}$$

By considering that

$$\dot{A}_\alpha = \frac{\partial A}{\partial t} + \left[\sum_{\beta=1}^{3} \dot{x}_\beta \frac{\partial A_\alpha}{\partial x_\beta}\right] \tag{3.53}$$

we find:

$$mc\frac{\mathrm{d}}{\mathrm{d}t}(\beta_\alpha \gamma) = q\left(\left[\sum_{\beta=1}^{3} \dot{x}_\beta \frac{\partial A_\beta}{\partial x_\alpha} - \sum_{\beta=1}^{3} \dot{x}_\beta \frac{\partial A_\alpha}{\partial x_\beta}\right] - \frac{\partial A}{\partial t} - \frac{\partial \Phi}{\partial x_\alpha}\right). \tag{3.54}$$

Finally, noting that $mc\frac{\mathrm{d}}{\mathrm{d}t}(\beta_\alpha \gamma) = F_\alpha$ and recalling the definition of electric and magnetic fields in terms of scalar and vector potentials, we can eventually write equation (3.54) in the Lorentz vector form:

$$\vec{F} = q\vec{E} + \vec{v} \times \vec{B}. \tag{3.55}$$

Exercise 3.3. Note that the equations of motion obtained in the previous exercises use time as an independent variable. Since we know where the RF cavities, magnets, etc. are located and we do not know the arrival time, it appears more convenient to work with the path variable (say z) rather than the time. Show that by identifying the Hamiltonian with p_z, one obtains:

$$H_1 = -\sqrt{\frac{(E - q\Phi)^2}{c^2} - (mc)^2 - (p_x - qA_x)^2 - (p_y - qA_y)^2} - qA_z \qquad (3.56)$$

and comment on the associated physical meaning.
Solution:
It should first be underlined that E in equation (3.56) is the total energy of the system and will be treated as a constant.

The Hamiltonian in equation (3.48) is obtained by solving equation (3.46) with respect to p_z, and the following correspondences are worth noting:

$$\begin{aligned} H_1 &\to -p_z \\ A_z &\to \Phi \\ z &\to -t \end{aligned} \qquad (3.57)$$

along with the following redefinition of the canonical conjugate variables:

$$(x, p_x), (y, p_y), (-t, H) \qquad (3.58)$$

which imply that:

$$\begin{aligned} \frac{d}{dz} p_x &= -\frac{\partial H_1}{\partial z} \\ \frac{dt}{dz} &= \frac{\partial H_1}{\partial H}. \end{aligned} \qquad (3.59)$$

The consistency of the motion equations obtained by means of the Hamiltonian (3.46) or (3.56) is easily established by noting that, on account of the second of equations (3.59), we obtain:

$$dt = \frac{\partial H_1}{\partial H} dz \qquad (3.60)$$

which allows us to write the Hamilton equations obtained from the Hamiltonian (3.46) in the form of (3.59), namely:

$$\dot{p}_x = -\frac{\partial H}{\partial x} \to \frac{d}{dz} p_x = -\frac{\partial H_1}{\partial H} \frac{\partial H}{\partial x} = -\frac{\partial H_1}{\partial x} \qquad (3.61)$$

while the second of equations (3.59) yields:

$$\frac{dz}{dt} = \frac{\partial H}{\partial H_1} = \frac{\partial H}{\partial p_z}. \qquad (3.62)$$

Exercise 3.4. The Hamiltonian (3.56) is not particularly useful in the treatment of problems related to the statistical properties of an electron beam transported through a magnetic channel. Find the physical (and mathematical) reasons to consider the following further form:

$$H_1 \to K = \frac{\delta}{\beta_0} - \sqrt{\left(\frac{1}{\beta_0} + \delta - \frac{q\Phi}{P_0 c}\right)^2 - (\pi_x - a_x)^2 - (\pi_y - a_y)^2 - \frac{(mc)^2}{P_0^2}} - a_z \qquad (3.63)$$

$$\delta = \frac{E}{P_0 c} - \frac{1}{\beta_0}, \quad \frac{mc}{P_0} = \frac{1}{\gamma_0 \beta_0}, \quad \gamma_0 = \frac{1}{\sqrt{1 - \beta_0^2}}, \quad \pi_{x,y} = \frac{p_{x,y}}{P_0}, \quad a_{x,y,z} = \frac{qA_{x,y,z}}{P_0}$$

where $P_0 = m\gamma_0\beta_0 c$ is the momentum of the reference particle.
Solution:
The introduction of P_0, exploited to normalize the kinematic variables defining the Hamiltonian H_1, introduces a new dynamical element which should be understood carefully.

The variable δ represents the 'energy deviation' from the reference particle; it is indeed exactly zero for particles with momentum P_0.

It is a given that if δ is considered to be a canonical momentum, the conjugated variable is:

$$s = \frac{z - \beta_0 c t}{\beta_0} \qquad (3.64)$$

which gives the distance from the particle governed by the Hamiltonian K with respect to its reference counterpart.

Exercise 3.5. Use the considerations developed in the previous exercise and interpret the physical meaning of $\frac{dz}{ds}$.
Solution:
The canonical variables of the Hamiltonian K are

$$(x, \pi_x), \ (y, \pi_y), \ (z, \delta) \qquad (3.65)$$

and s plays the role of the 'time variable.' We conclude, therefore, that (in the absence of interaction with external fields):

$$\frac{dz}{ds} = \frac{\partial K}{\partial \delta} = \frac{1}{\beta_0} - \frac{\frac{1}{\beta_0} + \delta}{\sqrt{\left(\frac{1}{\beta_0} + \delta\right)^2 - \pi_x^2 - \pi_y^2 - \frac{1}{(\beta_0\gamma_0)^2}}} \qquad (3.66)$$

which is a kind of restatement of equation (3.64).

Exercise 3.6. Provide (or at least try) a rigorous derivation of equation (3.63).
Solution:
A safe way to translate one Hamiltonian into another (physically equivalent), without any ambiguity in the definition of the canonical variables, is to use the canonical transformations. The relevant advantage of the transformed Hamiltonian

and canonical variables is the preservation of the 'canonical' form of the associated equations of motion.

The transition $H_1 \to K$ is obtained (see the bibliography at the end of the chapter) by means of the following generating function (of the second kind):

$$F_2(x, \pi_x, y, \pi_y, -t, \delta; z) = x\pi_x + y\pi_y + \left(\frac{z}{\beta_0} - ct\right)\left(\frac{1}{\beta_0} + \delta\right). \tag{3.67}$$

The transition from the old to the new Hamiltonian is then realized by the identities

$$K = H_1 + \frac{\partial F_2}{\partial z} \tag{3.68}$$

which yield equation (3.63), apart from an inessential constant.

Exercise 3.7. We are interested in the study of magneto-optic problems to treat the transverse components of electron beam evolution along (almost linear) linear pathways. The full Hamiltonian (3.63) is rather excessive for this task. In addition, let us note that, according to our assumption, the following inequalities hold:

$$\pi_{x,y}, \delta \ll 1, \gamma_0 \gg 1 \tag{3.69}$$

Accordingly, derive a form of Hamiltonian more useful for the specific problems under study.

Solution:

For simplicity, we consider the 'free Hamiltonian'

$$K = \frac{\delta}{\beta_0} - \sqrt{\left(\frac{1}{\beta_0} + \delta\right)^2 - \pi_x^2 - \pi_y^2 - \frac{1}{\beta_0\gamma^2}} \tag{3.70}$$

and note that $\pi_{x,y}$ are all quantities of order $\frac{1}{\gamma}$.

We can therefore eliminate the square root in equation (3.70) using a suitable Taylor expansion:

$$\begin{aligned}
K &= \frac{\delta}{\beta_0} - \left(\frac{1}{\beta_0} + \delta\right)\sqrt{1 - \frac{\pi_x^2 + \pi_y^2 + \frac{1}{\beta_0\gamma^2}}{\left(\frac{1}{\beta_0} + \delta\right)^2}} \simeq \frac{\delta}{\beta_0} - \left(\frac{1}{\beta_0} + \delta\right)\left(1 - \frac{1}{2}\frac{\pi_x^2 + \pi_y^2 + \frac{1}{\beta_0\gamma^2}}{\left(\frac{1}{\beta_0} + \delta\right)^2}\right) \\
&= \frac{\delta}{\beta_0} - \left(\frac{1}{\beta_0} + \delta\right) + \frac{1}{2}\frac{\pi_x^2 + \pi_y^2 + \frac{1}{\beta_0\gamma^2}}{\left(\frac{1}{\beta_0} + \delta\right)} \simeq \delta\left(\frac{1}{\beta_0} - 1\right) \\
&\quad - \frac{1}{\beta_0} + \frac{\beta_0}{2}\left(\pi_x^2 + \pi_y^2 + \frac{1}{\beta_0\gamma^2}\right)(1 - \beta_0\delta + \beta_0\delta^2) \\
&\simeq -1 + \frac{\delta}{2\gamma_0^2} + \frac{1}{2}\left(\pi_x^2 + \pi_y^2 + \frac{1}{\gamma^2}\right)(1 - \delta + \delta^2).
\end{aligned} \tag{3.71}$$

Neglecting all the terms larger than o (3.49) and the constant -1, we are left with (we reuse H to avoid any further tidying up of the notation):

$$H = \frac{1}{2}\left(\pi_x^2 + \pi_y^2 + \frac{\delta^2}{\gamma^2}\right) \tag{3.72}$$

which is interpreted as a nonrelativistic Hamiltonian that contains both transverse and longitudinal parts.

Exercise 3.8. Use the Hamiltonians (3.72) and (3.71) to derive the changes of the electron's transverse coordinates inside a quadrupole.
Solution:
From equation (3.63), after neglecting the longitudinal term (and using an expansion), we obtain:

$$H = \frac{1}{2}\left(\pi_x - a_x^2 + \pi_y - a_y^2 + \frac{\delta^2}{\gamma^2}\right) - a_z. \tag{3.73}$$

Bearing in mind that to the second order in the spatial coordinates, the vector potential components are:

$$a_x \simeq a_y = 0$$
$$a_z = -\frac{1}{2}\kappa(x^2 - y^2) \tag{3.74}$$

we end up with:

$$H = \frac{1}{2}\left(\pi_x^2 + \pi_y^2\right) + V_Q(x, y)$$
$$V_Q(x, y) = \frac{1}{2}\kappa_1(x^2 - y^2). \tag{3.75}$$

Exercise 3.9. Complete the previous exercise by including multipolar contributions (sextupoles, octupoles, …).
Hint:

$$H = \frac{1}{2}\left(\pi_x^2 + \pi_y^2\right) + V(x, y)$$
$$V(x, y) = V_Q(x, y) + V_S(x, y) + V_0(x, y)$$
$$V_S(x, y) = \frac{\kappa_2}{6}x^3 - 3x^2 y \tag{3.76}$$
$$V_0(x, y) = \frac{\kappa_2}{24}x^4 - 6x^2 y + y^4.$$

Exercise 3.10. Use Liouville's theorem to derive the transverse phase-space distribution of a relativistic electron beam moving in a transport magnetic channel.
Solution:
For simplicity, we consider the Hamiltonian (3.75), which does not imply any coupling between the x and y components.

We use $F(x, \pi_x)$ to denote the phase-space distribution. Liouville's theorem states that it is a conserved quantity, namely (s is the longitudinal variable, which in the present approximation is equivalent to $z \simeq ct$):

$$\frac{\mathrm{d}}{\mathrm{d}s} F = 0 \tag{3.77}$$

which implies that the phase-space distribution is described by the partial differential equation

$$\frac{\partial}{\partial s} F = -\pi_x \frac{\partial}{\partial x} F + \kappa_Q \, x \frac{\partial}{\partial \pi_x} F \tag{3.78}$$

$$F(x, \pi_x)\big|_{s=0} = F_0(x, \pi_x).$$

Exercise 3.11. Use equation (3.78) to establish a link with the matrix formalism developed in the previous parts of this chapter.
Solution:
The exact solution of equation (3.51) can be straightforwardly obtained by the use of the method of characteristics or any other method based on algebraic manipulation, such as Lie algebraic methods (see the bibliography at the end of the chapter).

The solution in equation (3.78) can be formally solved as follows:

$$\begin{aligned} F(x, \pi_x) &= \hat{U} F_0(x, \pi_x) \\ \hat{U} &= \mathrm{e}^{s \left(-\pi_x \frac{\partial}{\partial x} + V'_Q(x) \frac{\partial}{\partial \pi_x} \right)} \\ V'_Q(x) &= \frac{\partial}{\partial x} V_Q(x) = \kappa_Q x \end{aligned} \tag{3.79}$$

where $\hat{U}$ is the Liouville evolution operator. Although the problem can be solved exactly in this specific case, in order to make a more direct link with the matrix formalism, we note that we can write:

$$\hat{U} \simeq \mathrm{e}^{-\frac{\delta s}{2} \pi_x \frac{\partial}{\partial x}} \mathrm{e}^{\delta s \, V'_Q(x) \frac{\partial}{\partial \pi_x}} \mathrm{e}^{-\frac{\delta s}{2} \pi_x \frac{\partial}{\partial x}} F_0(x, \pi_x) \tag{3.80}$$

using three-step exponential disentanglement, where δs is a small advance in the forward direction.

Before going further, it is worth underscoring that the exponential operators appearing in equation (3.80) produce a shift in the variables of the function $F_0(x, \pi_x)$, as specified below:

$$e^{-\frac{\delta s}{2}\pi_x \frac{\partial}{\partial x}} F_0(x, \pi_x) = F_0\left(x - \frac{\delta s}{2}\pi_x, \pi_x\right)$$
$$e^{\delta s\, V'_Q(x)\frac{\partial}{\partial \pi_x}} F_0(x, \pi_x) = F_0\left(x, \pi_x + \delta V'_Q(x)\right).$$
(3.81)

The solution, based on the disentanglement in equation (3.80), involves three distinct steps:
1. Free propagation over a straight section of length s/2
2. A kick approximation of the force effect induced by the potential V
3. Free propagation over a straight section of length s/2

which are recognized as the thin lens approximation mentioned in the previous parts of the chapter.

We can therefore make the following identifications (we use the replacement $\pi_x \to x'$):

$$\begin{pmatrix} 1 & s \\ 0 & 1 \end{pmatrix} \Leftrightarrow e^{-s\, x' \frac{\partial}{\partial x}}$$
$$\begin{pmatrix} 1 & 0 \\ -k\, s & 1 \end{pmatrix} \Leftrightarrow e^{ksx' \frac{\partial}{\partial x}}$$
(3.82)

which eventually allows the following interpretation of the evolution operator and of the relevant disentangled form

$$e^{-\frac{s}{2} x' \frac{\partial}{\partial x}} e^{s\, V'_Q(x) \frac{\partial}{\partial x}} e^{-\frac{s}{2} x' \frac{\partial}{\partial x}} \Leftrightarrow \begin{pmatrix} 1 & \frac{s}{2} \\ 0 & 1 \end{pmatrix} \begin{pmatrix} 1 & 0 \\ -k\, s & 1 \end{pmatrix} \begin{pmatrix} 1 & \frac{s}{2} \\ 0 & 1 \end{pmatrix}$$
$$= \begin{pmatrix} 1 - k\frac{s^2}{2} & s \\ -k\, s & 1 - k\frac{s^2}{2} \end{pmatrix} + o(\delta s^3).$$
(3.83)

We have accordingly recovered the thin lens approximation within the framework of analytical mechanics.

Exercise 3.12. Note that the disentanglement in equation (3.80) is not unique and that at the order $o(\delta s^3)$, the alternative decoupling $e^{-\frac{\delta s}{2} x' \frac{\partial}{\partial x}} e^{\delta s\, V'(x) \frac{\partial}{\partial x}} e^{-\frac{\delta s}{2} x' \frac{\partial}{\partial x}}$ leads to the same conclusion.

Hint:
It can be directly confirmed that:

$$e^{\frac{s}{2}V'(x)\frac{\partial}{\partial x}}e^{-s\,x'\frac{\partial}{\partial x}}e^{\frac{s}{2}V'(x)\frac{\partial}{\partial x}} \Leftrightarrow \begin{pmatrix} 1 & 0 \\ -k\frac{s}{2} & 1 \end{pmatrix}\begin{pmatrix} 1 & s \\ 0 & 1 \end{pmatrix}\begin{pmatrix} 1 & 0 \\ -k\frac{s}{2} & 1 \end{pmatrix}$$

$$\simeq \begin{pmatrix} 1 - k\frac{s^2}{2} & s \\ -ks & 1 - k\frac{s^2}{2} \end{pmatrix} + o(\delta s^3). \tag{3.84}$$

Exercise 3.13. Find an analytical justification for the caveat 'at the order $o(\delta s^3)$' we added to the specification of the previous exercise.

Exercise 3.14. Show that for the case of a generic potential $V(x)$, the small-step solution of Liouville's equation can be written as:

$$e^{-\frac{\delta s}{2}\pi_x\frac{\partial}{\partial x}}e^{\delta s\,V'(x)\frac{\partial}{\partial \pi_x}}e^{-\frac{\delta s}{2}\pi_x\frac{\partial}{\partial x}}F_0(x,x') = F_0\left(x - \delta s\,x' - \frac{\delta s^2}{2}V'\left(x - \frac{\delta s}{2}x'\right),\, x' + \delta s\,V'\left(x - \frac{\delta s}{2}x'\right)\right). \tag{3.85}$$

Exercise 3.15. Prove that the small-step solution in equation (3.85) can be iterated to get the evolution on a larger distance S, as reported below:

$$\hat{U}^n F_0 x, x' = F_0\left(x_n, x'_n\right)$$

$$x_n = x_{n-1} - \delta s\,x'_{n-1} - \frac{\delta s^2}{2}V'\left(x_{n-1} - \frac{\delta s}{2}x'_{n-1}\right)$$

$$x'_n = x'_{n-1} + \delta s\,V'\left(x_{n-1} - \frac{\delta s}{2}x'_{n-1}\right) \tag{3.86}$$

$$n = \frac{S}{\delta s}.$$

Hint:
The problem is solved by an iterated application of the evolution operator, as shown in the example below for $n = 2$:

$$\hat{U}^2 F_0(x,x') = e^{-\frac{\delta s}{2}\pi_x\frac{\partial}{\partial x}}e^{\delta s\,V'(x)\frac{\partial}{\partial \pi_x}}e^{-\frac{\delta s}{2}\pi_x\frac{\partial}{\partial x}}F_0\left(x - \delta s\,x' - \frac{\delta s^2}{2}V'\left(x - \frac{\delta s}{2}x'\right),\, x' + \delta s\,V'\left(x - \frac{\delta s}{2}x'\right)\right) \tag{3.87}$$

and by the application of the translation operator as reported in equation (3.81).

Exercise 3.16. Show that (x_n, x'_n) are, at any step n, canonical variables.
Solution:
The proof is easily obtained by showing that the determinant of the Jacobian matrix of transformation corresponds to unit. The Jacobian matrix is denoted by

$$\hat{J} = \begin{pmatrix} \dfrac{\partial x_n}{\partial x_{n-1}} & \dfrac{\partial x'_n}{\partial x_{n-1}} \\ \dfrac{\partial x_n}{\partial x'_{n-1}} & \dfrac{\partial x'_n}{\partial x'_{n-1}} \end{pmatrix}. \tag{3.88}$$

It can therefore be confirmed by direct computation that:

$$\hat{J} = \begin{pmatrix} 1 - \dfrac{\delta s^2}{2} V''\left(x_{n-1} - \dfrac{\delta s}{2} x'_{n-1}\right) & -\delta s V''\left(x_{n-1} - \dfrac{\delta s}{2} x'_{n-1}\right) \\ -\delta s + \dfrac{\delta s^3}{4} V''\left(x_{n-1} - \dfrac{\delta s}{2} x'_{n-1}\right) & 1 - \dfrac{\delta s^2}{2} V''\left(x_{n-1} - \dfrac{\delta s}{2} x'_{n-1}\right) \end{pmatrix} \rightarrow \tag{3.89}$$

$$\rightarrow |\hat{J}| = 1.$$

Exercise 3.17. Reconcile the C–S formalism with the Hamilton–Liouville method.
Solution:
We make the assumption that the initial phase-space distribution is provided by the two-variable Gaussian:

$$F_0(x, x') = \dfrac{1}{2\pi\varepsilon} e^{-\frac{1}{2\varepsilon} Q(x, x')}$$

$$Q(x, x') = (x\ x') \begin{pmatrix} \gamma & \alpha \\ \alpha & \beta \end{pmatrix} \begin{pmatrix} x \\ x' \end{pmatrix}. \tag{3.90}$$

According to the previous discussion, the evolution of $F_0(x, x')$ along any transport element can be expressed through a corresponding translation operator.

In the case of propagation along a drift section, we find:

$$F(x, x'; s) = e^{-s x' \frac{\partial}{\partial x}} F_0(x, x') = \dfrac{1}{2\pi\varepsilon} e^{-\frac{1}{2\varepsilon} Q(x, x'; s)}. \tag{3.91}$$

The action of the exponential operator is readily obtained, as follows:

$$Q(x, x'; s) = (x\ x') \begin{pmatrix} 1 & 0 \\ -s & 1 \end{pmatrix} \begin{pmatrix} \gamma & \alpha \\ \alpha & \beta \end{pmatrix} \begin{pmatrix} 1 - s \\ 0 & 1 \end{pmatrix} \begin{pmatrix} x \\ x' \end{pmatrix}. \tag{3.92}$$

Reshuffling the right-hand side of equation (3.92) by transferring the s dependence to the Twiss coefficients, we end up with:

$$Q(x, x'; s) = (x\ x') \begin{pmatrix} \gamma(s) & \alpha(s) \\ \alpha(s) & \beta(s) \end{pmatrix} \begin{pmatrix} x \\ x' \end{pmatrix}, \qquad (3.93)$$

$$\gamma(s) = \gamma,\ \alpha(s) = \alpha - \gamma\ s,\ \beta(s) = \beta - 2\ \alpha\ s + \gamma\ s^2.$$

Exercise 3.18. Apply the same procedure to the case of the thin lens quadrupole.
Hint:

$$Q(x, x'; s) = (x\ x') \begin{pmatrix} 1 & ks \\ 0 & 1 \end{pmatrix} \begin{pmatrix} \gamma & \alpha \\ \alpha & \beta \end{pmatrix} \begin{pmatrix} 1 & 0 \\ k\ s & 1 \end{pmatrix} \begin{pmatrix} x \\ x' \end{pmatrix},$$

$$\gamma(s) = \gamma + 2\ \alpha\ k\ s + (k\ s)^2 \beta, \qquad (3.94)$$

$$\alpha(s) = \alpha + k\ s\ \beta,$$

$$\beta(s) = \beta.$$

Exercise 3.19. Consider the content of figure 3.18 and explain its physical meaning.
Hint:

The figure shows a focusing quadrupole with chromatic aberration. In other words, the magnet is not properly focusing the rays; the focal length depends indeed on the deviation of the electron's momentum deviation from that of the reference particle. The addition of two more focusing and defocusing quadrupoles restores correct focusing (see the next chapter for an appropriate discussion).

Exercise 3.20. Discuss how the sextupole provides the abovementioned corrections.
Hint:

See the next chapter, where the problem is discussed while also including the perspective of classical optics.

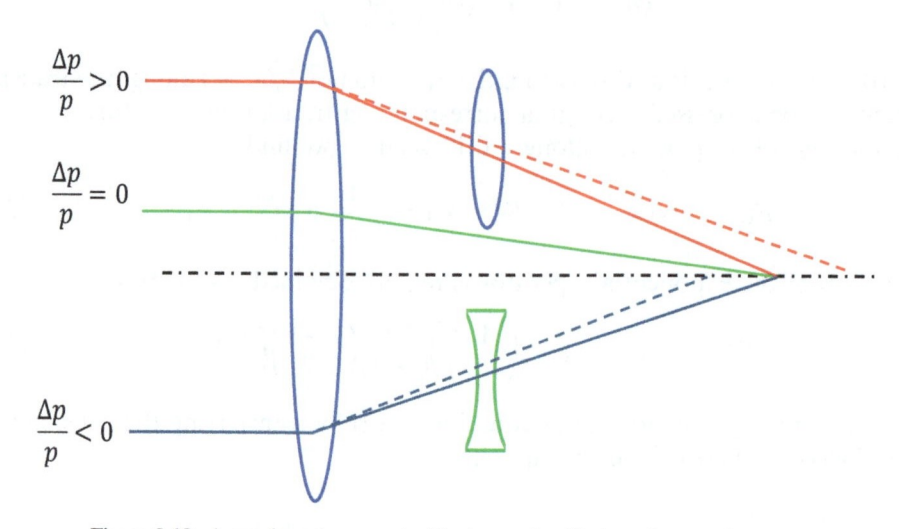

Figure 3.18. A quadrupole magnet with chromatic effects and correction terms.

References and further reading

For accelerators and charged beam transport, see the following:

Courant E D and Snyder H S 1958 Theory of the alternating-gradient synchrotron *Ann. Phys.* **3** 1–48

Penner S 1961 Calculations of properties of magnetic deflection systems *Rev. Sci. Instrum.* **32** 150–60

Brown K L, Belbeoch R and Bounin P 1964 First- and second-order magnetic optics matrix equations for the midplane of uniform-field wedge magnets *Rev. Sci. Instrum.* **35** 481–5

Brown K L 1965 *A General First- and Second-Order Theory of Beam Transport Optics and Its Application to the Design of High-energy Particle Spectrometers SLAC-PUB-0132* Stanford University https://www.slac.stanford.edu/pubs/slacpubs/0000/slac-pub-0132.pdf

Steffen K G 1965 *High Energy Beam Optics* (Interscience Monographs and Texts in Physics and Astronomy) (New York, NY:Interscience) https://books.google.it/books?id=xB9RAAAAMAAJ

For more recent contributions, see the following:

Reiser M 2008 *Theory and Design of Charged Particle Beams* 1st edn (New York:Wiley) doi: https://dx.doi.org/10.1002/9783527617623

Wiedemann H 2007 *Particle Accelerator Physics* 3rd edn (Berlin: Springer) doi: https://dx.doi.org/10.1007/978-3-540-49045-6

Steffen K 1990 Fundamentals of accelerator optics *CAS—CERN Accelerator School: SynchrotronRadiation and Free electron Lasers* ed Turner S (Geneva: CERN) https://cds.cern.ch/record/115996

Barletta W A 2010 Basic Properties of Particle Beams (Introduction to Accelerators). US Particle Accelerator School Lecture 4 https://uspas.fnal.gov/materials/09UNM/Unit_2_Lecture_4_Beam_Properties.pdf

Dattoli G et al 2017 *Charged Beam Dynamics, Particle Accelerators and Free Electron Lasers* (Bristol: IOP Publishing) doi: https://dx.doi.org/10.1088/978-0-7503-1239-4

Pivi M n.d. Introduction to Accelerator Optics. An Introduction to the USPAS'12 Course Storage and Damping Ring Design https://uspas.fnal.gov/materials/12MSU/DampingRings-Lecture0.pdf

Henderson S H J and Zhang Y 2009 Transverse beam optics. USPAS'12 Course, Part I. https://uspas.fnal.gov/materials/09VU/Lecture6.pdf

Ciocci F et al 2000 *Insertion Devices for Synchrotron Radiation and Free Electron Laser* (Series on Synchrotron Radiation Techniques and Applications 6) (Singapore:World Scientific) doi: https://dx.doi.org/10.1142/4066

For the use of sextupoles, see the following:

Reichel I 2012 Lattice Design II: Nonlinear Dynamics. USPAS'12 Course https://uspas.fnal.gov/materials/12MSU/DampingRings-Lecture3.pdf

Wiedemann H 2022 *Particle Accelerator Physics* (Cham: Springer) doi: https://dx.doi.org/10.1007/978-3-319-18317-6

For the use of Lie algebraic methods in beam transport, see the following:

Draft A J *et al* 1988 Lie algebraic treatment of linear and nonlinear Beam dynamics *Ann. Rev. Nucl. Part. Sci.* **38** 455–96

Dattoli G and Galli M and Ottaviani P L 1993 Anharmonic betatron motion in linearly polarized undulators *J. Appl. Phys.* **73** 7046–52

Quattromini M 1999 Time-dependent symplectic integrators and time-ordering contributions *Il Nuovo Cimento* A **112** 415–9

Backscattering Sources, Volume 1
Theoretical framework and Thomson backscattering sources
Alessandro Curcio, Giuseppe Dattoli and Emanuele Di Palma

Chapter 4

Optical beam transport

4.1 Introduction

In the previous chapter we provided the general tools and criteria for the design of devices dedicated to charged beam transport. We underscored that the treatment is greatly simplified by the use of the matrix formalism, allowing a straightforward manipulation of the beam's phase-space evolution within the magnetic elements constituting an actual transport line. We mentioned that closely similar methods are adopted in optical ray tracing. In this chapter we provide an analogous treatment for the design of optical beam lines, which are devices of paramount importance for the laser beam transport in Compton backscattering (CBS) devices. In the first part of the chapter we study light propagation from a solely geometric point of view and provide a comparison with the matter treated in previous chapter. In the second part we go a step further by introducing Gaussian wave optics and the associated phase space. Within this context we establish the optical Twiss parameters, characterize the statistical properties of the wave beams, and provide an accurate comparison with the electron case. The starting point for any discussion of light beam deflection is Snell's refraction law. In the Comments and exercises section we report a detailed discussion, along with the relevant derivation.

A ray crossing two media with different indices of refraction is bent at the interface between the two media. It forms an angle with respect to the normal that differs from the angle of incidence. The most elementary application of this phenomenon is a lens, which is an optical device that affects a ray beam as shown in figure 4.1, which displays convex (converging) and concave (diverging) systems. In this figure, a paraxial beam, namely a collection of rays moving parallel to the *optical axis*,[1] is:

    a) Deflected toward a common point on the axial region at the right-hand side of the lens in the converging case;

    b) Deflected as from a common point on the left-hand side in the diverging case.

---

[1] The optical axis is orthogonal to the lens and passes through its center. In the course of this chapter we will provide a more appropriate definition.

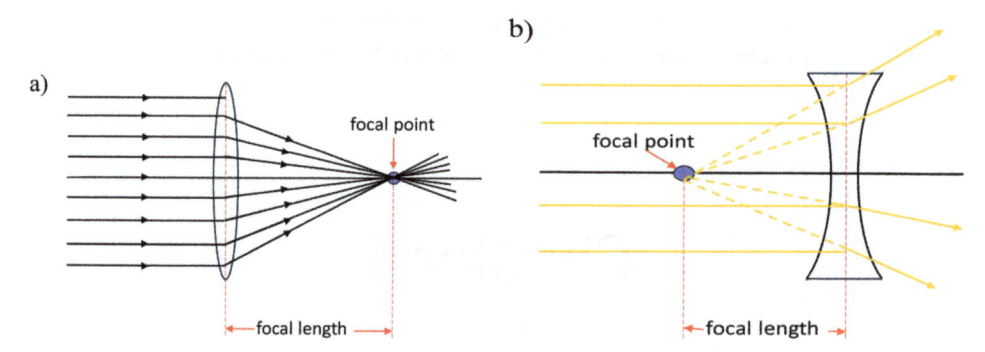

Figure 4.1. (a) Converging and (b) diverging lenses.

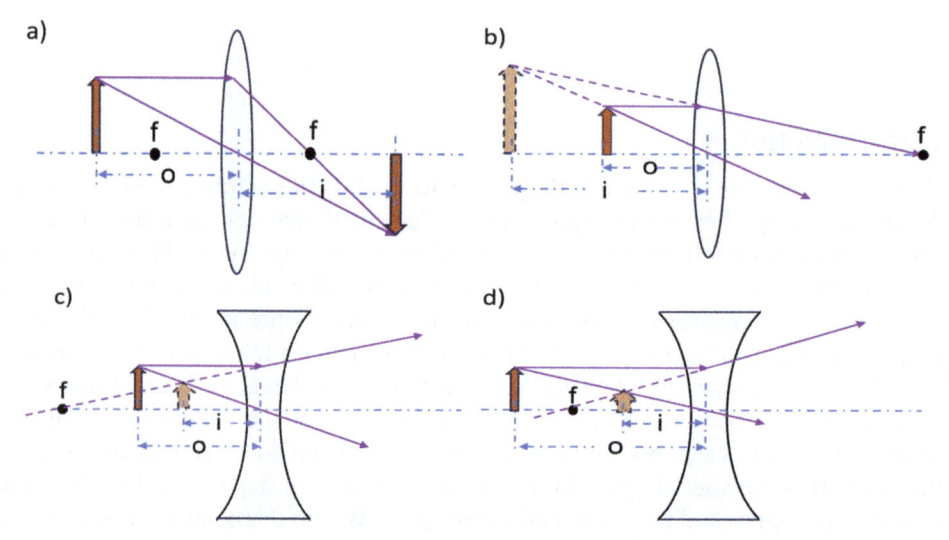

Figure 4.2. Images produced by lenses. (a) Converging real. (b) Converging virtual. (c) Diverging (virtual) with real object after focus. (d) Diverging (virtual) with real object before focus.

The common point is the *focus* of the lens. As in the case of the quadrupole, a lens is said to be 'thin' when it acts as a kick, namely when it produces an abrupt bend that is independent of its thickness. Any object placed at a certain distance from a lens forms an image (see figure 4.2), which can be either ***real or virtual***. in the first case, it is formed on its right-hand side, while in the latter, it is formed on its left. The image produced by a diverging lens is ***always*** virtual, whereas that produced by a convex lens can be virtual or real according to the position of the object with respect to the first focus. The distinction between real and virtual images is more subtle than might be thought and is further discussed in section 4.6 at the end of the chapter. The conventional rules for constructing lens images are given below (see also figure 4.3):
  1. Figures are drawn with the ray traveling from left to right;
  2. Distances are measured from a reference surface; for our purpose, that of the lens or any other refractive surface;

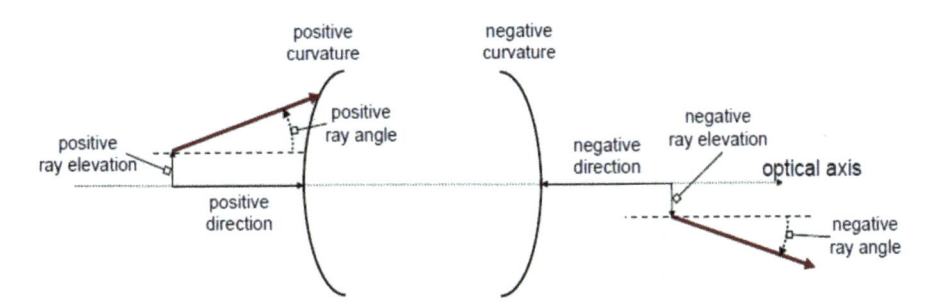

Figure 4.3. A visual summary of the sign convention.

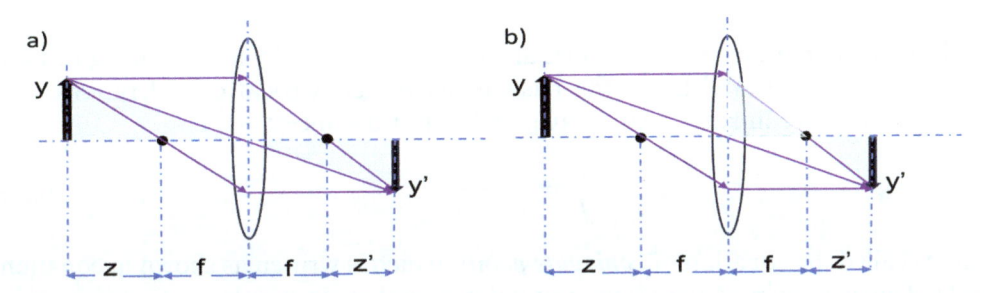

Figure 4.4. The geometric properties of lens image construction: (a) triangles on the left and the right of the lens, (b) triangles on the right-hand side only.

3. The focal length of a lens inducing a clockwise angular deflection (convergent) is positive;
4. The distance to a real object is positive;
5. The distance to a real image is positive;
6. Heights above the optical axis are positive;
7. Heights below the optical axis are negative;
8. Angular orientation is counterclockwise with respect to the optical axis.

Referring to figure 4.4, we define:
1. The lens focal length f
2. The distance from the object to the lens $o = f + z$
3. The distance from the image to the lens $i = f + z'$

We draw a ray tracing diagram for, say, an object placed at a distance $>2f$ from a convex lens (see figure 4.2(a)) by applying the following procedure:
1. We trace a segment from the top of the object and parallel to the optical axis;
2. We draw a second segment connecting the top of the object and the center of the lens; this ray proceeds without being deflected;
3. The intersection between the two segments defines the ray image (namely its height, position, and sign).

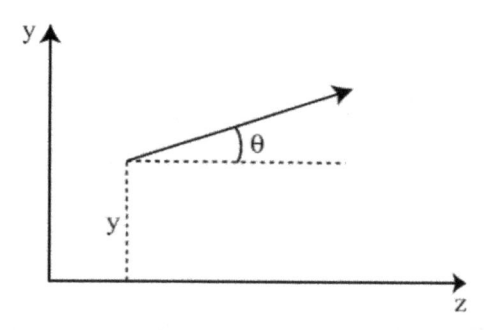

Figure 4.5. A geometrical representation of an optical ray in the paraxial approximation.

By keeping in mind the previous rules and conventions and by inspecting figures 4.4 and 4.5, we can establish, on the basis of naïve geometrical considerations, the following identity linking focal, image, and object distances:

$$\frac{1}{f} = \frac{1}{o} + \frac{1}{i}. \tag{4.1}$$

This is called the equation of **conjugate points**, which is written as shown in equation (4.1). This is known as the Gauss normal convention, in which the Cartesian sign rule is not taken into account (see the Comments and exercises section for further discussion). By recasting equation (4.1) in the form:

$$\frac{1}{f} = \frac{1}{z+f} + \frac{1}{z'+f} \tag{4.2}$$

we find the alternative form:

$$(z+f)(z'+f) = (z'+f)f + f(z+f) \Rightarrow$$
$$f^2 = z\,z' \tag{4.3}$$

which justifies the denomination of the conjugate points' equation. The relative magnification is provided by the ratio:

$$M = -\frac{h'}{h} = -\frac{i}{o} = -\frac{\frac{f}{o}}{1 - \frac{f}{o}}. \tag{4.4}$$

The previous relation is also just a consequence of the similarity of the triangles involved in the construction of the ray tracing image (see figure 4.6). Even though it was derived for a converging lens, for an object placed at a distance larger than twice the focal length, equations (4.1), (4.3), and (4.4) hold under all circumstances involving thin lenses. For further applications of the above and the use of the quoted rules, see the Comments and exercises section.

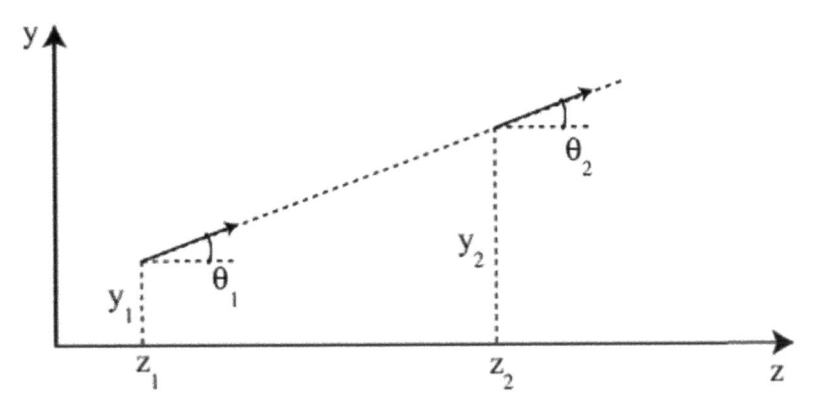

Figure 4.6. Free propagation of a ray beam.

4.2 The ray matrix method

The matrix formalism applied to electron beam transport can be translated almost unchanged to the case of optical beams. Here, we consider the **optical matrix conventions** according to **Yariv** and develop the relevant formalism according to the following rules:

1. A ray beam in the paraxial approximation is specified (see figure 4.5) by its angular divergence and by its height (namely the distance away from the optical axis). These can be embedded into a two-component vector denoted by:

$$\underline{r} = \begin{pmatrix} y \\ \theta \end{pmatrix} \tag{4.5}$$

   The angle θ is associated with the slope $\tan(\theta) = \Delta y/\Delta z$ which, due to the paraxial nature of the motion, reduces to $\Delta y/\Delta z \cong \theta$.

2. Elementary optical systems (lenses, mirrors, etc.) are mathematically realized by 2×2 matrices, which, acting on a ray vector, determine the variation of either the angular or the spatial part;

3. Any other complex optical system can be realized as a 2×2 matrix obtained as a suitable product of the 'minimal' matrices corresponding to the elementary systems.

Within this context, two further matrices play a central role. We consider the cases corresponding to: (a) ray propagation along a straight section and (b) a ray crossing a thin lens. The propagation of a ray in a homogeneous medium, namely a medium with a constant refractive index, corresponds to the following transformation (see figure 4.6):

$$\underline{r}_o = {}_d\hat{T}\, \underline{r}_i,$$
$${}_d\hat{T} = \begin{pmatrix} 1 & d \\ 0 & 1 \end{pmatrix}. \tag{4.6}$$

The subindices *i* and *o* stand for input and output, respectively, and *d* denotes the length of the straight section. Accordingly, we find:

$$\underline{r}_o = \begin{pmatrix} y_i + d\,\theta_i \\ \theta_i \end{pmatrix} \tag{4.7}$$

The ray beam propagation at the interface between two media with different indices of refraction is characterized by the matrix:

$$_n\hat{T} = \begin{pmatrix} 1 & 0 \\ 0 & \dfrac{n_1}{n_2} \end{pmatrix} \tag{4.8}$$

which is just a restatement of Snell's law for paraxial propagation. In the case of thin lenses, the ray height is left unchanged by the action of the optical element. The transfer matrix is therefore the same as for the quadrupole:

$$_L\hat{T} = \begin{pmatrix} 1 & 0 \\ k & 1 \end{pmatrix}. \tag{4.9}$$

The non-diagonal matrix element k can be specified by the definition of a lens itself, namely an optical device which bends all the parallel (to the lens axis) rays to the focus. An incoming 'parallel' ray beam is accordingly specified by the two-component vector $\underline{p}_i = \begin{pmatrix} h \\ 0 \end{pmatrix}$. After a ray has crossed the lens and a distance d, the ray vector is transformed into:[2]

$$\underline{p}_o = \begin{pmatrix} 1 & d \\ 0 & 1 \end{pmatrix}\left[\begin{pmatrix} 1 & 0 \\ k & 1 \end{pmatrix}\begin{pmatrix} h \\ 0 \end{pmatrix}\right] = \begin{pmatrix} 1+d\,k & d \\ k & 1 \end{pmatrix}\begin{pmatrix} h \\ 0 \end{pmatrix} = \begin{pmatrix} (1+d\,k)\,h \\ k\,h \end{pmatrix}. \tag{4.10}$$

If, after being bent by the lens, the ray propagates for a distance equal to the lens' focal length (namely $d = f$), then the output ray height vanishes. The k matrix element in equation (4.9) is therefore specified by the condition $1 + f\,k = 0$, thus yielding the expression:

$$_L\hat{T} = \begin{pmatrix} 1 & 0 \\ \mp \dfrac{1}{f} & 1 \end{pmatrix} \tag{4.11}$$

for the lens matrix operator, where the + sign denotes a defocusing lens. The output ray is accordingly specified by:

$$\underline{p}_o = \begin{pmatrix} 0 \\ -\dfrac{h}{f} \end{pmatrix}. \tag{4.12}$$

---

[2] The order of the matrices is due to the fact that the straight section operator acts on the vector modified by the lens. The beam indeed crosses the lens first and then propagates freely. We will carefully reconsider the problem of optical operator order later in this book.

The inverse ($_o\hat{T}^{-1}$) of a matrix $_o\hat{T}$, representing an optical device, is understood to be another optical element that changes the orientation of a ray crossing $_o\hat{T}$. Examples are better than words; we recall that the inverse of a matrix is:

$$\begin{pmatrix} a & b \\ c & d \end{pmatrix}^{-1} = \frac{1}{\Delta}\begin{pmatrix} d & -b \\ -c & a \end{pmatrix}, \quad (4.13)$$
$$\Delta = ad - bc$$

where Δ is the matrix determinant. Taking the inverse of the matrix on the right-hand side of equation (4.10), we find:

$$\underline{p}_i = \begin{pmatrix} 1 & 0 \\ -\frac{1}{f} & 1 \end{pmatrix}\left[\begin{pmatrix} 1 & -f \\ 0 & 1 \end{pmatrix}\begin{pmatrix} 0 \\ -\frac{h}{f} \end{pmatrix}\right] = \begin{pmatrix} 1 & -f \\ -\frac{1}{f} & 0 \end{pmatrix}\begin{pmatrix} 0 \\ -\frac{h}{f} \end{pmatrix} = \begin{pmatrix} h \\ 0 \end{pmatrix}. \quad (4.14)$$

We note that for all the optical elements considered so far, we have $\Delta = 1$. This statement does not hold for matrices such as those in equation (4.8), which involve rays crossing media with different indices of refraction. We invite the reader to consider the physical meaning of the previous statement. Having defined the rules above, we note that the combination of different optical elements is, as also discussed in the previous chapter, equivalent to straightforward (albeit sometimes cumbersome) matrix algebra. Before entering into more specific details, let us stress again that the matrix corresponding to the optical devices shown in figure 4.7 should be composed of the product of the operators[3] realizing the individual subsystems, using the order given in the figure. In other words, the last matrix on the left should be that corresponding to the last optical element crossed by the ray.

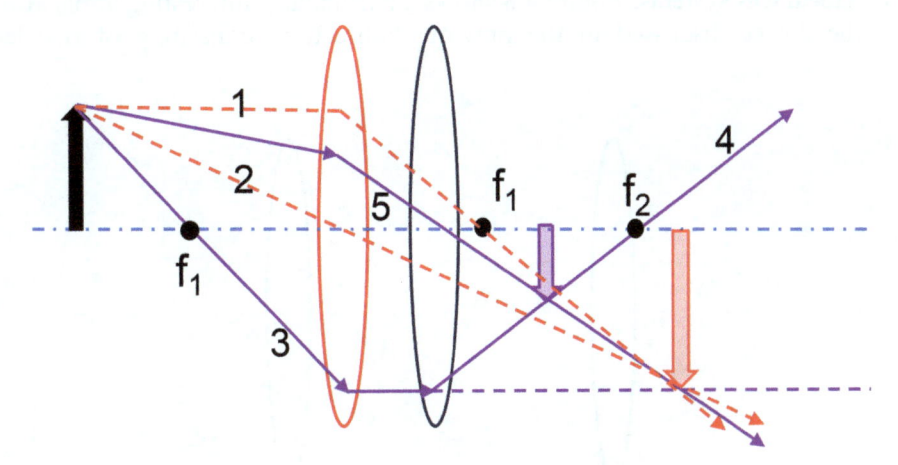

Figure 4.7. Lens doublet ray tracing and image formation.

---

[3] We refer either to a matrix or to an operator because the action of an optical element can be realized using e.g. differential operators as well, as we will see in the following chapters.

1. **Consecutive thin lenses.** According to the Cartesian convention, the ray travels from left to right. Assuming that the ray crosses the lenses in the sequence 1, 2, ..., we find:

$$_L\hat{T} = {}_{L_n}\hat{T}\,{}_{L_{n-1}}\hat{T} \cdots {}_{L_1}\hat{T} = \begin{pmatrix} 1 & 0 \\ -\dfrac{1}{f} & 1 \end{pmatrix}$$

$$\frac{1}{f} = \sum_{i=1}^{n} \frac{1}{f_i}.$$

(4.15)

The geometrical content relevant to the case of two successive lenses is displayed in figure 4.7, where we have shown the relevant ray tracing for the determination of the image formation.

2. **Lens–straight section–lens.** We consider an optical device composed of a convex lens, a straight section, and a second converging lens. By just applying the rules stipulated so far, we obtain the following transfer matrix describing the ray transport system:

$$\hat{T} = \begin{pmatrix} 1 & 0 \\ -\dfrac{1}{f_2} & 1 \end{pmatrix} \begin{pmatrix} 1 & d \\ 0 & 1 \end{pmatrix} \begin{pmatrix} 1 & 0 \\ -\dfrac{1}{f_1} & 1 \end{pmatrix} = \begin{pmatrix} 1 - \dfrac{d}{f_1} & d \\ -\left(\dfrac{1}{f_2} + \dfrac{1}{f_1} - \dfrac{d}{f_1 f_2}\right) & 1 - \dfrac{d}{f_2} \end{pmatrix}.$$

(4.16)

In section 4.6, we further comment on equation (4.16) and show that it implicitly contains the Gullstrand equation.

3. **Afocal lens systems.** Figure 4.8 shows a particularly interesting application of the device discussed in the previous subsection, consisting of two lenses

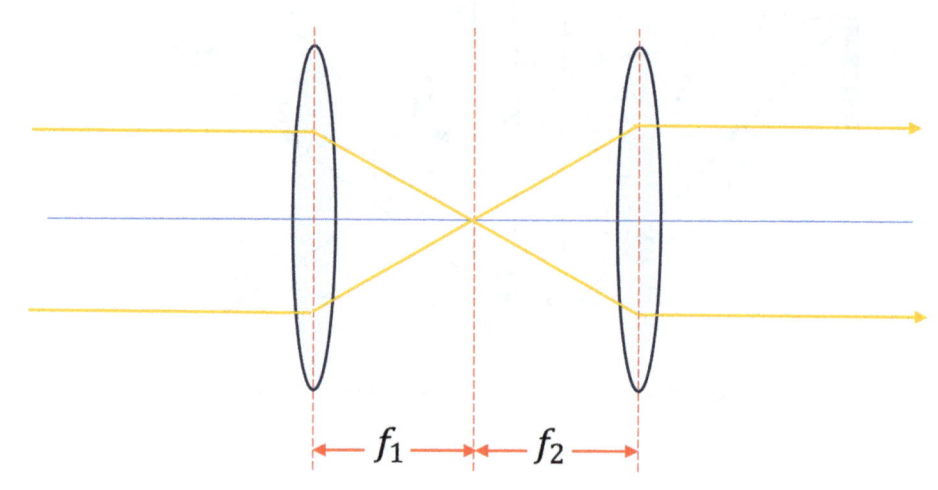

Figure 4.8. The geometry of an afocal optical device.

separated by a drift section $d = f_1 + f_2$. This lens system is said to be **afocal** in the sense that it transforms parallel rays into another (inverted) pair of parallel rays. We find indeed that:

$$\begin{pmatrix} y_o \\ \theta_o \end{pmatrix} = \begin{pmatrix} -\frac{f_2}{f_1} & f_1 + f_2 \\ 0 & -\frac{f_1}{f_2} \end{pmatrix} \begin{pmatrix} y_i \\ 0 \end{pmatrix} = \begin{pmatrix} -\frac{f_2}{f_1} y_i \\ 0 \end{pmatrix}. \tag{4.17}$$

It is worth stressing that the system acts as a telescope if $f_2/f_1 > 1$. Further comments on telescopic systems can be found in the exercises at the end of this chapter.

4.3 Matrix optics and lens images

In the previous sections we have introduced the notion of real and virtual images and made a naïve distinction by specifying that, in the case of converging lenses, such images are formed at the right or the left of the focal point, respectively. We will reconsider the problem of image formation within the context of the matrix formalism. According to the discussion of the previous sections, a parallel ray passing through a lens and a drift section transforms as follows:

$$\begin{pmatrix} y_o \\ \theta_o \end{pmatrix} = \begin{pmatrix} 1 & d \\ 0 & 1 \end{pmatrix} \begin{pmatrix} 1 & 0 \\ -\frac{1}{f_1} & 1 \end{pmatrix} \begin{pmatrix} h \\ 0 \end{pmatrix} = \begin{pmatrix} 1 - \frac{d}{f_1} & d \\ -\frac{1}{f_1} & 1 \end{pmatrix} \begin{pmatrix} h \\ 0 \end{pmatrix} = \begin{pmatrix} \left(1 - \frac{d}{f_1}\right) h \\ -\frac{h}{f_1} \end{pmatrix}. \tag{4.18}$$

If we use the equation of the conjugate points and replace d with the distance to the object image calculated from the equation of conjugated points, namely $d = \frac{f_1 o}{o - f_1}$, where o is the position of the object facing the lens,[4] we find:

$$\begin{pmatrix} y_o \\ \theta_o \end{pmatrix} = \begin{pmatrix} -Mh \\ -\frac{h}{f_1} \end{pmatrix}, \tag{4.19}$$

$$M = \frac{f_1}{o - f_1}.$$

This corresponds to an upside-down image of the object with a size magnification $M > 1$ when the object is placed at a distance $f_1 < o < 2f_1$ and $M < 1$ in the case $o > 2f_1$. If the object is placed at $o < f_1$, the image is upright for $M > 1$ and lies on the left of the first focus; note indeed that $i < 0$, and $|i| < f_1$. Let us now consider a

---

[4] Note that no matrix transport is necessary for propagation from the object position to the lens, because a parallel ray vector is left unaltered by such propagation.

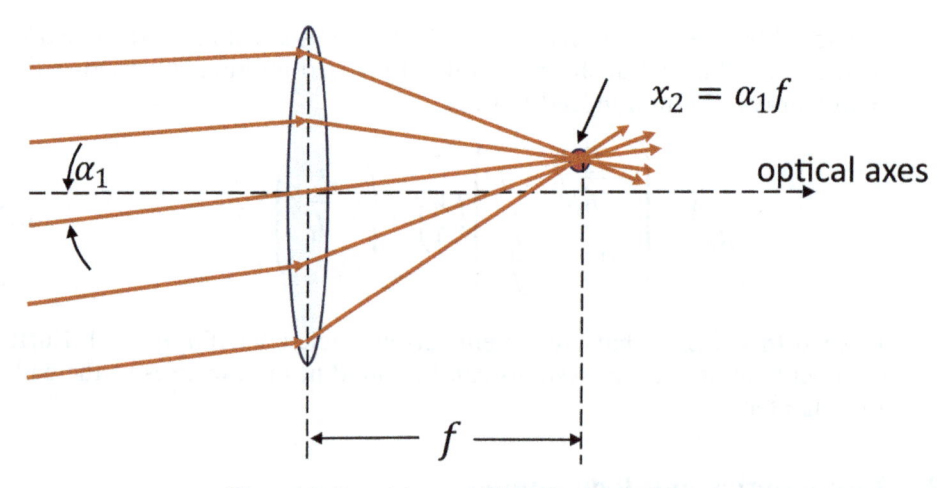

Figure 4.9. Paraxial propagation of non-parallel rays.

converging lens. From the equation of conjugate rays, after replacing f_1 with $-f_1$ in equation (4.19), we find that the image position, regardless of the object position, is always negative. Therefore, according to the sign convention, it is placed on the left of the lens and is virtual. So far, we have considered rays strictly parallel to the optical axis, but we can ask: what happens to an optical bundle like that shown in figure 4.9, where a collection of parallel rays with a slight inclination with respect to the optical axis crosses a converging lens? By applying the same formalism as before, we obtain:

$$\begin{pmatrix} y_o \\ \theta_o \end{pmatrix} = \begin{pmatrix} 1 & f_1 \\ 0 & 1 \end{pmatrix} \begin{pmatrix} 1 & 0 \\ -\frac{1}{f_1} & 1 \end{pmatrix} \begin{pmatrix} h \\ \alpha \end{pmatrix} = \begin{pmatrix} 0 & f_1 \\ -\frac{1}{f_1} & 1 \end{pmatrix} \begin{pmatrix} h \\ \alpha \end{pmatrix} = \begin{pmatrix} f_1 \alpha \\ -\frac{h}{f_1} + \alpha, \end{pmatrix} \qquad (4.20)$$

which ensures that the focal point is not located on the lens' optical axis but is slightly shifted above it if the ray divergence is positive. The geometrical locus on which the rays converge for different beam divergences is the *foci line*. Even though we have treated the lens image formation in different formal contexts, we have not yet specified the meaning of real and virtual images. The main difference between the two is that the first only can be seen on a screen, although it is possible to see both. In the case of a simple magnifier, the eye refracts the rays from the enlarged image onto the retina. We will comment on this last point in section 4.6 dedicated to exercises.

4.4 A phase-space formalism for optical wave transport

Even though not explicitly stated in the previous discussion, the matrix formalism for combinations of linear optical systems is equivalent to a particular case of Liouville's theorem. This is a rather naïve statement, except that we have implicitly assumed that phase space is a physical quantity characterizing both charged and optical beams. In this book we have reversed the paradigm; we discussed optical ray tracing after our discussion of charged beam transport. This choice was not casual.

The Courant–Snyder formalism is indeed very effective in providing an understanding of the statistical properties of the propagated electron beam, and the same effectiveness can be translated to Gaussian optics. This correspondence becomes an effective design tool for devices, as it anticipates the overlap of electrons and photon beams (as in the subject of this book). The ABCD matrix governs the evolution of the parameters specifying the development of an optical beam (namely its height and divergence). It can be written as

$$\begin{pmatrix} y_o \\ \theta_o \end{pmatrix} = \begin{pmatrix} A & B \\ C & D \end{pmatrix} \begin{pmatrix} y_i \\ \theta_i \end{pmatrix} \tag{4.21}$$

which applies to all optical systems discussed so far. We have seen that electron ray tracing is an approximation of Hamilton mechanics and that geometric optics is a limit of wave optics. The propagation of a Gaussian beam (a collection of rays) emerging from a source with a **waist** W_0 is characterized by the evolution of its spot size $W(z)$ shown in figure 4.10.

The beam intensity distribution can indeed be written as:

$$\begin{aligned} I(r, z) &= I_0 e^{-\frac{2r^2}{w(z)^2}}, \\ I_0 &= \frac{2P}{\pi w(z)^2}, \\ w(z) &= w_0 \sqrt{1 + \left(\frac{\lambda_0 z}{\pi w_0^2}\right)^2} \end{aligned} \tag{4.22}$$

where r is the distance from the propagation axis and P is the total power carried by the wave. If we introduce the quantity:

$$z_R = \frac{\pi w_0^2}{\lambda_0} \tag{4.23}$$

the waist evolution can be rewritten in the form:

$$w(z) = w_0 \sqrt{1 + \left(\frac{z}{z_R}\right)^2}. \tag{4.24}$$

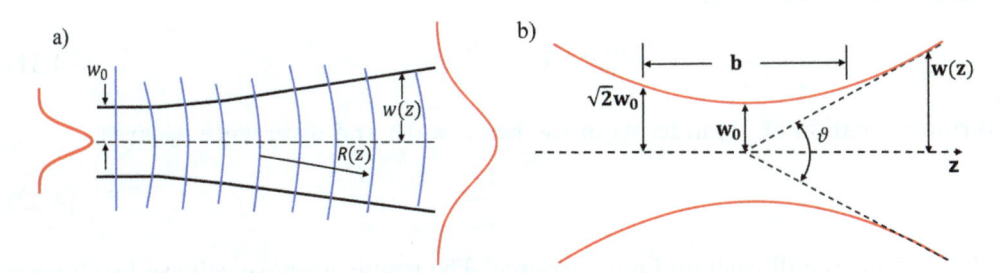

Figure 4.10. (a) The evolution of a Gaussian beam's spot size and (b) the definition of beam divergence (the evolutions of spot size and divergence are invariant for propagation axis reflection).

This clarifies the physical meaning of the parameter in equation (4.23): it is the distance (called the Rayleigh distance) at which the size of the waist is increased by a factor of $\sqrt{2}$. Let us now consider the evolution of the waist from points 1 and 2 on the axis z and write:

$$w_2^2 = w_1^2 + \left(\frac{\lambda_0}{4\pi}\right)^2 \frac{L^2}{w_1^2}. \tag{4.25}$$

Let us now assume that the waist can be written using the same formula we adopted in the case of the electron beam, namely $W^2 = \beta \varepsilon_{\text{ph}}$. The quantity ε_{ph} is not yet specified and plays the role of *optical* emittance. In terms of the Twiss parameter β, equation (4.25) can be written as:

$$\beta_2^2 = \beta_1^2 + \frac{L^2}{\beta_1^2} \tag{4.26}$$

where we have assumed:

$$\varepsilon_{\text{ph}} = \frac{\lambda_0}{4\pi}. \tag{4.27}$$

Regarding the analogy between equation (4.26) and the corresponding equation for the electron beam case, we note that it essentially consists of free propagation over a distance L from the waist position. Bearing in mind that (at the waist) $\alpha_1 = 0$ and that $\beta_1 = 1/\gamma_1$, we obtain:

$$\beta_2^2 = \beta_1^2 + (L\gamma_1^2). \tag{4.28}$$

Let us now clarify the meaning of equation (4.27). We start from the Heisenberg principle for the case of a single photon:

$$\Delta x \Delta p_x = \frac{\hbar}{2} \tag{4.29}$$

write the x component of the photon momentum as:

$$p_x = \hbar k x' = \frac{2\pi \hbar}{\lambda_0} x' \tag{4.30}$$

and eventually write (4.29) as:

$$\Delta x \Delta x' = \frac{\lambda_0}{4\pi}. \tag{4.31}$$

Writing equation (4.31) in terms of the beam waist and divergence θ_d gives:

$$w_0 \theta_d = \frac{\lambda_0}{4\pi} \tag{4.32}$$

which specifies a diffraction-limited source. The results we have achieved so far are extremely important and can be summarized as reported below:

1. A ray beam can be treated using the phase-space concepts adopted for e-beam propagation.
2. The optical beam phase-space area is given by equation (4.27).
3. The formalism has provided a correspondence between emittance, wavelength, and the Planck constant.

The last point has noticeable conceptual consequences. For example, the design of fourth-generation diffraction-limited x-ray sources requires the following condition to apply to electron and photon ray emittances: $\varepsilon_e \leqslant \varepsilon_{ph}$. Finding a parallelism between $\bar{\lambda}$ and $\hbar$ opens further speculations, which are discussed in the next section.

4.5 Gaussian beams and the formal quantum theory of light rays

Leontovich and Fock showed that Maxwell's wave equation describing the propagation of the electromagnetic field reduces, under the assumption of paraxial motion, to a Schrödinger-type equation. The analogy with quantum mechanics was reinforced by Gloge and Marcuse, who arrived at the same conclusion starting from the canonical quantization paradigm associated with the $\bar{\lambda}\lambda$and$\hbar$ correspondence (for further details, see the Comments and exercises section and the bibliography at the end of the chapter). The parabolic equation determining the evolution of a two-dimensional Gaussian beam can indeed be written as:

$$i k \, \partial_z F(x, y, z) = -\frac{1}{2}\big[\partial_x^2 + \partial_y^2\big] F(x, y, z). \tag{4.33}$$

Its solutions are obtained using the standard methods, resulting in, for example:

$$F(x, y, z) = e^{\frac{i}{2k}[\partial_x^2 + \partial_y^2]} F(x, y, 0). \tag{4.34}$$

Factoring both the evolution operator and the initial distribution function, we can set:

$$\begin{aligned} F(x, y, z) &= \Psi(x, z)\,\Psi(y, z) \\ \Psi(\xi, z) &= e^{\frac{i}{2k}\partial_\xi^2}\Psi(\xi, 0). \end{aligned} \tag{4.35}$$

Using a Gaussian as the initial function, we end up with the z-dependent solution:

$$\begin{aligned} \Psi(\xi, z) &= \frac{N}{wz} e^{-\frac{\xi^2}{wz^2}} e^{+i\frac{k}{2}\frac{\xi^2}{R(z)} + ikz - i\zeta(z)}, \\ R(z) &= z\left[1 + \left(\frac{z_R}{z}\right)^2\right], \quad \zeta(z) = \frac{1}{2} tg^{-1}\!\left(\frac{z}{z_R}\right) \end{aligned} \tag{4.36}$$

where the radius $R(z)$ accounts for the wave front evolution and $\zeta(z)$ is known as the Gouy phase (see the Comments and exercises section). Comparing this with equation (4.22), we conclude that:

$$I(r, z) = |F(x, y, z)|^2$$
$$x^2 + y^2 = r^2. \tag{4.37}$$

The phase term $i\,kz$ takes account of the wave's propagation in the z-direction. In order to use the phase-space formalism, we need to apply a distribution function in the canonical coordinates. This can be achieved using the so-called Wigner transform, which can be viewed as a generalization of the ordinary Fourier transform, namely:

$$W(\xi, p_\xi, z) = \frac{1}{2\pi\,\varepsilon_p} \int_{-\infty}^{+\infty} e^{-\frac{ip_\xi \delta}{\varepsilon_p}} \Psi\left(\xi + \frac{\delta}{2}, z\right) \Psi^*\left(\xi - \frac{\delta}{2}, z\right) d\delta. \tag{4.38}$$

In more precise terms, the Wigner function W is a quasi probability distribution (sometimes, it is not positively defined everywhere) and reduces to the Liouville function in the 'classical' limit, namely when $\varepsilon_p \to 0$ (when the Planck constant is assumed to be vanishing in quantum mechanics). The equation satisfied by the Wigner function is known as the von Neumann equation (for further details, see the Comments and exercises section) and reads:

$$\frac{\partial}{\partial z} W = \left[-p_x \frac{\partial}{\partial x} - p_y \frac{\partial}{\partial y} + \frac{1}{i\varepsilon_p} \left(V\left(x + i\frac{\varepsilon_p}{2}\frac{\partial}{\partial p_x}, y + i\frac{\varepsilon_p}{2}\frac{\partial}{\partial p_y}\right) - V\left(x - i\frac{\varepsilon_p}{2}\frac{\partial}{\partial p_x}, y - i\frac{\varepsilon_p}{2}\frac{\partial}{\partial p_y}\right) \right) \right] W \tag{4.39}$$

and (as easily checked—see the Comments and exercises section) reduces to the ordinary Liouville function for vanishing ε_p:

$$\lim_{\varepsilon_p \to 0} \frac{1}{i\varepsilon_p} \left(V\left(\xi + i\frac{\varepsilon_p}{2}\frac{\partial}{\partial p_\xi}\right) - V\left(\xi - i\frac{\varepsilon_p}{2}\frac{\partial}{\partial p_\xi}\right) \right) = V'(\xi)\frac{\partial}{\partial p_\xi}. \tag{4.40}$$

More generally, the following expansion holds:

$$\frac{1}{i\varepsilon_p}\left(V\left(\xi + i\frac{\varepsilon_p}{2}\frac{\partial}{\partial p_\xi}\right) - V\left(\xi - i\frac{\varepsilon_p}{2}\frac{\partial}{\partial p_\xi}\right)\right) = \frac{1}{\varepsilon_p}\sin\left(\hat{o}\frac{\partial}{\partial \xi}\right)V(\xi) = \frac{1}{\varepsilon_p}\sum_{n=0}^{\infty}\frac{\hat{o}^{2n+1}}{(2n+1)!}V^{(2n+1)}(\xi)$$
$$\hat{o} = \frac{\varepsilon_p}{2}\frac{\partial}{\partial p_\xi}. \tag{4.41}$$

It is now important to understand that if the potential is a quadratic function of the spatial coordinates, the phase-space evolution is ruled by the same Liouville equation that is exploited in the case of electron beams crossing a quadrupole-like device. Accordingly, the same conclusion holds for light beams passing through lens-like optical systems. Limiting ourselves to this case, we can use the same mathematical technicalities adopted for the charged beam phase-space evolution for radiation beams passing through optical guiding devices. This statement has important consequences from the practical point of view regarding, for example, the design of radiation sources, in which the interaction between electron and optical beams plays a central role. This is essential for synchrotron radiation sources, free-electron laser devices, and for CBS machines. The concepts we have sketched so far have been formulated and developed within the context of the study of the phenomena emerging from the necessity of defining the electron and optical beam brightnesses on the same footing. This point of view is crucial in designing a guide that brings the beams to the overlap point and in defining the conditions for the optimum transfer of brightness from the electrons to the light beams. In the following chapters we will apply the methods we have described in chapters 3 and 4 to the design of an actual CBS device. The reader is referred to the Comments and exercises section for further details of the theoretical and practical aspects of the previously developed concepts. The literature concerning the topics summarized in the last two chapters is immense; it ranges from classical to quantum optics and accelerator physics. We have just scraped the surface; it is therefore mandatory to refer the reader to selected (but exhaustive) reviews reported in the bibliography at the end of the chapter. We believe, however, that the method of presentation we have chosen is sufficient for the reader to acquire the tools required for autonomous development of the design of electron/optical transport systems.

4.6 Comments and exercises

4.6.1 Exercise

Exercise 4.1. Derive Snell's law, which links the reflection and refraction of a light beam crossing two media with different indices of refraction:

$$\frac{\sin(\theta_1)}{\sin(\theta_2)} = \frac{n_2}{n_1}. \tag{4.42}$$

Solution:

We assume that an optical ray moves in a straight line and that its strategy when going from a point A to another B, while crossing two media with different refractive indices (see figure 4.11), is that of choosing the most convenient path to minimize the transit time. The ray moves from A to B in a time given by:

$$\begin{aligned} T_{AB} &= T_{AO} + T_{OB} \\ T_{AO} &= \frac{\sqrt{a^2 + x^2}}{v_1}, \\ T_{OB} &= \frac{\sqrt{b^2 + (d-x)^2}}{v_2}. \end{aligned} \tag{4.43}$$

The search for the path that is traversed in the minimum time is determined by the condition:

$$\frac{\mathrm{d}}{\mathrm{d}x} T_{AB} = 0 \Rightarrow \frac{x}{v_1 \sqrt{a^2 + x^2}} = \frac{d-x}{v_2 \sqrt{b^2 + (d-x)^2}}. \tag{4.44}$$

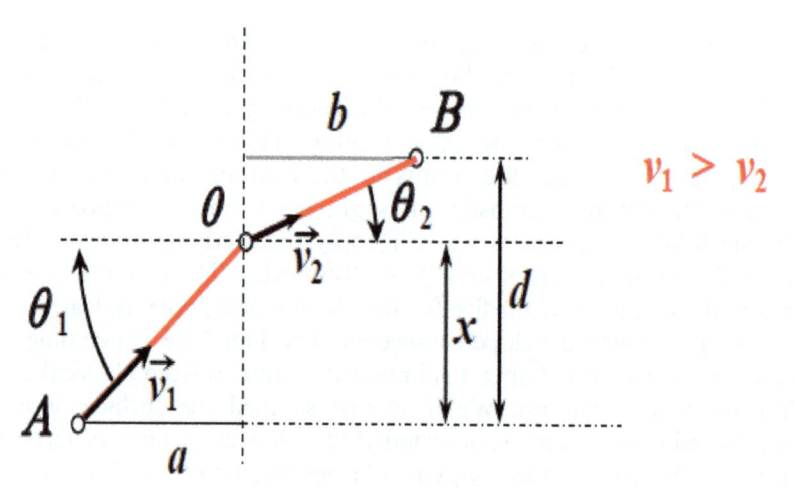

Figure 4.11. The geometry of Snell's law.

Furthermore, since:

$$\frac{x}{\sqrt{a^2 + x^2}} = \sin(\theta_1),$$

$$\frac{d - x}{\sqrt{b^2 + (d - x)^2}} = \sin(\theta_2) \qquad (4.45)$$

we can determine the relation:

$$\frac{\sin(\theta_1)}{\sin(\theta_2)} = \frac{v_1}{v_2} \qquad (4.46)$$

which is recognized as Snell's law. Furthermore, taking into account that the velocity of light in a medium is linked to its refractive index by:

$$v = \frac{c}{n} \qquad (4.47)$$

we eventually end up with:

$$\frac{\sin(\theta_1)}{\sin(\theta_2)} = \frac{n_2}{n_1}. \qquad (4.48)$$

Exercise 4.2. Express the focal power in terms of the lens' constitutive elements.
  Solution:
  A thin lens' focal power depends on the inner/outer $n_{L,o}$ refractive indices and on the radii of curvature $R_{1,2}$ of the lens surfaces, namely

$$P = \frac{n_L - n_0}{n_0}\left(\frac{1}{R_1} - \frac{1}{R_2}\right) \qquad (4.49)$$

$$P = f^{-1}.$$

The relation reported in equation (4.49) is known as **lens-maker formula** and expresses the condition that the power of a thin lens is (approximately) given by the sum of the powers of its individual surfaces[5]; this is the bare essence of the thin lens approximation.

Exercise 4.3. Guess the focal length and power of a thick lens.
Solution:
A thick lens can be modeled as the combination of two thin lenses arranged as in figure (4.55). Therefore, using the ray matrix rule reported in equation (4.15), we obtain:

$$f = \frac{f_1 f_2}{f_1 + f_2 - \dfrac{d}{n}} \tag{4.50}$$

and, regarding the thick lens power, according to equation (4.49) we find the so-called **Gullstrand equation**:

$$P = P_1 + P_2 - P_1 P_2 \frac{d}{n}. \tag{4.51}$$

From the physical point of view, the action of a thick lens on a ray beam can be viewed as the result of three successive independent steps (see figure 4.12 where we compare thin and thick lenses), as reported below:

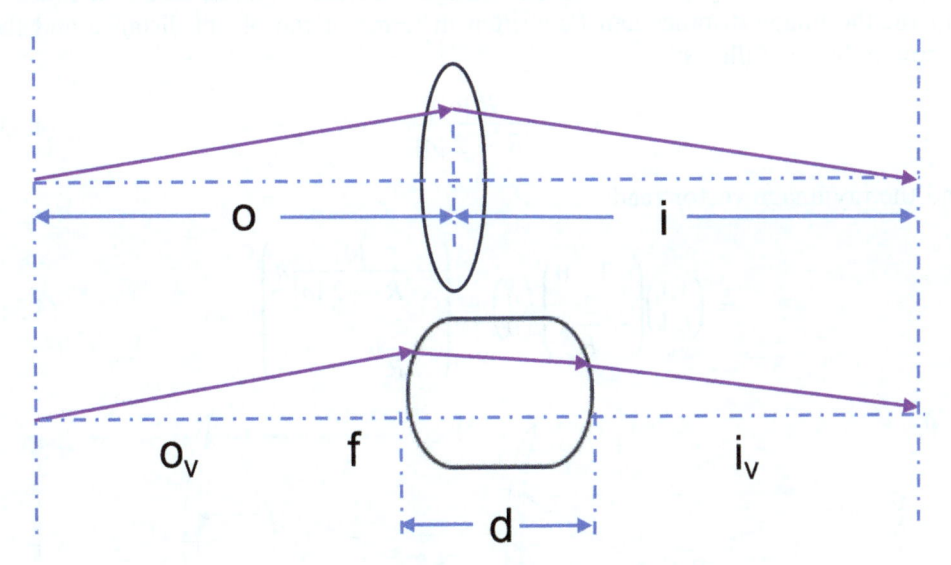

Figure 4.12. Thin and thick lenses.

---

[5] The focal power, which has the dimension of the inverse of a length, is also called the dioptric power and is denoted by D. A lens that has a focal power of $+4\,D$ in its front curve and a focal power of $-6\,D$ at the back has a total power of $-2\,D$.

1. A thin lens
2. Free propagation
3. A thin lens

Exercise 4.4. Use the matrix formalism to describe the ray propagation in mirrors.
Solution:
According the previous discussion, concave spherical mirrors with radius R can be treated as convex lenses. The associated operator reads

$$_M\hat{T} = \begin{pmatrix} 1 & 0 \\ -\dfrac{2}{R} & 1 \end{pmatrix}. \tag{4.52}$$

Regarding the sign convention, it should be noted that:
1. The object distance is positive if the object is in front of the mirror (a real object) but negative if it is behind the mirror (a virtual object)
2. The image distance is positive if the image is in front of the mirror (a real image) but negative if the image is behind the mirror (a virtual image)
3. The magnification is positive for upright images but negative for inverted images (figure 4.13).

The equation for conjugate points is still valid. Therefore, in the case of a concave mirror, the image distance can be written in terms of the object distance and the mirror radius as follows:

$$i = -\frac{R|o|}{R - 2|o|} \tag{4.53}$$

and the ray image vector reads

$$\underline{i} = \begin{pmatrix} 1 & i \\ 0 & 1 \end{pmatrix} \begin{pmatrix} 1 & 0 \\ -\dfrac{2}{R} & 1 \end{pmatrix} \begin{pmatrix} h \\ 0 \end{pmatrix} = \begin{pmatrix} -\dfrac{|o|}{R - 2|o|}h \\ -\dfrac{2}{R}h \end{pmatrix}. \tag{4.54}$$

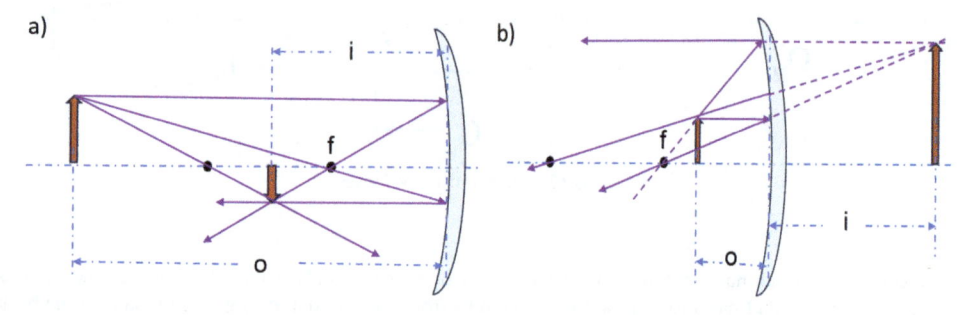

Figure 4.13. The geometry of a concave spherical mirror: (a) a real image, $|o| > R/2$; (b) a virtual image, $|o| < R/2$.

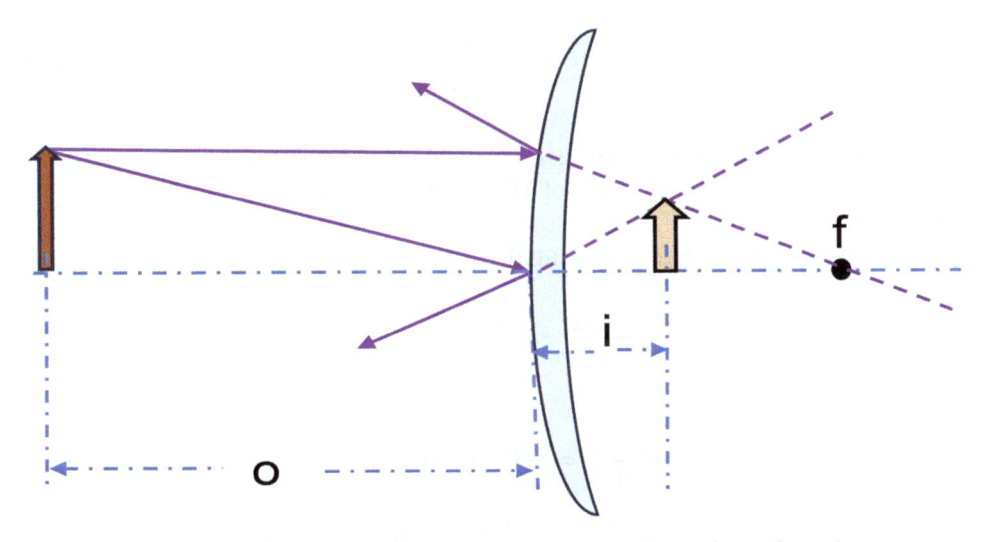

Figure 4.14. The geometry of a convex mirror and the relevant image formation.

For a convex mirror (figure 4.14), the transfer operator is that of a diverging lens; therefore, the previous equations (4.53, 4.54) should be written as

$$i = -\frac{R\,|o|}{R + 2\,|o|}$$

$$\underline{i} = \begin{pmatrix} \frac{|o|}{2\,|o| + R}h \\ \frac{2}{R}h \end{pmatrix}. \tag{4.55}$$

Exercise 4.5. Use the methods of canonical quantization and the discussion in this chapter to derive the paraxial wave propagation equation reported in equation (4.64).

Solution:

In view of the established correspondence between quantum mechanics and paraxial wave propagation, we can also consider the following definition of the ray angular divergence operator:

$$\hat{\theta}_\xi = -i\bar{\lambda}\frac{\partial}{\partial \xi} \tag{4.56}$$

which yields the following commutation bracket

$$\left[\hat{\xi}, \hat{\theta}_\xi\right] = i\bar{\lambda}. \tag{4.57}$$

The associated (unidimensional) free propagation Schrödinger equation is written as:

$$i\bar{\lambda}\frac{\partial}{\partial z}\Psi(\xi, z) = -\frac{1}{2}\bar{\lambda}^2\frac{\partial^2}{\partial \xi^2}\Psi(\xi, z). \tag{4.58}$$

Exercise 4.6. Use the previously established correspondences to derive the free propagation ray equation in terms of Heisenberg's equations of motion.
Solution:
The Hamiltonian operator associated with equation (4.58) is written:

$$\hat{H} = \frac{1}{2}\hat{\theta}_\xi^2. \tag{4.59}$$

Therefore:

$$\begin{aligned} i\bar{\lambda}\frac{d}{dz}\hat{\xi} &= [\hat{H}, \xi] = i\bar{\lambda}\hat{\theta}_\xi \\ i\bar{\lambda}\frac{d}{dz}\hat{\theta}_\xi &= [\hat{H}, \hat{\theta}_\xi] = 0. \end{aligned} \tag{4.60}$$

Show that the relevant solution is the operator form of the ray free propagation 'law.'

Exercise 4.7. Write the solution of equation (4.58) using the following initial condition:

$$\Psi(\xi, 0) = \frac{N}{w_0}e^{-\frac{\xi^2}{w_0}}e^{+i\Phi} \tag{4.61}$$

where Φ is an arbitrary phase. Write the obtained solution in the form given in equation (4.36).
Solution:
According to the previous discussion, we find:

$$\Psi(\xi, z) = \frac{N}{w_0}e^{+i\Phi}e^{\frac{i\bar{\lambda}z}{2}\frac{\partial^2}{\partial \xi^2}}e^{-\frac{\xi^2}{w_0}}. \tag{4.62}$$

The use of the Glaisher operational rule (see the bibliography)

$$e^{k\frac{\partial^2}{\partial x^2}}e^{-x^2} = \frac{1}{\sqrt{1+4k}}e^{-\frac{x^2}{1+4k}} \tag{4.63}$$

yields:

$$\Psi(\xi, z) = \frac{N}{w_0\sqrt{1+2i\frac{z\bar{\lambda}}{w_0^2}}}e^{+i\Phi}e^{-\frac{\xi^2}{w_0^2\left(1+2i\frac{z\bar{\lambda}}{w_0^2}\right)}} \tag{4.64}$$

which should be further elaborated to end up with the solution reported in equation (4.58). We therefore note that:

$$w_0\sqrt{1 + 4\mathrm{i}\frac{z\bar{\lambda}}{w_0^2}} = \sqrt{w(z)}\,\mathrm{e}^{\frac{\mathrm{i}}{2}\zeta z} \tag{4.65}$$

$$\frac{\xi^2}{w_0^2\left(1 + 2\mathrm{i}\frac{z\bar{\lambda}}{w_0^2}\right)} = \frac{\xi^2}{w(z)^2} - \frac{\mathrm{i}k}{2}\frac{\xi^2}{R(z)}. \tag{4.66}$$

Upon replacing $\mathrm{e}^{+\mathrm{i}\Phi}$ with $\mathrm{e}^{+\mathrm{i}kz}$, the equivalence between solutions (4.64) and (4.36) can be established using operational methods, which are further discussed in the following. We should, however, comment on the replacement of the phase Φ with kz: we must accordingly remember that Ψ represents a wave propagating in the z-direction. This emerges as the paraxial approximation of the Helmholtz equation and actually, the complete function should be understood as $\Psi(\xi, z) = \Psi_s(\xi, z)\mathrm{e}^{\mathrm{i}kz}$, where Ψ_s denotes the solution of equation (4.58). The important physical content of this result is that the propagating wave is no longer plane, and a curvature of the phase front emerges with a radius of curvature $R(z)$.

Exercise 4.8. Explain how the effect of a lens can be included in this quantum-like formalism.
Solution:
Taking into account what we have learned from the electron phase-space formalism, we can assume that the Heisenberg equation ruling the evolution of a beam crossing a focusing lens can be written as:

$$\begin{aligned}\mathrm{i}\,\bar{\lambda}\frac{\mathrm{d}}{\mathrm{d}z}\hat{\xi} &= [\hat{H}, \xi] = \mathrm{i}\bar{\lambda}\hat{\theta}_\xi \\ \mathrm{i}\,\bar{\lambda}\frac{\mathrm{d}}{\mathrm{d}z}\hat{\theta}_\xi &= [\hat{H}, \xi] = -\mathrm{i}\bar{\lambda}\kappa\hat{\xi}.\end{aligned} \tag{4.67}$$

Therefore, the associated Hamiltonian is written:

$$\hat{H} = \frac{1}{2}\left(\hat{\theta}_x^2 + \kappa\hat{\xi}^2\right). \tag{4.68}$$

It is evident that the solution of the of the associated Schrödinger equation can be written as:

$$\Psi(\xi, z) = \frac{N}{w_0}\mathrm{e}^{+\mathrm{i}\Phi}\mathrm{e}^{\frac{\mathrm{i}\bar{\lambda}z}{2}\frac{\partial^2}{\partial\xi^2} - \mathrm{i}\frac{z\kappa}{2\bar{\lambda}}\hat{\xi}^2}\,\mathrm{e}^{-\frac{\xi^2}{w_0}}. \tag{4.69}$$

A naïve disentanglement of the evolution operator yields:

$$\mathrm{e}^{\frac{\mathrm{i}\bar{\lambda}z}{2}\left(\frac{\partial^2}{\partial\xi^2} + \kappa\xi^2\right)} \simeq \mathrm{e}^{-\frac{\mathrm{i}z}{2\bar{\lambda}}(\kappa\xi^2)}\mathrm{e}^{\frac{\mathrm{i}\bar{\lambda}z}{2}\left(\frac{\partial^2}{\partial\xi^2}\right)}. \tag{4.70}$$

which is easily seen to correspond to equation (4.64). It is important to understand that the introduction of the complex radius $q(z)$ is not merely formal. It is the key parameter that describes the properties of the Gaussian beam. As shown below, it is therefore the most suitable element to form a bridge with the matrix method.

Exercise 4.12. Show that

$$e^{\frac{i\lambda z_2}{2}\left[\frac{\partial^2}{\partial x^2}+\frac{\partial^2}{\partial y^2}\right]}\Psi(x, y, z_1) = \Psi(x, y, z_1 + z_2). \tag{4.74}$$

Solution:
Note that:

$$\Psi(x, y, z_1) = e^{\frac{i\lambda z_1}{2}\left[\frac{\partial^2}{\partial x^2}+\frac{\partial^2}{\partial y^2}\right]}\Psi(x, y, 0). \tag{4.75}$$

Therefore:

$$e^{\frac{i\lambda z_2}{2}\left[\frac{\partial^2}{\partial x^2}+\frac{\partial^2}{\partial y^2}\right]}\Psi(x, y, z_1) = e^{\frac{i\lambda(z_1+z_2)}{2}\left[\frac{\partial^2}{\partial x^2}+\frac{\partial^2}{\partial y^2}\right]}\Psi(x, y, 0) = \Psi(x, y, z_1 + z_2) \tag{4.76}$$

which is propagation along a drift section expressed as a *waveform*.

Exercise 4.13. Explain the meaning of the complex radius at $z = 0$.
Solution:

$$\frac{1}{q(0)} = \frac{1}{R(0)} - i\frac{\lambda_0}{\pi w^2 0} \tag{4.77}$$
$$R(0) = \infty, \; w0 = w_0.$$

Therefore

$$q(0) = i\frac{\pi w_0^2}{\lambda_0} = iZ_R. \tag{4.78}$$

The parameter is an imaginary number at $z = 0$; this location corresponds to the focal plane.[6]

Exercise 4.14. Show that

$$\frac{1}{q(z)} = \frac{1}{z + iZ_R}. \tag{4.79}$$

---

[6] The plane perpendicular to the axis of a lens or mirror and passing through the focal point.

Hint:
The proof is straightforward, and

$$\frac{1}{q(z)} = \frac{z - iZ_R}{z^2 + Z_R^2} = \frac{1}{z\left(1 + \left(\frac{Z_R}{z}\right)^2\right)} - i\frac{1}{Z_R\left(1 + \left(\frac{z}{Z_R}\right)^2\right)}. \quad (4.80)$$

Exercise 4.15. Show that the complex parameter $q_1(z)$ of a Gaussian beam passing through an optical device described by the matrix in equation (4.21) is modified as reported below:

$$q_2 = \frac{Aq_1 + B}{Cq_1 + D}. \quad (4.81)$$

Solution:
We suggest a check rather than a formal proof. Therefore, we first consider the case of free propagation: $A = 1$, $B = d$, $C = 0$, $D = 1$; accordingly, we find:

$$q_2 = q_1 + d. \quad (4.82)$$

In the case of a thin lens, we go back to the 'quantum point of view' reported in equation (4.71) and find:

$$e^{-\frac{i\xi^2}{2\lambda f}}\frac{1}{q(z)}e^{-ik\frac{\xi^2}{2}\frac{1}{q(z)}} = \frac{1}{q(z)}e^{-ik\frac{\xi^2}{2}\frac{1}{q(z)} - \frac{i\xi^2}{2\lambda f}}$$

$$= \frac{1}{q(z)}e^{-ik\frac{\xi^2}{2}\left(\frac{f + q(z)}{q(z)f}\right)}. \quad (4.83)$$

We therefore recover the prescription (4.81), where the parameters of the thin lens are $A = 1$, $B = 0$, $C = -1/f$, $D = 1$, and indeed we find (after having specified the sign of f) that

$$\frac{1}{q_2} = \frac{-|f| + q_1}{q_1(-|f|)}$$

$$q_2 = \frac{q_1}{-\frac{q_1}{|f|} + 1}. \quad (4.84)$$

It should be noted that, in line with with equation (4.71), we have written the first exponential in equation (4.81) by assuming that $e^{-\frac{i\xi^2}{2\lambda f}} = e^{-\frac{iz_2}{2k}k\xi^2}$. This is a consequence of the thin lens approximation. Indeed, as z_2 is the thickness of the lens, we can set $kz_2 = 1/f$. For a comparison with the quadrupole case, see equations (3.18) and (3.19).

Exercise 4.16. In figure 4.18, we show a lens placed at the waist of a Gaussian beam. Find the location and size of the new waist.
Solution:

where the photon beam emittance is $\bar{\lambda}/2$. Let us now introduce the following parameters:

$$\gamma = \frac{1}{Z_R} \quad ; \quad \beta = Z_R\left[1 + \left(\frac{z}{Z_R}\right)^2\right] \quad ; \quad \alpha = \frac{z}{Z_R} \quad (4.92)$$

$$\gamma\beta - \alpha^2 = 1.$$

The obvious conclusion is that the Wigner function coincides, in this specific case, with the Liouville function. If we keep $z = 0$ in equation (4.92), we obtain the input Twiss parameters $\gamma = \frac{1}{Z_R}$, $\beta = Z_R$, and $\alpha = 0$; accordingly, the result of equations (4.92) is a reformulation of the result obtained by the matrix formalism, which yields the Twiss parameters at z in the form:

$$\begin{pmatrix} \beta \\ \alpha \\ \gamma \end{pmatrix}_z = \begin{pmatrix} 1 & 2z & z^2 \\ 0 & 1 & z \\ 0 & 0 & 1 \end{pmatrix} \begin{pmatrix} \beta \\ 0 \\ \gamma \end{pmatrix}_{z=0}. \quad (4.93)$$

Exercise 4.18. Apply the considerations given in exercise 15 to the case of a thin lens.
Solution:
We consider the action of a thin lens on the wave function at $z = 0$:

$$\Psi_L(\xi) = N e^{-\chi\xi^2 - \frac{i\xi^2}{2\lambda f}} \quad (4.94)$$

and get:

$$W_L(\xi, \theta_\xi) = \frac{2}{\pi\bar{\lambda}} e^{-\frac{1}{\bar{\lambda}}\left[\left(\frac{Z_R}{f^2} + \frac{1}{Z_R}\right)\xi^2 + \frac{4Z_R}{f}\theta\xi + 4Z_R\theta^2\right]} = \frac{2}{\pi\bar{\lambda}} e^{-\frac{1}{\bar{\lambda}}\gamma_L\xi^2 + 4\alpha_L\theta\xi + 4\beta_L\theta^2}. \quad (4.95)$$

The use of the following identity for the Twiss coefficients of a beam passing through a thin lens:

$$\begin{pmatrix} \beta \\ \alpha \\ \gamma \end{pmatrix}_L = \begin{pmatrix} 1 & 0 & 0 \\ \frac{1}{f} & 1 & 0 \\ \frac{1}{f^2} & \frac{2}{f} & 1 \end{pmatrix} \begin{pmatrix} \beta \\ 0 \\ \gamma \end{pmatrix}_0 = \begin{pmatrix} 1 & 0 & 0 \\ \frac{1}{f} & 1 & 0 \\ \frac{1}{f^2} & \frac{2}{f} & 1 \end{pmatrix} \begin{pmatrix} Z_R \\ 0 \\ \frac{1}{Z_R} \end{pmatrix} \quad (4.96)$$

allows the same conclusion as before.

Exercise 4.19. Evaluate the transport of the Wigner distribution along the sequence drift–lens–drift.

Solution:
The smart way to solve this problem is to calculate the matrix product corresponding to a sequence composed of a drift of length L_1, a thin lens, and finally a drift of length L_2 using equations (4.93) and (4.96).

$$W(\xi, \theta_\xi, L_1 + \text{lens} + L_2) = \frac{2}{\pi \bar{\lambda}} e^{-\frac{1}{\bar{\lambda}}\left[\gamma_f \xi^2 + 4\beta_f \theta_\xi^2 + 4\alpha_f \xi \theta_\xi\right]} \qquad (4.97)$$

where:

$$\beta_f = \left(1 - \frac{2L_2}{f} + \frac{L_2^2}{f^2}\right) Z_R + \left[L_2^2 + L_1^2\left(1 - \frac{2L_2}{f} + \frac{L_2^2}{f^2}\right) - L_1\left(\frac{2L_2^2}{f} - 2L_2\right)\right]\frac{1}{Z_R}$$

$$\alpha_f = \left(\frac{1}{f} - \frac{L_2}{f^2}\right) Z_R + \left[L_1^2\left(\frac{1}{f} - \frac{L_2}{f^2}\right) - L_2 - L_1\left(1 - \frac{2L_2}{f}\right)\right]\frac{1}{Z_R} \qquad (4.98)$$

$$\gamma_f = \frac{Z_R}{f^2} + \left(1 - \frac{2L_1}{f} + \frac{L_1^2}{f^2}\right)\frac{1}{Z_R}.$$

It is easy to verify that $\gamma_f \beta_f - \alpha_f^2 = 1$ and that for $L_1 = L_2 = 2f$, $\beta_f = Z_R = \beta_0$, which corresponds to 1:1 imaging.

Exercise 4.20. Show that:

$$\int_{-\infty}^{+\infty} W(\xi, \theta_\xi, z)\, d\theta_\xi = |\Psi(\xi, z)|^2$$
$$\int_{-\infty}^{+\infty} W(\xi, \theta_\xi, z)\, d\xi = |\tilde{\Psi}(\theta_\xi, z)|^2 \qquad (4.99)$$

where $\tilde{\Psi}(\theta_\xi, z)$ is the Fourier transform of $\Psi(\xi, z)$, namely:

$$\tilde{\Psi}(\theta_\xi, z) = \frac{1}{\sqrt{2\pi}} \int_{-\infty}^{+\infty} \Psi(\xi, z) e^{-\theta_\xi \xi}\, d\xi \qquad (4.100)$$

and explain the relevant physical meaning.

Hint:
The Wigner distribution defined in equation (4.38) is linked to the product of wave functions through an integral transform (called the Wigner–Weyl transform). It is a generalization of the ordinary Fourier transform and represents a phase-space quasi probability distribution.[8] Integration over the momentum variable θ_ξ yields the space probability density, while integration over the space variable yields the distribution in momentum space.

---

[8] 'Quasi' is added here because it is not always defined as positive.

Exercise 4.21. Outline the derivation of equation (4.39).
Solution:

Let us consider the definition of the Wigner function in equation (4.38) and keep the partial derivative of both sides with respect to time, namely:

$$\frac{\partial}{\partial z} W(\xi, p_\xi, z) = \frac{1}{2\pi \bar{\lambda}} \int_{-\infty}^{+\infty} e^{-\frac{i p_\xi \delta}{\bar{\lambda}}} \left[\Psi^*\left(\xi - \frac{\delta}{2}, z\right) \frac{\partial}{\partial z} \Psi\left(\xi + \frac{\delta}{2}, z\right) + \Psi\left(\xi + \frac{\delta}{2}, z\right) \frac{\partial}{\partial z} \Psi^*\left(\xi - \frac{\delta}{2}, z\right) \right] d\delta \quad (4.101)$$

where:

$$\frac{\partial}{\partial z} \Psi \xi, z = \left[-\frac{\bar{\lambda}}{2i} \frac{\partial^2}{\partial \xi^2} + \frac{V(\xi)}{i} \right] \Psi \xi, z. \quad (4.102)$$

The derivation of the von Neumann equation is now rather cumbersome. We therefore follow a procedure and highlight the main steps. We use the Schrödinger equation (4.102) inserted into (4.101) to write:

$$\frac{\partial}{\partial z} W(\xi, p_\xi, z) = \frac{\partial}{\partial z} [W_K(\xi, p_\xi, z) + W_P(\xi, p_\xi, z)] \quad (4.103)$$

where K and P denote the kinetic and potential parts, respectively, namely:

$$\frac{\partial}{\partial z} W_K(\xi, p_\xi, z) = \frac{1}{4\pi i \bar{\lambda}} \int_{-\infty}^{+\infty} e^{-\frac{i p_\xi \delta}{\bar{\lambda}}} \left[\Psi\left(\xi + \frac{\delta}{2}, z\right) \frac{\partial^2}{\partial \xi^2} \Psi^*\left(\xi - \frac{\delta}{2}, z\right) - \Psi^*\left(\xi - \frac{\delta}{2}, z\right) \frac{\partial^2}{\partial \xi^2} \Psi\left(\xi + \frac{\delta}{2}, z\right) \right] d\delta$$

$$\frac{\partial}{\partial z} W_U(\xi, p_\xi, z) = \frac{2\pi}{i\bar{\lambda}^2} \int_{-\infty}^{+\infty} e^{-\frac{i p_\xi \delta}{\bar{\lambda}}} \left[U\left(\xi + \frac{\delta}{2}\right) - U\left(\xi - \frac{\delta}{2}\right) \right] \Psi\left(\xi + \frac{\delta}{2}, z\right) \Psi^*\left(\xi - \frac{\delta}{2}, z\right) d\delta. \quad (4.104)$$

We can now write the right-hand side of the second equation of (4.104) and proceed by expanding the potential to the second order in δ. We accordingly note that:

$$U\left(\xi + \frac{\delta}{2}\right) - U\left(\xi - \frac{\delta}{2}\right) = 2\sin\left(\frac{\delta}{2} \frac{\partial}{\partial \xi}\right) U(\xi) \cong \delta U'(\xi) - \frac{\delta^3}{24} U'''(\xi). \quad (4.105)$$

Plugging this expansion into the second of equations (4.104), we find:

$$\int_{-\infty}^{+\infty} e^{-\frac{i p_\xi \delta}{\bar{\lambda}}} \Psi\left(\xi + \frac{\delta}{2}, z\right) \Psi^*\left(\xi - \frac{\delta}{2}, z\right) \left[U\left(\xi + \frac{\delta}{2}\right) - U\left(\xi - \frac{\delta}{2}\right) \right] d\delta \cong$$

$$\cong U'(\xi) \int_{-\infty}^{+\infty} \delta \, e^{-\frac{i p_\xi \delta}{\bar{\lambda}}} \Psi\left(\xi + \frac{\delta}{2}, z\right) \Psi^*\left(\xi - \frac{\delta}{2}, z\right) d\delta - \frac{\delta^3}{24} U'''(\xi) \quad (4.106)$$

$$\int_{-\infty}^{+\infty} \delta^3 e^{-\frac{i p_\xi \delta}{\bar{\lambda}}} \Psi\left(\xi + \frac{\delta}{2}, z\right) \Psi^*\left(\xi - \frac{\delta}{2}, z\right) d\delta.$$

The integrals containing δ, δ^3 can be now treated by noting that:

$$\int_{-\infty}^{+\infty} \delta^n \, e^{-\frac{ip_\xi \delta}{\lambda}} f(\xi, \delta) \, d\delta = i^n \frac{\partial^n}{\partial p_\xi^n} \phi(\xi, p_\xi)$$

$$\phi(\xi, p_\xi) = \int_{-\infty}^{+\infty} e^{-\frac{ip_\xi \delta}{\lambda}} f(\xi, \delta) \, d\delta. \tag{4.107}$$

Therefore, we can write:

$$\frac{\partial}{\partial z} W_U(\xi, p_\xi, z) \cong i \left[U'(\xi) \frac{\partial}{\partial p_\xi} W(\xi, p_\xi) - U''(\xi) \frac{\partial^3}{\partial p_\xi^3} W(\xi, p_\xi) \right]. \tag{4.108}$$

If we include the higher-order terms of expansion, we can easily recover the result shown in equation (4.39). Regarding the kinetic part of equations (4.104), we first note that:

$$\Psi\!\left(\xi + \frac{\delta}{2}, z\right) \frac{\partial^2}{\partial \xi^2} \Psi^*\!\left(\xi - \frac{\delta}{2}, z\right) = -2\Psi\!\left(\xi + \frac{\delta}{2}, z\right) \frac{\partial^2}{\partial \xi \partial \delta} \Psi^*\!\left(\xi - \frac{\delta}{2}, z\right)$$

$$\Psi^*\!\left(\xi - \frac{\delta}{2}, z\right) \frac{\partial^2}{\partial \xi^2} \Psi\!\left(\xi + \frac{\delta}{2}, z\right) = 2\Psi^*\!\left(\xi + \frac{\delta}{2}, z\right) \frac{\partial^2}{\partial \xi \partial \delta} \Psi\!\left(\xi - \frac{\delta}{2}, z\right). \tag{4.109}$$

Integration by parts yields the following for the first term in the previous equation:

$$-2 \int_{-\infty}^{+\infty} e^{-\frac{ip_\xi \delta}{\lambda}} \left[\Psi\!\left(\xi + \frac{\delta}{2}, z\right) \frac{\partial^2}{\partial \xi \partial \delta} \Psi^*\!\left(\xi - \frac{\delta}{2}, z\right) \right] d\delta$$

$$= -2ip_\xi \int_{-\infty}^{+\infty} e^{-\frac{ip_\xi \delta}{\lambda}} \left[\Psi\!\left(\xi + \frac{\delta}{2}, z\right) \frac{\partial}{\partial \xi} \Psi^*\!\left(\xi - \frac{\delta}{2}, z\right) \right] d\delta \tag{4.110}$$

$$+ \bar{\lambda} \int_{-\infty}^{+\infty} e^{-\frac{ip_\xi \delta}{\lambda}} \left[\left(\frac{\partial}{\partial \xi} \Psi\!\left(\xi + \frac{\delta}{2}, z\right)\right) \left(\frac{\partial}{\partial \xi} \Psi^*\!\left(\xi - \frac{\delta}{2}, z\right)\right) \right] d\delta$$

and for the second:

$$2 \int_{-\infty}^{+\infty} e^{-\frac{ip_\xi \delta}{\lambda}} \left[\Psi^*\!\left(\xi + \frac{\delta}{2}, z\right) \frac{\partial^2}{\partial \xi \partial \delta} \Psi\!\left(\xi - \frac{\delta}{2}, z\right) \right] d\delta$$

$$= -2ip_\xi \int_{-\infty}^{+\infty} e^{-\frac{ip_\xi \delta}{\lambda}} \left[\Psi^*\!\left(\xi - \frac{\delta}{2}, z\right) \frac{\partial}{\partial \xi} \Psi\!\left(\xi + \frac{\delta}{2}, z\right) \right] d\delta + \tag{4.111}$$

$$- \bar{\lambda} \int_{-\infty}^{+\infty} e^{-\frac{ip_\xi \delta}{\lambda}} \left[\left(\frac{\partial}{\partial \xi} \Psi\!\left(\xi + \frac{\delta}{2}, z\right)\right) \left(\frac{\partial}{\partial \xi} \Psi^*\!\left(\xi - \frac{\delta}{2}, z\right)\right) \right] d\delta.$$

In conclusion, we arrive at:

$$\frac{\partial}{\partial z} W_K(\xi, p_\xi, z) = -p_\xi \frac{\partial}{\partial \xi} W_K(\xi, p_\xi, z). \tag{4.112}$$

4.6.2 Quadratic forms

In the last two chapters we have discussed the statistical properties and the transport of electron and photon beam distributions. We have set up the formalism to treat phase-space evolution, and the matrix method has been shown to be a powerful tool for analysis. In the exercises which follow, we will take a step further in this direction using quadratic forms, which allow the derivation of cumbersome multidimensional Gaussian integrals.

These results are propaedeutic to the analysis of the next chapter, in which we treat the photon and electron superposition at the interaction point.

The procedure we have followed is based on technicalities associated with convolutions of 2D Gaussian distributions, which were derived using different assumptions and, in some cases, specific approximations allowing more physically meaningful results.

In the previous chapters we mentioned different mathematical tools used in the treatment of laser and electron beam transports. We have underscored the importance of the matrix formalism, which significantly simplifies the computation of the phase-space statistical quantities.

The possibilities offered by these techniques can be pushed further by the incorporation of methods employing quadratic forms.

The following exercises are helpful in defining the associated rules.

Exercise 4.22. Prove the identity given below:

$$a x^2 + 2b x y + c y^2 = \underline{z}^T \hat{M}^{-1} \underline{z}$$
$$\underline{z} = \begin{pmatrix} x \\ y \end{pmatrix}, \quad \hat{M}^{-1} = \begin{pmatrix} a & b \\ b & c \end{pmatrix}. \tag{4.113}$$

Hint:

$$a x^2 + 2b x y + c y^2 = \underline{z}^T \hat{M}^{-1} \underline{z}$$
$$\underline{z} = \begin{pmatrix} x \\ y \end{pmatrix}, \quad \hat{M}^{-1} = \begin{pmatrix} a & b \\ b & c \end{pmatrix}. \tag{4.114}$$

It is sufficient to note that: $\underline{z}^T = (x \ y)$, so that $\underline{z}^T \hat{M}^{-1} \underline{z} = (x \ y) \begin{pmatrix} a & b \\ b & c \end{pmatrix} \begin{pmatrix} x \\ y \end{pmatrix}$....

Exercise 4.23. Prove that a two-variable distribution can be written as:

$$P_G(x, y) = \frac{1}{2\pi |\hat{M}|} e^{-\frac{1}{2} \underline{z}^T \hat{M}^{-1} \underline{z}},$$
$$|\hat{M}| = \det(\hat{M}) > 0. \tag{4.115}$$

Hint:
The proof is achieved by noting that:

$$\int_{-\infty}^{+\infty} dy \int_{-\infty}^{+\infty} dx P_G(x, y) = 1. \qquad (4.116)$$

This result can be achieved by the use of 'brute force,' namely by explicitly deriving the integral in equation (4.116):

$$\frac{1}{2\pi |\hat{M}|^{1/2}} \int_{-\infty}^{+\infty} dy \int_{-\infty}^{+\infty} dy e^{-\frac{(ax^2+2bxy+cy^2)}{2}} = \frac{1}{2\pi |\hat{M}|^{1/2}} \int_{-\infty}^{+\infty} dy e^{-\frac{cy^2}{2}} \int_{-\infty}^{+\infty} dx e^{-\frac{(ax^2+2bxy)}{2}}$$

$$\int_{-\infty}^{+\infty} dx e^{-\frac{(ax^2+2bxy)}{2}} = \sqrt{\frac{2\pi}{a}} e^{\frac{(by)^2}{2a}}$$

$$\frac{1}{2\pi |\hat{M}|^{1/2}} \int_{-\infty}^{+\infty} dy e^{-\frac{cy^2}{2}} \int_{-\infty}^{+\infty} dx e^{-\frac{(ax^2+2bxy)}{2}} = \frac{1}{\sqrt{\frac{2\pi}{a}} |\hat{M}|^{1/2}} \int_{-\infty}^{+\infty} dy e^{-\frac{1}{2}\left(\frac{ac-b^2}{a}\right)y^2}$$

$$= \frac{1}{|\hat{M}|^{1/2} \sqrt{ac-b^2}}$$

$$|\hat{M}|^{1/2} \sqrt{ac-b^2} = 1.$$

We invite the reader to prove the last identity. From the previous results, we can infer a practical rule which will be exploited in the following.

If we compare the 1D and 2D Gaussian integrals:

$$\int_{-\infty}^{+\infty} e^{-ax^2} dx = \int_{-\infty}^{+\infty} e^{-xax} dx = \sqrt{\pi} \, |a^{-1}|^{1/2}$$

$$\int_{-\infty}^{+\infty} e^{-\underline{z}^T \hat{N} \underline{z}} d\underline{z} = \pi \, |\hat{N}^{-1}|^{1/2}$$

we can conclude that the two results are formally equivalent on account of the equivalence:

$$\begin{aligned} a &\to \hat{N} \\ |a^{-1}| &\to |\hat{N}^{-1}| \\ |\hat{N}^{-1}| &= \frac{1}{|\hat{N}|}. \end{aligned} \qquad (4.117)$$

Exercise 4.24. Use the given analogy to evaluate the convolution:

$$\int_{-\infty}^{+\infty} e^{-(\underline{z}-\underline{\xi})^T \hat{N}\underline{z} - \underline{\xi}-\underline{\xi}^T \hat{M}\underline{\xi}} d\underline{z}.$$

Solution:
To proceed, we consider the 1D counterpart:

$$\int_{-\infty}^{+\infty} e^{-a(x-\xi)^2 - b\xi^2} d\underline{z} = \sqrt{\frac{\pi}{a+b}} e^{-xa^{-1} + b^{-1-1}x}.$$

The use of the correspondences in equation (4.117) allows the conclusion:

$$\int_{-\infty}^{+\infty} e^{-(\underline{z}-\underline{\xi})^T \hat{N}\underline{z} - \underline{\xi}-\underline{\xi}^T \hat{M}\underline{\xi}} d\underline{z} = \pi \, |\hat{N} + \hat{M}|^{-\frac{1}{2}} e^{-\underline{z}^T \hat{N}^{-1} + \hat{M}^{-1-1}\underline{z}}. \qquad (4.118)$$

Note that:
$$|\hat{N} + \hat{M}| \neq |\hat{N}| + |\hat{M}|.$$

Exercise 4.25. Apply the outlined technique to prove that:
$$\int_{-\infty}^{+\infty} e^{-\underline{z}^T \hat{N} \underline{z} - \underline{z}^T \hat{M} \underline{\alpha}} d\underline{z} = \pi \, |\hat{N}|^{-\frac{1}{2}} e^{\frac{1}{4}\underline{\alpha}^T \hat{M}^T \hat{N}^{-1} \hat{M} \underline{\alpha}}$$
$$\underline{\alpha} = \begin{pmatrix} a \\ b \end{pmatrix}, \; a, b \equiv \text{constants}$$
(4.119)

Hint:
Note that in the 1D case, one gets:
$$\int_{-\infty}^{+\infty} e^{-xa\,x - xb} d\underline{z} = \sqrt{\frac{\pi}{|a|}} e^{\frac{ba - \frac{1}{2}b}{4}}$$
(4.120)

and apply the previously outlined correspondences.

Exercise 4.26. Use the previous results to define the distribution resulting from the convolution of the electron and laser beam phase-space distributions and interpret the associated physical meaning.

Hint:

$$\hat{N}^{-1} = \frac{1}{\sqrt{\varepsilon_e}} \begin{pmatrix} \gamma_e & \alpha_e \\ \alpha_e & \beta_e \end{pmatrix}, \qquad \hat{M}^{-1} = \frac{1}{\sqrt{\varepsilon_{ph}}} \begin{pmatrix} \gamma_{ph} & \alpha_{ph} \\ \alpha_{ph} & \beta_{ph} \end{pmatrix}$$

$$\hat{N}^{-1} + \hat{M}^{-1} = \begin{pmatrix} \dfrac{\gamma_e}{\sqrt{\varepsilon_e}} + \dfrac{\gamma_{ph}}{\sqrt{\varepsilon_{ph}}} & \dfrac{\alpha_e}{\sqrt{\varepsilon_e}} + \dfrac{\alpha_{ph}}{\sqrt{\varepsilon_{ph}}} \\ \dfrac{\alpha_e}{\sqrt{\varepsilon_e}} + \dfrac{\alpha_{ph}}{\sqrt{\varepsilon_{ph}}} & \dfrac{\beta_e}{\sqrt{\varepsilon_e}} + \dfrac{\beta_{ph}}{\sqrt{\varepsilon_{ph}}} \end{pmatrix}$$

$$\left(\frac{\gamma_e}{\sqrt{\varepsilon_e}} + \frac{\gamma_{ph}}{\sqrt{\varepsilon_{ph}}}\right)\left(\frac{\beta_e}{\sqrt{\varepsilon_e}} + \frac{\beta_{ph}}{\sqrt{\varepsilon_{ph}}}\right) - \left(\frac{\alpha_e}{\sqrt{\varepsilon_e}} + \frac{\alpha_{ph}}{\sqrt{\varepsilon_{ph}}}\right)^2$$
$$= \frac{\gamma_e}{\sqrt{\varepsilon_e}} \frac{\beta_e}{\sqrt{\varepsilon_e}} + \frac{\gamma_e}{\sqrt{\varepsilon_e}} \frac{\beta_{ph}}{\sqrt{\varepsilon_{ph}}} + \frac{\gamma_{ph}}{\sqrt{\varepsilon_{ph}}} \frac{\beta_e}{\sqrt{\varepsilon_e}} + \frac{\gamma_{ph}}{\sqrt{\varepsilon_{ph}}} \frac{\beta_{ph}}{\sqrt{\varepsilon_{ph}}}$$
$$- \left(\left(\frac{\alpha_e}{\sqrt{\varepsilon_e}}\right)^2 + 2 \frac{\alpha_e}{\sqrt{\varepsilon_e}} \frac{\alpha_{ph}}{\sqrt{\varepsilon_{ph}}} + \left(\frac{\alpha_{ph}}{\sqrt{\varepsilon_{ph}}}\right)^2\right)$$
$$= \frac{1}{\varepsilon_e} + \frac{1}{\varepsilon_{ph}} + \frac{1}{\sqrt{\varepsilon_e \varepsilon_{ph}}} \gamma_e \beta_{ph} + \gamma_{ph} \beta_e - 2\alpha_e \alpha_{ph}.$$

4.6.3 Gravitational lenses

The possibility that an optical ray may be deflected by a gravitational field was known since the beginning of the nineteenth century. In 1801 Soldner used Newton's theory to predict the deflection angle (see figure 4.19)

$$\alpha_S = \frac{2GM}{v^2 b}. \tag{4.121}$$

In 1911 Einstein discussed the same effect on the basis of the equivalence principle and the use of a non-Euclidean metric. Einstein's prediction is structurally equivalent to equation (4.121) and indeed reads (where the subscripts S stands for Soldner):

$$\alpha_S = \frac{2GM}{c^2 b}. \tag{4.122}$$

A question often raised in the scientific literature (see the bibliography at the end of the chapter) is whether there is a similarity between the physical mechanisms underlying the two deflections. Although the tools needed to resolve this question involve the concepts and technicalities of general relativity, we would just like to scratch the surface to offer a broader perspective on the topics discussed so far. Therefore, we note that, as loosely stressed above, Snell's law of light diffraction is a consequence of Fermat's principle, namely

$$\delta S = \delta \int_A^B n_r \sqrt{x'^2 + y'^2 + z'^2}\, ds = 0 \tag{4.123}$$

which provided the basis for the Hamilton principle:

$$\delta S = \delta \int_A^B L(x, y, z, x', y', z')\, ds \tag{4.124}$$

$$L(x, y, z, x', y', z') = n(x, y, z)\sqrt{x'^2 + y'^2 + z'^2}$$

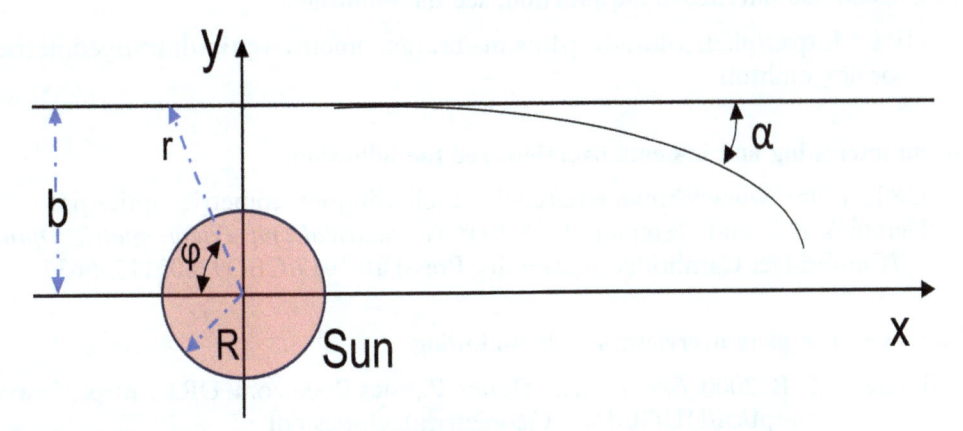

Figure 4.19. Sketch of the gravitational light bending, where b is the distance of the light ray from the Sun axes, r is the vector radius from the center of the Sun following the ray position, α is the deflection angle.

where L is the **optical Lagrangian**. Owing to the generality of the principle itself, it can be exploited as the starting point to look for the deep root of the two bendings. The formalism of general relativity yields the following for the gravitational refraction index (see the bibliography):

$$n = \left(1 + \frac{GM}{2rc^2}\right)^3 \left(1 - \frac{GM}{2rc^2}\right)^{-1}. \qquad (4.125)$$

Exercise 4.27. Assuming that $\frac{GM}{2rc^2}$ is a small quantity, expand the right-hand side of equation (4.125).
Hint:
At the first order we find

$$n \cong \left(1 + 3\frac{GM}{2rc^2}\right)\left(1 + \frac{GM}{2rc^2}\right) = 1 + 2\frac{GM}{rc^2} \qquad (4.126)$$

hence $\Delta n = 2\frac{GM}{bc^2}$. This justifies the result in equation (4.122). We invite the reader to repeat the same procedure to derive the Soldner equation (4.121).

Further reading

For a friendly and practical introduction, see the following:

URL: https://www.geeksforgeeks.org/difference-between-concave-convex-lens/#difference-between-convex-and-concave-lens
URL: https://www.geeksforgeeks.org/concave-and-convex-mirrors/
URL: https://www.labxchange.org/library/items/lb:LabXchange:d170f874-fb91-39ec-8fe8-d3733c4680f4:html:1

For a visual and interactive introduction, see the following:

URL: https://phet.colorado.edu/sims/html/geometric-optics/latest/geometric-optics_en.html

For an interesting and in-depth overview, see the following:

URL: https://www.lehman.edu/faculty/anchordoqui/Geometric_optics.pdf
Dereniak E L and Dereniak T D 2008 *Geometrical and Trigonometric Optics* (Cambridge: Cambridge University Press) 10.1017/CBO9780511755637

For a more complete overview, see the following:

Huggins E R 2000 *Geometrical Optics* Physics 2000.com URL: https://www.fisica.net/optica/HUGGINS_Geometrical_Optics.pdf

For a good reference manual, see the following:

Dubey A K n.d. *Lectures on geometrical optics.* URL: https://vixra.org/pdf/1608.0067v2.pdf.

For a detailed account, see the following:

Goodman D S 2010 General principles of geometrical optics *Handbook of Optics: Volume I—Geometrical and Physical Optics, Polarized Light, Components and Instruments* 3rd edn ed Bass M (New York:McGraw-Hill Professional) https://www.accessengineeringlibrary.com/content/book/9780071498890/chapter/chapter1

For an introduction to matrix optics, see the following:

Mondal P P 2022 Ray and matrix optics: a simple theory of light *Classical and Quantum Optics* (New York, NY:AIP Publishing LLC) doi: https://dx.doi.org/10.1063/9780735424517_001

Gerrard A and Burch J M 1994 *Introduction to Matrix Methods in Optics* (New York, NY:Dover Books) https://books.google.it/books?id=naUSNojPwOgC

Trebini R 2015 Geometrical Optics Lecture Notes Georgia Inst. of Technology, Optics I course. URL: https://www.studocu.com/en-us/document/georgia-institute-of-technology/optics-i/lecture-slides-lecture-15-geometrical-optics-1/745938

Gillespie A 2013 *Comparison between charged-particle (CP) and light/laser (L) optics* https://indico.cern.ch/event/266133/attachments/474621/656940/Gillespie_Aachen_CP_vs_P_optics.pdf.

For the backbone of the knowledge in classical optics, see the following:

Luneburg R K 2021 *Mathematical Theory of Optics* (Berkeley, CA: University of California Press) pp 1–2 https://www.ucpress.edu/book/9780520328259/mathematical-theory-of-optics

Kline M and Kay I W 1965 *Electromagnetic Theory and Geometrical Optics* (Pure and Applied Mathematics: A Series of Texts and Monographs) (New York, NY:Interscience) https://books.google.it/books?id=UD8v1Q-rqdAC.

Born M and Wolf E 1980 *Principles of Optics* 6th edn (Oxford: Pergamon) 10.1016/B978-0-08-026482-0.50014-1 https://wwwsciencedirect.com/science/article/pii/B9780080264820500141

Marcuse D 1982 *Light Transmission Optics* (Princeton, NJ: Van Nostrand Reinhold) https://books.google.it/books?id=3yRRAAAAMAAJ.

For waves, paraxial optics, and Gaussian beams, see the following:

Leontovich M A 1944 On a method of solving the problem of propagation of electromagnetic waves near the surface of the Earth *Izv. Akad. Nauk SSSR Ser. Fiz* **8** 16–22

Leontovich M A and Fock V A 1946 The solution of the problem of the diffraction of electromagnetic waves around the Earth by the method of the parabolic equation *Acad. Sci. USSR. J. Phys* **10** 13–24

Fock V A 1965 *Electromagnetic Diffraction and Propagation Problems* (Oxford: Pergamon)

Lax M, Louisell W H and McKnight W B 1975 From Maxwell to paraxial wave optics *Phys. Rev. A* **11** 1365–70

Siegman A E 1986 *Lasers* (Information and Interdisciplinary Subjects Series) (Mill Valley, CA: University Science Books) doi: https://books.google.it/books?id=1BZVwUZLTkAC

Yariv A 1989 *Quantum Electronics* 3rd edn (New York:Wiley) doi: https://books.google.it/books?id=UTWg1VIkNuMC

Pampaloni F and Enderlein J 2004 Gaussian, Hermite–Gaussian, and Laguerre–Gaussian beams: A primer arxiv arXiv:physics/0410021

Solimeno S, Crosignani B and Di Porto P 1986 *Guiding, Diffraction, and Confinement of Optical Radiation* (New York, NY: Academic Press) doi: https://dx.doi.org/10.1016/B978-0-12-654340-7.X5001-4

Bélanger P A 1991 Beam propagation and the ABCD ray matrices *Opt. Lett.* **16** 196–8

Stace C n.d. *Using ABCD Matrices to find the position of a Gaussian beam waist.* URL: https://lumoptica.com/file_download/5/ABCD+matrix+and+Gaussian+beam+waist.pdf

URL: https://www.hunter.cuny.edu/physics/courses/physics852/repository/files/1Waveandbeamoptics.pdf.

Vaveliuk P, Ruiz B and Lencina A 2007 Limits of the paraxial approximation in laser beams *Opt. Lett.* **32** 927–9

Feng S and Winful H G 2001 Physical origin of the Gouy phase shift *Opt. Lett.* **26** 485–7

Babusci D et al. 2019 *Mathematical Methods for Physicists* (Singapore:World Scientific). doi: https://dx.doi.org/10.1142/11315

Dattoli S K G and Ricci P E 2008 On Crofton-Glaisher type relations and derivation of generating functions for Hermite polynomials including the multi-index case *Integ. Trans. Spec. Funct.* **19** 1–9

Dattoli G, Gallardo J C and Torre A 1988 An algebraic view to the operatorial ordering and its applications to optics *La Rivista del Nuovo Cimento (1978–1999)* **11** 1–79

For optics and quantum mechanics, see the following:

Gloge D C and Marcuse D 1969 Formal quantum theory of light rays *J. Opt. Soc. Am.* **59** 1629–31

Stoler D 1981 Operator methods in physical optics *J. Opt. Soc. Am.* **71** 334–41

Dattoli G, Solimeno S and Torre A 1986 Algebraic time-ordering techniques and harmonic oscillator with time-dependent frequency *Phys. Rev. A* **34** 2646–53

Dattoli G, Solimeno S and Torre A 1987 Algebraic view of the optical propagation in a nonhomogeneous medium *Phys. Rev. A* **35** 1668–72

Dattoli G *et al.* 1992 Towards a wave theory of charged-beam propagation *IlNuovo Cimento D* **14** 271–8

Konrad T Quantum mechanics and classical optics: new ways to combine classical and quantum methods *J. Phys.* **2448** 012005

For the Wigner distribution, see the following:

Bastiaans M J 1979 Wigner distribution function and its application to first-order optics *J. Opt. Soc. Am.* **69** 1710–6

Simon R and Sudarshan E C G and Mukunda N Gaussian-Wigner distributions in quantum mechanics and optics *Phys. Rev. A* **36** 3868–80

Case W B 2008 Wigner functions and Weyl transforms for pedestrians *Am. J. Phys.* **76** 937–46

Brogaard A 2015 Wigner function formalism in quantum mechanics *Bachelor project in Physics* University of Copenhagen https://nbi.ku.dk/english/theses/bachelor-theses/jon-brogaard/Jon_Brogaard_Bachelorthesis_2015.pdf

Chapter 5

Beam–beam interactions

5.1 Introduction

In the previous chapters we have provided the *theoretical minimum* required to understand the design and optimization of Compton backscattering (CBS)- based photon sources. In the present and following chapters, we embed the previously outlined notions to provide criteria of practical interest, with the objective of supplying tools to enable the preliminary design (namely the search for the working points) of complex systems. We organize the discussion on two different levels: the first, which is more 'phenomenological,' aims to define the effect of the beam characteristics on the spectral line. It uses a point of view which was developed in the past, within the context of the study of the spectral properties of undulator radiation and the associated distortion induced by the beam energy spread and spatial/angular distributions. This preliminary discussion is intended as a gentle introduction to the more substantive treatment of the second part, which is largely based on the results obtained in chapter 2. This choice has a twofold motivation: it clarifies the use of well-known techniques based on pivotal parameters, which have played a significant role in the study of undulators and free-electron lasers (FELs) and incorporates a more suitable treatment of CBS radiation. We have already mentioned that CBS and FEL sources are complex devices; this statement may be largely misleading. According to the common vulgate, a system is complex if it is made of a large number of parts that interact with each other. It goes without saying that these systems are difficult to model because of the *intrinsic nonlinearities, adaptation, feedback loops*…. These kinds of behaviors seem to be extraneous to the types of devices we are discussing. It is, however, a common experience of scientists working on *laser, accelerator,* and *fusion plasma* machines that these devices exhibit all the issues we have outlined above. The *emergence*[1] of instabilities that plague their operation is a further manifestation of their complexity.

---

[1] The term 'emergence' refers to a behavior that occurs when the parts of a multicomponent device interact in a way that does not characterize their own separate behaviors.

The overlap of laser and accelerator physics which gave rise to FEL development triggered an understanding of the interaction between these systems and indicated new avenues that allow the use of an emerging instability in one system to 'cure' those occurring in the other. FEL (and CBS) devices are complex systems. In this chapter, we do not examine the relevant complexity in terms of the previously quoted behaviors; rather, we consider the engineering entanglement of the composing parts. These devices embed different physical environments, characterized by a large number of specific parameters. The relevant theoretical description, which is necessarily non trivial, requires massive numerical simulations to explore the associated phenomenology and the design details. The use of scaling formulae has played an important role in the design of FEL devices operating in the self-amplified spontaneous emission (SASE) and oscillator configurations and has recently been employed to examine the feasibility of FELs driven by plasma-accelerated beams. The logical steps underlying the reported strategy are summarized in figure 5.1. Specifically, scaling/empirical formulae are used to determine the parameter space that fixes the working point, and home-made or commercially available software is used to refine the first evaluation and eventually to benchmark the first step. CBS x-ray devices aim to provide photon sources with energies of more than tens of keV and adequate luminosity. In this review, we collect different ideas related to compact light sources previously provided by several authors and present the basic tools for

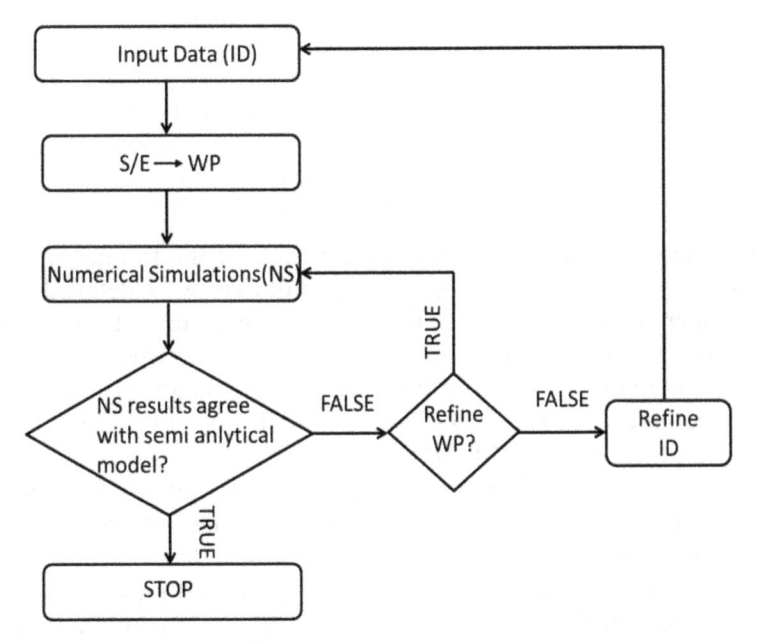

Figure 5.1. An FEL design strategy flowchart: (1) the input data (beam and undulator parameters). (2) S/E → WP = the use of a scaling/empirical formula to find the working point. (3) Numerical simulations (NS) = use of numerical software to refine the details of the preliminary definition of the space parameters. (4) Loop the process, changing the input data and/or the working point, until a benchmark test against a semianalytical model succeeds.

fixing the working point of an actual CBS-based device. The elementary physical mechanism underlying devices of this type is the head-on collision of a (sufficiently) energetic beam with a (sufficiently) intense and coherent optical source. The latter may also be provided by an undulator field, which is seen by the electrons as a coherent ensemble of photons. We summarize the essential tools of the forthcoming discussion below:

1. The energy of the scattered photons in a head-on collision process is given by:

$$E_s = \frac{(1 - \beta \cos(\phi_1))\hbar\omega_0}{(1 - \beta \cos(\phi_2)) + \frac{E_0}{E_e}1 - \cos(\theta)}. \tag{5.1}$$

2. The assumption that backscattering occurs ($\theta_f \simeq 0$, $\theta_i \simeq \pi$; ultrarelativistic $\beta \simeq 1 - 1/2\gamma^2$ or classical $\hbar\omega_0/E_e = \bar{\lambda}_c/\gamma\,\bar{\lambda}_0 \ll 1$) allows us to reduce the previous equation to $\omega_s \simeq 4\gamma^2\omega_0$.

3. The inclusion of the transverse motion effects (due to the spatial distribution of the laser electric field) yields:

$$\omega_s \simeq \frac{4\gamma^2\omega_0}{1 + \frac{a_0^2}{2}} \quad ; \quad a_0 = \frac{|e||\vec{A}_0|}{m_0 c} \tag{5.2}$$

where $\vec{E}_0$ is the electric field associated with the laser wave. In practical units (namely in terms of the laser intensity, wavelength...), the strength parameter a_0 is written as:

$$a_0 \simeq 8.57 \times 10^{-10} \lambda_0[\mu\text{m}]\sqrt{I_0[\text{W cm}^{-2}]}. \tag{5.3}$$

The physical mechanisms underlying the emergence of the frequency shift and the analogy with the emissions of undulator/wiggler devices were detailed in chapter 2. As already underscored in chapter 2, the laser strength parameter is a key quantity determining the regime of operation of the scattering process itself. In order to understand the numbers that may be involved in an actual experiment, we note that for an infrared laser ($\lambda_0 \simeq 1\,\mu$m) characterized by $a_0 \simeq 0.1$, the corresponding intensity is slightly larger than 10^{16} W cm^{-2}.

In a CBS process such as that shown in figure 5.2, a single-photon/single-electron head-on collision occurs between the N_e (non-interacting) electrons composing the electron bunch and the N_0 photons contributing to the laser field. The rate $\dot{N}_X$ of x-ray photons emerging from the interaction point (IP) per second after the scattering depends on the scattering cross section and on another quantity, the parameter L_0, which accounts for the number of electrons and photons and their overlap. We therefore set:

$$\dot{N}_X \simeq \sigma_T L_0 \tag{5.4}$$

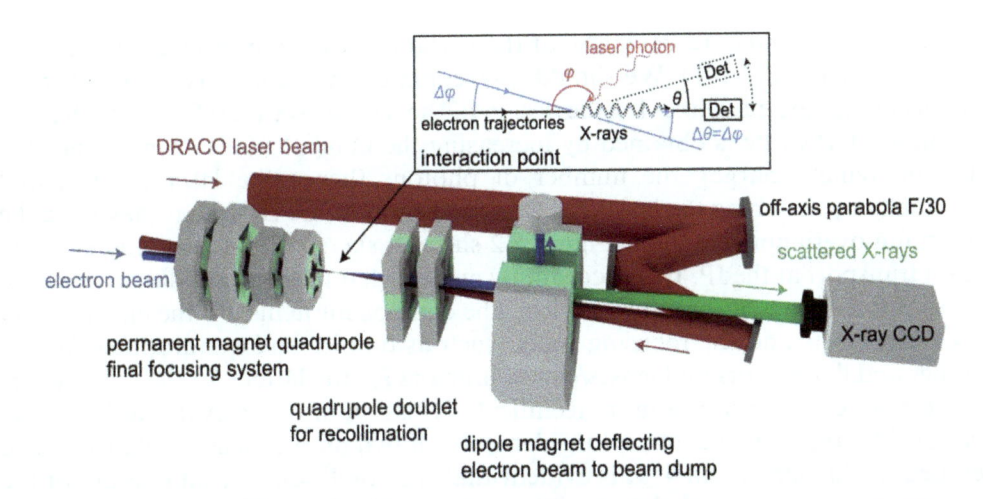

Figure 5.2. The IP of an experimental configuration. Reprinted by permission from Springer Nature Customer Service Centre GmbH: Krämer *et al* 2018 Making spectral shape measurements in inverse Compton scattering a tool for advanced diagnostic applications *Sci. Rep.* **8** 1398, CC BY 4.0.

where σ_T is the Thomson cross section and:

$$L_0 \simeq \frac{f_{\text{rep}} N_e N_0}{2\pi \Sigma_{e,0}^2}, \tag{5.5}$$

$$\Sigma_{e,0}^2 = \sigma_e^2 + \sigma_0^2$$

where f_{rep} denotes the repetition rate (rep. rate) of the scattering events, $\sigma_{e,0}$ the transverse root-mean-square (rms) dimensions of the electron/laser beams, and $N_{e,0}$ the relevant numbers of electrons and photons. The brightness of the source is defined as the rate of emitted photons per unit area and unit angle (for more general considerations related to the definition of the brightness of x-ray synchrotron sources, see the Comments and exercises section):

$$B \equiv \frac{\dot{N}_X}{4\pi^2 \Sigma_{e,0}^2 <\theta>_x <\theta>_y} = \frac{1}{4\pi^2} \frac{256 \gamma_0^2}{81\pi^2} \frac{\dot{N}_X}{\Sigma_{e,0}^2} \simeq 0.008 \times \frac{\gamma_0^2 \dot{N}_X}{\Sigma_{e,0}^2} \tag{5.6}$$

where the divergence of the CBS radiation in the linear regime is given by equation (2.108) when the electron divergence is negligible. Furthermore, the spectral brightness is defined as:

$$B_s = \frac{B}{\frac{\Delta\omega}{4\gamma_0^2 \omega_0}[0.1\%]} = \frac{B}{10^3 \frac{\Delta\omega}{4\gamma_0^2 \omega_0}}. \tag{5.7}$$

If the full spectrum is collected, the total bandwidth $\Delta\omega$ can be calculated using equation (2.101) (see the Comments and exercises section):

$$\Delta\omega = \sqrt{\frac{8}{5}} \gamma_0^2 \omega_0 \tag{5.8}$$

where $\Delta\omega$ is the standard deviation of the variable ω over the spectral distribution given by equation (2.101). We should now comment on equations (5.4) and (5.5), which contain the guidelines for the design of the device. We therefore note that the maximum x-ray rate is obtained by increasing the number of electrons (hence the electron bunch charge), the number of photons (hence the laser power), and eventually by matching the laser and electron beams at the IP so that they have the minimum overlapping section. Figure 5.2 shows a sketch of the laser and electron beam transport at the IP. It is accordingly evident that the relevant design is one of the sensitive points of the entire device. The component items are the electron and laser transports and the focusing tools (such as dipoles and quadrupoles for the former and flat mirrors and lenses/curved mirrors for the latter). In the second part of the chapter we present a more technical discussion of the relevant design; here, we consider the relevant conceptual elements, drawn on the basis of what we have learned in chapters 3 and 4. It is evident that the conditions for an overlap of the two beam waists require the matching of the associated phase spaces. In this preliminary analysis we assume that the two beams at the IP are characterized by suitable Twiss parameters (later in this chapter, we discuss definite solutions for their realization). The last equation (5.5) is less naïve than it may appear. We therefore modify it to get a more suitable expression for our purposes. Therefore, we set:

$$\dot{N}_X \simeq \kappa \frac{f_{\text{rep}}}{2\pi} \frac{\widetilde{E}_0 \bar{Q}_e \lambda_0}{1+d}$$

$$\kappa = \frac{8}{3} \frac{\pi\, r_0}{\hbar I_A} \quad ; \quad Q_e = N_e |e|$$

$$I_A \equiv \text{Alfven current} \qquad (5.9)$$

$$\bar{Q}_e = \frac{Q_e}{2\pi\sigma_e^2} \equiv \text{electron charge density of Gaussian beam}$$

$$\widetilde{E}_0 = N_0 E_0 \quad ; \quad d = \left(\frac{\sigma_0}{\sigma_d}\right)^2.$$

The design parameters appearing in equation (5.9) are the bunch charge, the laser energy, the transverse cross sections of the laser/electron beams, and the collisions' rep. rate. Regarding the design strategy, we cannot define, at least on the basis of the elements provided so far, any effective optimization criteria except those of maximizing the values of the bunch charge and the laser energy and minimizing the size of the overlapping waist at the IP. Regarding the collisions' rep. rate, we will see in chapter 6 that it can be enhanced by the combined use of recirculated beams and of laser pulse stacking in high-Q cavities. Albeit fairly elementary, the discussion outlined so far is sufficient to get preliminary indications regarding the working point of an x-ray CBS source. We can therefore sketch some design information with the tools at hand. Considering, therefore, the hard x-ray region, for example in the range around 10 keV, we note that it is achievable with a laser operating at a wavelength of 1 μm and an electron beam with an energy of 25 MeV. Equation (5.9) allows us to specify the amount of charge density required to obtain a specific x-ray rate. If we assume that the laser and electron rms beam transverse sections are matched ($d = 1$), we find:

$$\bar{Q}_e = \frac{4\pi}{\kappa f_{rep} \tilde{E}_0 \lambda_0} \dot{N}_X. \quad (5.10)$$

Table 5.1 lists the results of a 'simulation' of the x-ray rate attainable by the use of laser and electron beams with 'reasonable' parameters. Notable elements of the table are the small geometrical[2] emittance of the electron beam and the low beta value at the IP. The surface density corresponding to the parameters in table 5.1 is evaluated to be $\bar{Q}_e \simeq 0.009$ Cm^{-2}—a number within the current capabilities of small-scale linacs. Bearing in mind that (see the Comments and exercises section):

$$Q_e = \frac{8\pi^2 \beta_e \varepsilon_n}{\kappa f_{rep} \tilde{E}_0 \gamma \lambda_0} \dot{N}_X \quad (5.11)$$

we can conclude that the amount of charge required to achieve the rate in table 5.1 is ~115 pC. This is close to the values seen experimentally. In the exercises at the end of the chapter we present further useful handling of the previous equations to achieve further awareness of how the design parameters and simple formulae should be combined. We have now completed the first step in providing the necessary notions to combine theoretical/phenomenological tools to size a CBS x-ray source. However, albeit reasonable, the discussion misses important features that are considered useful in an actual experimental environment:

1. The effect of electron and 'optical' beams' (laser and x-ray) longitudinal and transverse distributions;
2. The line broadening induced by the electron beam's energy spread and emittances.

Table 5.1. Reference parameters for a CBS x-ray source.

$\dot{N}_X$ [s^{-1}]	10^8
f_{rep} [Hz]	10^3
$\tilde{E}_0$ [J]	10^{-2}
λ_0 [m]	10^{-6}
γ	50
β_e [m]	1
ε_n [mm · mmrad]	0.1

---

[2] By geometrical emittance, we mean $\varepsilon = \varepsilon_n/\gamma$, where ε_n is called the 'normalized emittance.'

In the following sections, we complete our discussion using the previously sketched outline. The analysis we develop in the following part of the chapter relies upon the consideration of pure common sense by taking account of the following:
1. Accurate modeling of the electron beam–photon beam overlap;
2. The impact of the beam qualities on the spectral quality of the emitted radiation;
3. In the forthcoming part of the chapter we provide a (tentatively) accurate description of points (1) and (2) and develop a toy model as a preliminary support for CBS source design.

5.2 Spectral broadening in CBS devices: on-axis contributions from energy spread and emittance

We have underscored that the mechanisms underlying the CBS processes can be treated using the same procedure as that adopted for the derivation of bremsstrahlung radiation in magnetic undulators. An important aspect of the study of the relevant spectral properties is an understanding of the effects determining line spectral broadening, which are related to the beam distributions and the transport conditions. This section is based on a simplified analysis of the spectral properties of CBS radiation in the linear regime of interaction, inspired by our previous experience of undulator/FEL physics. This point of view, necessarily simplified, will be corroborated by the study developed in the second part of the chapter. A strong motivation for this analysis was originally determined by the role that the beam distributions and transport play in the design of FEL devices, since they are the main factors determining the reduction of the gain and thus the limitations of their performance. These effects, known as inhomogeneous broadening, cause an increase in the spectral width and a reduction of its peak. Furthermore, they are responsible for a frequency shift and a more significant suppression of the higher-order harmonics. In order to underscore their importance, we should like to mention the important role they have played as a diagnostic tool for the qualities of the electron beam itself. The concept of inhomogeneous broadening is easily understood by looking at the equation defining the frequency of the on-axis Compton backscattered photons, as shown below:

$$\omega_s = \frac{4\gamma^2 \omega_0}{1 + (\gamma\sqrt{x'^2 + y'^2})^2 + \frac{a^2(x, y, \zeta + z; -z)}{2}} \qquad (5.12)$$

where, for small-divergence electron beams, x', $y' \simeq p_{x,y}/p_z$. The physical meaning of equation (5.12) should be interpreted by considering it along with equation (2.57), where we introduced the dependence on the initial transverse momenta $p_{x,y}(0)$ into the study of the CBS spectral properties. The term under the square root is, therefore, the frequency shift induced by electron trajectories that are initially unaligned with the axis of symmetry. It should also be mentioned that equation (5.12) omits the angular part $(\gamma\theta)^2$ because it is assumed to hold for observation angles very close to the propagation axis. This assumption will be relaxed later. Real-life CBS sources are operated with nonideal electron beams, namely an ensemble of electrons with different kinematic conditions. The scattering is indeed due to electrons which have a non-vanishing transverse position and/or follow trajectories forming an angle with respect to the laser propagation axis. The radiated spectrum is therefore the result of emissions produced by electrons with slightly different energies, entrance angles, and positions (see figure 5.3) relative to the reference beam particle. The dependence on the initial electron transverse coordinates is due to the spatial profile of the laser mode intensity, which in turn defines the equivalent strength parameter $a(x, y, \zeta + z; -z)$. The dependence of a upon the coordinate z is mostly due to the fact that laser beam intensity in the interaction region follows an hourglass shape, since the laser is focused toward the electron beam and naturally diffracts after the focal plane. The dependence upon $\zeta + z$ and $-z$ will shortly be clarified in this chapter; it is related to the counterpropagation of the electron and photon beams. Indeed, according to chapters 2 and 4, the spatiotemporal distribution of the laser pulse is given by:

$$a^2(x, y, \zeta + z; -z) = a_0^2 e^{-2\frac{x^2+y^2}{w^2(z)}} \text{rect}\left(\frac{\zeta + z}{L}\right) = a_0^2 e^{-2\frac{x^2+y^2}{w_0^2\left(1+\frac{z^2}{z_R^2}\right)}} \text{rect}\left(\frac{\zeta + z}{L}\right). \quad (5.13)$$

For a rectangular pulse of width L (here, the function that represents this pulse shape is rect(x)), the hourglass effect is only important when the so-called Rayleigh length ($z_R = \pi w_0^2/\lambda_0$ for a Gaussian beam) is shorter than $L/2$. Each electron scatters a photon with a relative frequency shift compared to the nominal frequency ω_s found

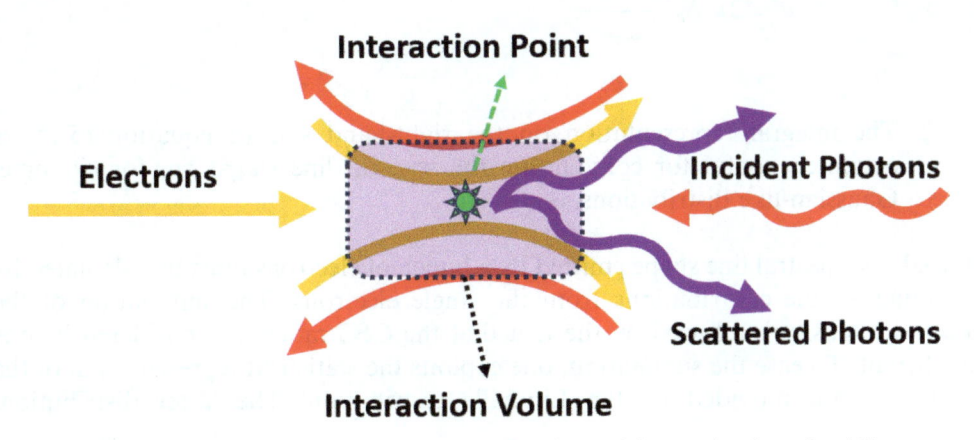

Figure 5.3. The geometry of Compton backscattering and the associated angular shifts.

in equation (5.2). These shifts are given by (see the Comments and exercises section for the derivation):

$$\left(\frac{\delta\omega}{\omega}\right)_s = \left(\frac{\delta\omega}{\omega}\right)_{\delta_e} + \left(\frac{\delta\omega}{\omega}\right)_{\theta_e} + \left(\frac{\delta\omega}{\omega}\right)_{\delta_0} + \left(\frac{\delta\omega}{\omega}\right)_r$$

$$\left(\frac{\delta\omega}{\omega}\right)_{\delta_e} = -2\frac{\delta\gamma}{\gamma_0} = -2\frac{\gamma - \gamma_0}{\gamma_0} = -2\delta_e$$

$$\left(\frac{\delta\omega}{\omega}\right)_{\theta_e} = \frac{\gamma_0^2(x'^2 + y'^2)}{1 + \frac{a_0^2}{2}} \tag{5.14}$$

$$\left(\frac{\delta\omega}{\omega}\right)_{\delta_0} = -\frac{\delta\omega}{\bar{\omega}_0} = -\frac{\omega - \omega_0}{\bar{\omega}_0} = -\delta_0$$

$$\left(\frac{\delta\omega}{\omega}\right)_r = -\frac{a_0^2(x^2 + y^2)}{\left(1 + \frac{a_0^2}{2}\right)w_0^2}$$

where $\bar{\omega}_0$ is the spectral barycenter of the laser pulse. The reference particle is assumed to be characterized by $\delta_e = 0$, $x = y = 0$, $x' = y' = 0$. According to equations (5.14), the effect of the beam distributions on the CBS spectrum is inferred by averaging the relevant spatial and angular distributions. We now give an idea of how the relevant calculations should be carried out, noting that:

1. The spectral line shape is reproduced by a sinc-square function (see equation (2.83)), namely:

$$S(\nu) = \left[\frac{\sin\left(\frac{\nu}{2}\right)}{\left(\frac{\nu}{2}\right)}\right]^2 = 2Re\left[\int_0^1 (1 - \kappa)e^{-i\nu\kappa}d\kappa\right] \tag{5.15}$$

$$\nu = 2\pi\, N_c \frac{\omega - \omega_s}{\omega_s}.$$

2. The integral representation on the right-hand side of equation (5.15) is extremely useful for convoluting the spectral line shape on, for example, Gaussian-like distributions.

Indeed, the spectral line shape emitted by a bunch of electrons must be calculated by summing all the contributions from the single electrons. The summation of the spectral intensity is allowed by the fact that the CBS radiation considered here is incoherent. To ease the summation, one exploits the statistical representation of the electron beam provided by the Liouville distribution. The latter distribution, denoted by f, accounts for the six-dimensional phase-space electron beam distribution and can be split as follows:

$$f(x, x'; y, y'; \delta_e, \zeta - z; z) = f_x(x, x'; z)f_y(y, y'; z)f_\zeta(\zeta - z)f_{\delta_e}(\delta_e)$$

$$f_\eta(\eta, \eta'; z) = \frac{N_e}{2\pi\varepsilon_\eta}\exp\left(-\frac{\gamma_{e\eta}(z)\eta^2 + 2\alpha_{e\eta}(z)\eta\eta' + \beta_{e\eta}(z)\eta'^2}{2\varepsilon_\eta}\right), \qquad \eta = x, y \qquad (5.16)$$

$$f_\zeta(\zeta - z) = \frac{N_e}{\sqrt{2\pi}\,\sigma_\zeta}\exp\left(-\frac{(\zeta - z)^2}{2\sigma_\zeta^2}\right), \qquad f_{\delta_e}(\delta_e) = \frac{1}{\sqrt{2\pi}\,\sigma_{\delta_e}}\exp\left(-\frac{\delta_e^2}{2\sigma_{\delta_e}^2}\right)$$

where the symbol (e) recalls the fact that the Twiss parameters in equation (5.16) refer to the electron beam, with the number of electrons in the incident beam given by N_e. It is evident that the transverse phase-space distribution $f_\eta(\eta, \eta'; z)$ is expressed in terms of the beam Twiss parameters, while the remaining energy/time distributions are uncorrelated Gaussians. The Wigner function W accounts for the six-dimensional phase-space photon beam distribution and, according to equations (4.91) and (4.92), it is:

$$W(x, x'; y, y'; \delta_0, \zeta + z; -z) = W_x(x, x'; z)W_y(y, y'; z)W_\zeta(\zeta + z)W_{\delta_0}(\delta_0)$$

$$W_\eta(\eta, \eta', z) = \frac{2}{\pi\bar{\lambda}}e^{-\frac{1}{\bar{\lambda}}\gamma_{0\eta}(z)\eta^2 + 4\beta_{0\eta}(z)\eta'^2 + 4\alpha_{0\eta}(z)\eta\eta'} \qquad (5.17)$$

$$W_\zeta(\zeta + z) = \frac{1}{L}\text{rect}\left(\frac{\zeta + z}{L}\right), \qquad W_{\delta_0}(\delta_0) = N_c\,\text{sinc}^2\pi\delta_0 N_c$$

where the symbol (0) refers to the incident photon beam and $\delta_0 = (\omega - \omega_0)/\omega_0$. In line with the approach taken in the previous chapters (in particular chapter 2), we have considered a rectangular pulse of length $L = 2\pi c N_c/\omega_0$ for the incident photon beam. The inclusion of the propagation information via the dependence $\zeta \mp z$ for the electron beam and the photon beam, respectively, is reminiscent of the head-on geometry of the electron–photon collision. The coordinates $\zeta_{e,0} = z \mp ct$ are comoving with the beams' barycenter (accounting for the pulse shape) and z is the coordinate of propagation (accounting for the propagation of the beams toward the IP at $z = 0$, seen from the reference frame). According to equation (2.96), the spectral intensity is proportional to a_0^2. In an adiabatic picture, where the variations of the laser envelopes are slow compared to the oscillation frequency of the electrons in the laser's electric field, we can replace $a_0^2 \rightarrow a^2(x, y, \zeta + z; -z)$. This phenomenologically accounts for the evolution of the laser during the interaction. A fully consistent way to account for such dynamics would have been to study the electron motion under the action of a non-plane wave, which is analytically impossible (figure 5.4). Therefore, the inhomogeneously broadened CBS spectral line is calculated by summing up the contributions of the single electrons described by the Liouville distribution f and also by summing up the contributions from the evolving laser beam included in a^2. Furthermore, the dynamics studied in chapter 2 assumed a monochromatic plane wave, while in reality, the incident laser is never so. In order to include the effect of a finite laser spectrum, we also consider an average over the incident laser photon energies, using the partial Wigner distribution W_{δ_0}. Taking into account that the scattered CBS line is shifted for any particle other than the reference

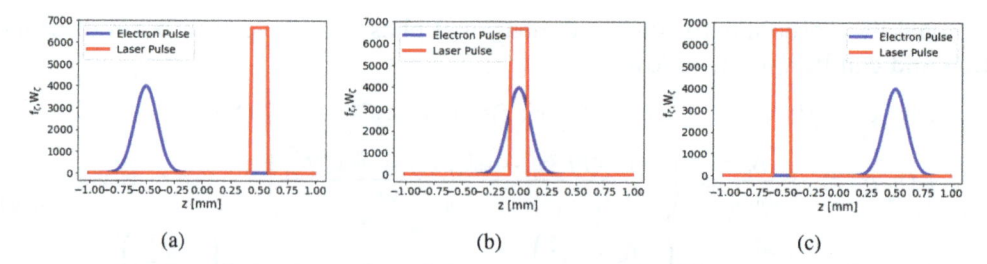

Figure 5.4. From (a) to (c): different stages of the electron–laser pulse collision according to equations (5.16) and (5.17). These figures graphically display the meaning of the dependence $\zeta \mp z$ in the distribution functions for the electron beam and the photon beam, respectively.

particle leads to a sort of convolution between the natural spectral line of equation (5.15) and the product distribution given by $fa^2 W_{\delta_0}$:

$$\langle S(\nu) \rangle = \int_V S(\nu + \delta\nu) f(x, x'; y, y'; \delta_e, \zeta - z; z) a^2(x, y, \zeta + z; -z) W_{\delta_0}(\delta_0) d^7 V$$

$$\delta\nu = 2\pi N_c \left(\frac{\delta\omega}{\omega}\right)_s, \qquad d^7 V = dxdydzdx'dy'd\delta_e d\delta_0.$$
(5.18)

The explicit evaluation of equation (5.18) is outlined in the Comments and exercises section. The only technique required is the derivation of multidimensional Gaussian integrals, which according to the integral representation in equation (5.15) yields the following expression for $\langle S(\nu) \rangle$:

$$\langle S(\nu) \rangle = 2 \int dz K(z) Re \left[\int_0^1 (1 - \kappa) \frac{e^{-i\nu\kappa - \frac{1}{2} 4\pi N_c \sigma_{\delta_e} \kappa^2}}{\sqrt{R_x(\kappa, z) R_y(\kappa, z)}} d\kappa \right]$$

$$R_\eta(\kappa, z) = \left[1 + \frac{4\sigma_\eta^2}{w_0^2} + 4i\pi\gamma_0^2 \left(\sigma_\eta'^2 + \frac{4\varepsilon_\eta^2}{w_0^2} \right) N_c \kappa + \frac{4\pi N_c a_0^2 \kappa}{\left(1 + \frac{a_0^2}{2}\right) w_0^2} \left(4\pi N_c \gamma_0^2 \varepsilon_\eta^2 \kappa - i\sigma_\eta^2 \right) \right]$$
(5.19)

$$K(z) = \frac{1}{2L} \left[\text{erf}\left(\frac{4z + L}{2\sqrt{2}\,\sigma_\zeta}\right) - \text{erf}\left(\frac{4z - L}{2\sqrt{2}\,\sigma_\zeta}\right) \right]$$

which is useful to infer the spectral line distortion induced by beam distributions with non-negligible energy spread and emittances. The rms electron bunch length is σ_ζ, which is assumed to be constant. The first factor in equation (5.19), involving error functions, takes into account the temporal overlap of the two incident beams, which affects the efficiency of the collision process and reduces the peak of the radiated spectral intensity. It is worth noting that the second term defining R_η generalizes the denominator of the luminosity introduced in equation (5.5), which takes account of the loss of collision efficiency due to a limited spatial overlap of the two incident beams (in fact: $\sigma_0 = w_0/2$). In the case of linear Thomson backscattering ($a_0 \ll 1$), the last term defining R_η is negligible. Equation (5.19) is useful to determine the line broadening and peak reduction in terms of the electron beam qualities and transport parameters. Moreover, for less-than-perfect head-on collisions, the formulas we have derived still hold true, although ω_s is appropriately rescaled by a factor of

$1 - \beta_0 \cos \phi_1$, introduced in (2.57) (where for a head-on interaction, $\phi_1 = \pi$). It should be underlined that the guidance of the electron beam to the IP is far from being a secondary detail but an important aspect of the optimization of CBS devices. Figure 5.5 shows the geometrical overlap of the beam at the IP for a less-than-perfect head-on collision. We have also noted that increasing the CBS brightness requires small electron beam transverse sections and therefore small Twiss beta values, which, in turn, cause an increase in the inhomogeneous broadening parameters. It should be noted that the Twiss parameters depend on z, i.e. they do evolve during the collision process; therefore, the spectral line also evolves in such a way that the measured spectrum is an average over all configurations of the beams' phase space during their interaction. Figure 5.6 yields an idea of the impact of the beam distributions on the spectral distribution. Figure 5.6 yields an idea of the dependence of the spectral line on the energy spread. In the case of a symmetric energy distribution, no spectral peak shift is exhibited and the most significant effects are the

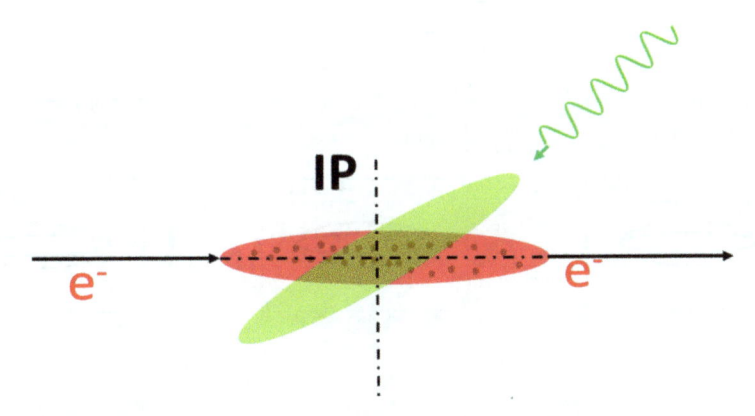

Figure 5.5. Electrons (red) and photons (green) overlapping at the IP.

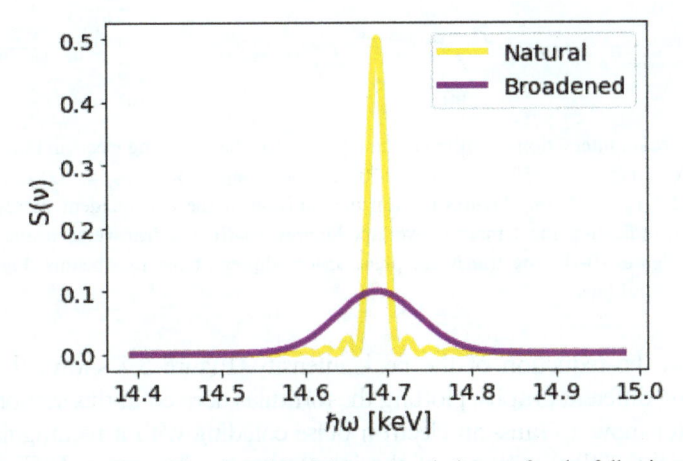

Figure 5.6. The broadened spectral line (purple) of the scattered photons for the following electron and laser parameters: $\lambda_0 = 0.8~\mu m$, $N_c = 187$, $\gamma = 50$, $\sigma_{\delta_e} = 5 \times 10^{-3}$, $\sigma_{0x} = \sigma_{0y} = \sigma_{ex} = \sigma_{ey} = 1~mm$, $L = 150~\mu m$, $\sigma_\zeta = 100~\mu m$, and $\gamma_0 \varepsilon_\eta = 1.5~\mu m$. The natural spectral line (yellow) corresponding to negligible energy spread and emittance of the two incident beams.

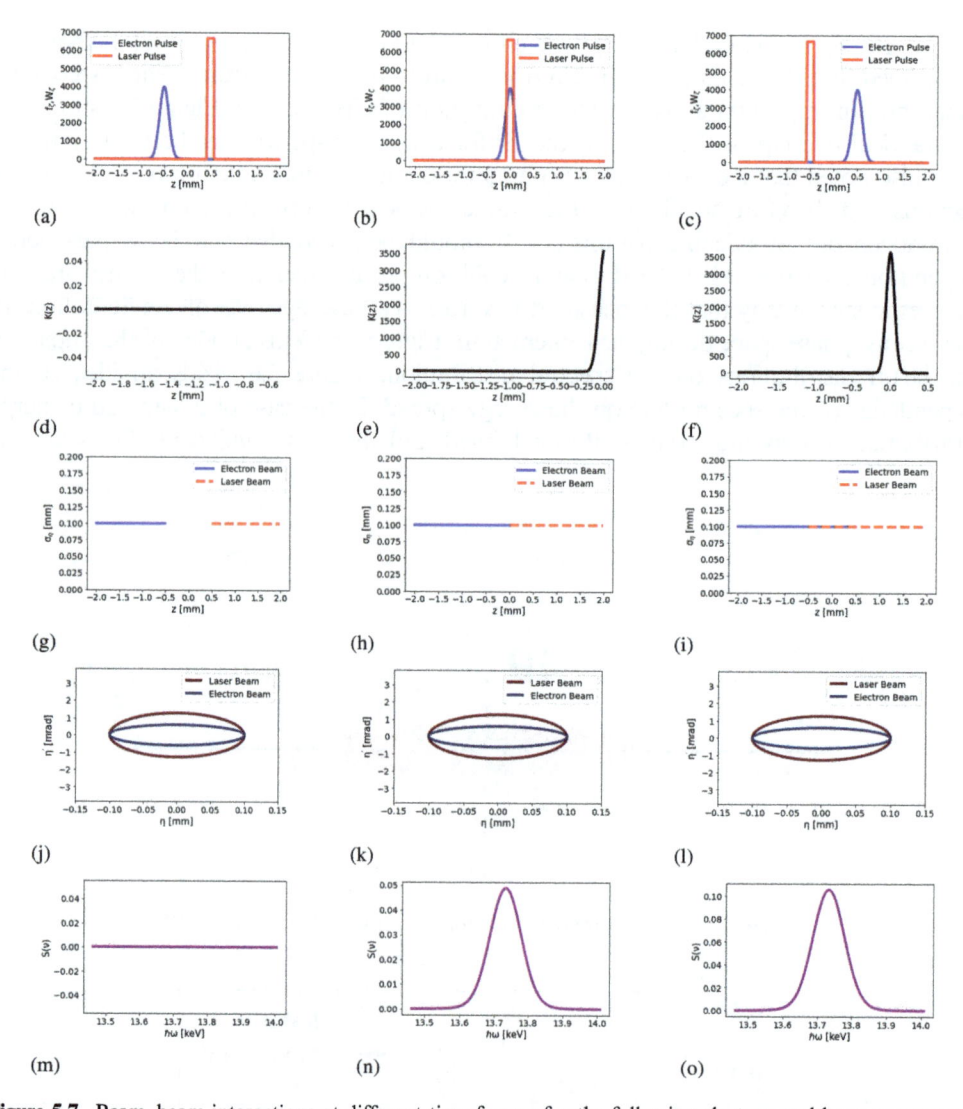

Figure 5.7. Beam–beam interactions at different time frames for the following electron and laser parameters: $\lambda_0 = 0.8\ \mu\text{m}$, $N_c = 175$, $\gamma = 50$, $\sigma_{\delta_e} = 5 \times 10^{-3}$, $\sigma_{0x} = \sigma_{0y} = \sigma_{ex} = \sigma_{ey} = 100\ \mu\text{m}$, $L = 150\ \mu\text{m}$, $\sigma_\zeta = 100\ \mu\text{m}$, and $\gamma_0 \varepsilon_\eta = 1.5\ \mu\text{m}$. **Figures (a)–(c)**: the collision of the two incident pulses. **Figures (d)–(f)**: the function $K(z)$, indicating the temporal overlap. **Figures (g)–(i)**: the transverse beam sizes of the two colliding beams. **Figures (j)–(l)**: the transverse phase-space ellipses of the two beams. **Figures (m)–(o)**: the broadened CBS spectral line.

broadening and the reduction of the peak intensity. Figure 5.7 shows the beam–beam interaction using different graphs, plotting the formulas derived in this section. Specifically, figures 5.7(a)–(c) show a Gaussian electron pulse colliding with a rectangular laser pulse. From left to right, different stages of the interaction are represented. Figures 5.7(d)–(f) show the function $K(z)$, defined in equation (5.19) and account for the temporal overlap between the incident pulses. $K(z)$ is zero before the collision, starts growing during the collision, reaches a maximum, and finally decreases to zero when the two pulses have passed each other and the interaction is finished. Figures 5.7(g)–(i) represent the transverse beam sizes of the two colliding beams. The electron beam travels from the left and the photon beam from the right. In this case, diffraction is negligible; therefore, the beam sizes stay constant. This is also visible in figures 5.7(j)–(l), which depict the transverse phase-space ellipses of the two colliding beams. The absence of rotation of the ellipses corresponds to an absence of diffraction, i.e. to a constant value of the beam size during the interaction. Figures 5.7(m)–(o) show the scattered CBS spectral line. A broadened line of Gaussian shape is observed, mainly due to the Gaussian shape of the electron energy spread distribution. From left to right the intensity of the line increases, following the evolution of the interaction. Figure 5.8 shows the beam–beam interaction for the case in which diffraction is not negligible. Specifically, figures 5.8(a)–(c) and (d)–(f) are the same as in figure 5.7, since diffraction does not affect the temporal overlap. Figures 5.8(g)–(i) represent the transverse beam sizes of the two colliding beams. The transverse beam sizes are focused while approaching the IP. As in figure 5.7, the electron beam travels from the left and the photon beam travels from the right. For the parameters chosen for the simulation, the two beam envelopes do not follow the same trend, but they do reach the same value of transverse beam size at the IP. After the interaction, due to diffraction, the beam sizes defocus. Within the pulse described by $K(z)$, which fixes the scale of the temporal overlap, the transverse beam sizes do evolve, affecting the efficiency of the interaction. Figures 5.8(j)–(l) represent a rotation of the ellipses for the two incident beams. Before the interaction, the two ellipses show a negative correlation, indicating that the beams are subjected to focusing while traveling towards the IP. The beam ellipses reach a zero correlation in the phase space at the IP, where the waist of the beam size is realized. After the interaction, the beams diverge, leading to a rotation of the ellipses, which acquire a positive correlation. The effect of diffraction is to cause a significant decrease in the intensity of the scattered radiation line, as depicted in figures 5.8(m)–(o). In the diffractive case, the broadened line is also deformed, acquiring a long low-energy tail. This corresponds not only to a broadening effect but also to a shift, since the spectral barycenter moves toward lower frequencies.

Before closing this section, we underscore that the line broadening is specified as the rms of the spectral frequency. We therefore set:

$$\Delta\omega = \sqrt{\langle\omega^2\rangle - \langle\omega\rangle^2}$$
$$\langle\omega^m\rangle = \left(\frac{1}{2\pi N_c}\right)^m \langle\nu^m\rangle. \tag{5.20}$$

The total broadening also includes the natural broadening $(\Delta\omega)_0$ given by $4\sqrt{12}\,\gamma_0^2\omega_0/2\pi N_c$. In conclusion, for a linear CBS radiation source, we define:

$$\left(\frac{\Delta\omega}{\omega}\right)_s \simeq \sqrt{\left(\frac{\Delta\omega}{\omega}\right)_0^2 + \left(\frac{\Delta\omega}{\omega}\right)_{\delta_e}^2 + \left(\frac{\Delta\omega}{\omega}\right)_{\theta_e}^2 + \left(\frac{\Delta\omega}{\omega}\right)_{\delta_0}^2}$$
$$= \left(\frac{\Delta\omega}{\omega}\right)_0 \sqrt{1 + \frac{4\pi^2}{3}N_c^2\sigma_{\delta_e}^2 + \frac{2\pi^2}{3}N_c^2\gamma_0^4\sigma_{x'}^{\prime 4} + \frac{2\pi^2}{3}\gamma_0^4\sigma_{y'}^{\prime 4}} \tag{5.21}$$
$$\equiv \left(\frac{\Delta\omega}{\omega}\right)_0 \sqrt{1 + \mu_{\delta_e}^2 + \mu_{x'}^2 + \mu_{y'}^2}, \qquad \left(\frac{\Delta\omega}{\omega}\right)_0 = \frac{\sqrt{3}}{\pi N_c}.$$

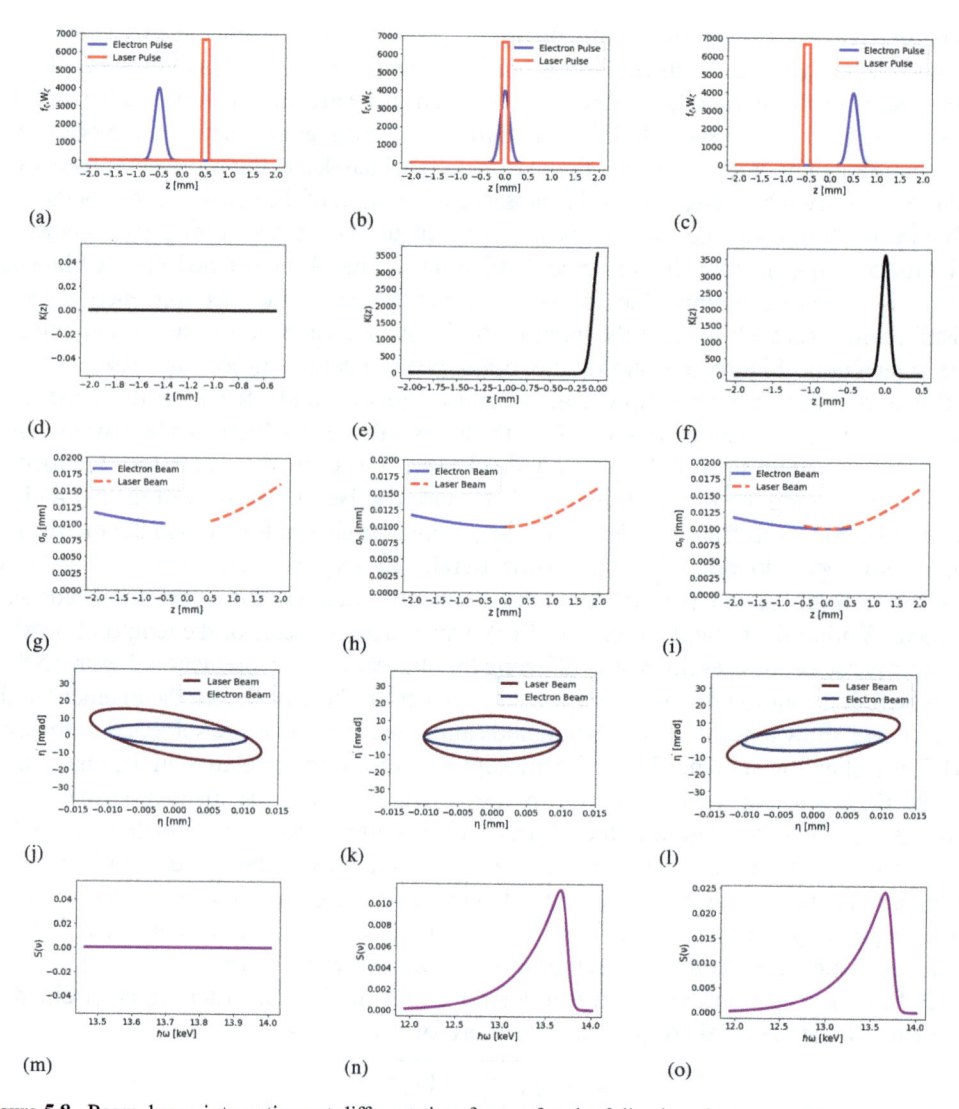

Figure 5.8. Beam–beam interactions at different time frames for the following electron and laser parameters: $\lambda_0 = 0.8$ μm, $N_c = 175$, $\gamma = 50$, $\sigma_{\delta_e} = 5 \times 10^{-3}$, $\sigma_{0x} = \sigma_{0y} = \sigma_{ex} = \sigma_{ey} = 10$ μm, $L = 150$ μm, $\sigma_\zeta = 100$ μm, and $\gamma_0 \varepsilon_\eta = 1.5$ μm. **Figures (a)–(c)**: the collision of the two incident pulses. **Figures (d)–(f)**: the function $K(z)$, indicating the temporal overlap. **Figures (g)–(i)**: the transverse beam sizes of the two colliding beams. **Figures (j)–(l)**: the transverse phase-space ellipses of the two beams. **Figures (m)–(o)**: the broadened and shifted CBS spectral line.

It is evident that the inhomogeneous parameters are a direct measure of the line broadening itself, which is an 'attribute' of the spectrum at fixed observation angles. An important point to be clarified is how to integrate over the solid angle of the detector collecting the scattered photons. This is a point of paramount importance, because the collection angles are important in specifying the spectral coherence of the scattered photon beam. The discussion of this section is biased by previous studies of FELs in which the effect of the undulator's spectral broadening was defined at θ angles of approximately zero. The next section deals with the importance of the off-axis contribution in the study of CBS devices.

5.3 Spectral broadening in CBS devices: off-axis contributions from the divergence of scattered photons

In order to provide a more effective understanding of CBS spectral line broadening, we should like to mention the following more general expression for the Compton backscattered frequency in the linear regime (see chapter 2):

$$\omega_s \simeq \frac{4\gamma^2 \omega_0}{1 + \gamma^2 \tilde{\theta}^2} \qquad (5.22)$$

where $\tilde{\theta}^2 \equiv \theta^2 + \theta_e^2 - 2\theta\theta_e \cos(\phi - \phi_e)$. The electron polar angle is defined in terms of the initial transverse momenta, i.e. $\theta_e = \sqrt{p_x(0)^2 + p_y^2(0)}/p_z(0)$, while the azimuthal angle is $\phi_e = \arctan(p_y(0)/p_x(0))$. The angular part of the off-axis contribution depends both on the observation angle and on the transverse component of the electrons' momenta. For symmetric electron angular distributions, it is possible to neglect the term $\propto \cos(\phi - \phi_e)$, resulting in a zero contribution to the broadening. Therefore, it is possible to rewrite equation (5.22) as:

$$\omega_s(\theta) \simeq \frac{4\gamma^2 \omega_0}{1 + \gamma^2 (\theta^2 + \theta_e^2)}. \qquad (5.23)$$

Here, we consider the weak-field case and leave the nonlinear corrections to the exercises at the end of the chapter. It is important to understand the importance of inhomogeneous broadening from the perspective of the measured spectrum per unit angle. To this aim, we revisit equation (2.96):

$$\frac{d^2 N}{d\omega d\theta} \simeq N(\theta) \operatorname{sinc}^2\left[\frac{\nu\theta}{2}\right]$$
$$N(\theta) = \frac{\alpha \omega a_0^2 N_c^2 \theta}{128\pi^3 \gamma_0^2 \omega_0^2}\left(1 + \frac{\theta^4 \omega^2}{8\omega_0^2} - \frac{\theta^2 \omega}{2\omega_0}\right) \qquad (5.24)$$
$$\nu\theta = \frac{\omega - \omega_s(\theta)}{\omega_s(\theta)}.$$

The inclusion of the inhomogeneous broadening leads to the expression:

$$\frac{d^2 N}{d\omega d\theta} \simeq N(\theta) \langle S(\nu(\theta)) \rangle \qquad (5.25)$$

where:

$$\langle S(\nu(\theta))\rangle = 2\int dz K(z) Re\left[\int_0^1 (1-\kappa)\frac{e^{-i\nu(\theta)\kappa - \frac{1}{2}4\pi N_c \sigma_{\delta_e}\kappa^2}}{\sqrt{R_x(\kappa,z)R_y(\kappa,z)}}d\kappa\right]. \quad (5.26)$$

The shift induced by non-axial detection of the radiation spectrum can be calculated using a similar approach to that of equation (5.14) for a shift related to the initial angles of the electron trajectories, since θ and θ_e appear symmetrically in equation (5.23). Therefore:

$$(\delta\nu)_\theta = 2\pi N_c \left(\frac{\delta\omega}{\omega}\right)_\theta = 2\pi N_c \gamma_0^2 \theta^2. \quad (5.27)$$

Recalling equation (2.96), the relevant integral that must be solved to evaluate the broadening due to a finite observation angle is given by:

$$\int_0^{\theta_{max}} d\theta\theta\left(1 + \frac{\theta^4\omega^2}{8\omega_0^2} - \frac{\theta^2\omega}{2\omega_0}\right)e^{-i(\delta\nu)_\theta\kappa}. \quad (5.28)$$

Equation (5.28) can be solved analytically, giving a cumbersome expression. In most of cases of interest, the radiation cone is collimated to reduce the chromatic spread, or in any case limited by apertures. This means that one can consider an expansion to the first orders in θ_{max} for equation (5.28). This approximation leads to:

$$\int_0^{\theta_{max}} d\theta\theta\left(1 + \frac{\theta^4\omega^2}{8\omega_0^2} - \frac{\theta^2\omega}{2\omega_0}\right)e^{-i(\delta\nu)_\theta\kappa} \simeq \frac{\theta_{max}^2}{2\sqrt{1 + \frac{\omega}{2\omega_0}\theta_{max}^2 + 2i\pi N_c\gamma_0^2\theta_{max}^2\kappa}}. \quad (5.29)$$

A more important aspect of the finite observation cone is line broadening. This can be calculated as follows:

$$\left(\frac{\Delta\omega}{\omega}\right)_\theta^2 = \frac{\int_0^{\theta_{max}} d\theta\theta^5\gamma_0^4\left(1 + \frac{\theta^4\omega^2}{8\omega_0^2} - \frac{\omega}{2\omega_0}\theta^2\right)}{\int_0^{\theta_{max}} d\theta\theta\left(1 + \frac{\theta^4\omega^2}{8\omega_0^2} - \frac{\omega}{2\omega_0}\theta^2\right)}$$
$$- \frac{\left[\int_0^{\theta_{max}} d\theta\theta^3\gamma_0^2\left(1 + \frac{\theta^4\omega^2}{8\omega_0^2} - \frac{\omega}{2\omega_0}\theta^2\right)\right]^2}{\left[\int_0^{\theta_{max}} d\theta\theta\left(1 + \frac{\theta^4\omega^2}{8\omega_0^2} - \frac{\omega}{2\omega_0}\theta^2\right)\right]^2} \simeq \frac{\gamma_0^4\theta_{max}^4}{12}. \quad (5.30)$$

We underscore that inhomogeneous broadening is responsible for the reduction of both photon flux and monochromaticity and hence of brightness. It is important to note that the collection of photons over a given angular aperture of the detector dilutes the source's monochromaticity. According to the derivation shown in chapter 2, we can write the spectral–angular distribution of photons scattered by one electron as:

$$\frac{d^2N}{d\omega d\theta} = \frac{\pi\alpha\omega a_0^2 N_c^2 \theta}{8\gamma_0^2 \omega_0^2}\left(1 + \frac{\theta^4\omega^2}{8\omega_0^2} - \frac{\theta^2\omega}{2\omega_0}\right)\text{sinc}^2\left[\left(\Psi_1 - \frac{\omega_0}{c}\right)\frac{\zeta}{2}\right] \quad (5.31)$$

where Ψ_1 depends upon θ and ω. Using the integral representation of the sinc function, introduced in equation (5.19), and using the result at equation (5.29), we end up with:

$$\frac{dN}{d\omega} \simeq \frac{\pi\alpha\omega a_0^2 N_c^2 \theta_{\max}^2}{16\gamma_0^2 \omega_0^2}\text{sinc}^2\left[\left(\Psi_1 - \frac{\omega_0}{c}\right)\frac{\zeta}{2}\right] \simeq \frac{\pi\alpha\omega a_0^2 N_c^2 \theta_{\max}^2}{16\gamma_0^2 \omega_0^2}\frac{\omega_0}{N_c}4\gamma_0^2\delta\left(\omega - 4\gamma_0^2\omega_0\right) \quad (5.32)$$

where, for the last passage, we have rewritten equation (2.104) as:

$$\text{sinc}^2\left(\frac{\nu}{2}\right) \simeq \frac{\omega_0}{N_c}4\gamma_0^2\delta\left(\omega - 4\gamma_0^2\omega_0\right) = \frac{\pi\omega_0}{\sqrt{3}}4\gamma_0^2\left(\frac{\Delta\omega}{\omega}\right)_0\delta\left(\omega - 4\gamma_0^2\omega_0\right). \quad (5.33)$$

The integration of equation (5.32) over the frequencies is trivial, yielding:

$$N \simeq \pi\alpha a_0^2 N_c \gamma_0^2 \theta_{\max}^2. \quad (5.34)$$

Equation (5.34) expresses the number of photons emitted by a single electron within a finite solid angle $\pi\theta_{\max}^2$ centered around the axis z. To obtain the number of photons emitted by a bunch of electrons, we develop the heuristic derivation shown below. Indeed, the radiation spectrum emitted by a bunch of electrons can be expressed by replacing the sinc function with equations (5.26) and (5.29):

$$\frac{dN}{d\omega} \simeq \frac{N_e \pi\alpha\omega a_0^2 N_c^2 \theta_{\max}^2}{8\gamma_0^2 \omega_0^2}\int dz K(z) Re\left[\int_0^1 (1-\kappa)\frac{e^{-i\nu(\omega)\kappa - \frac{1}{2}4\pi N_c \sigma_{\delta_e}\kappa^2}}{\sqrt{R_x(\kappa,z)R_y(\kappa,z)R_\theta(\kappa,\omega)}}d\kappa\right] \quad (5.35)$$

$$R_\theta(\kappa,\omega) = 1 + \frac{\omega}{2\omega_0}\theta_{\max}^2 + 2i\pi N_c \gamma_0^2 \theta_{\max}^2 \kappa.$$

Using a suitable modification of equation (2.104) in order to include the homogeneous broadening, it is possible to calculate the total photon number. Therefore, analogously to equation (5.33), by replacing the sinc function with the broadened line and considering linear interactions ($a_0 \ll 1$), one obtains:

$$\int dz K(z) Re\left[\int_0^1 (1-\kappa)\frac{e^{-i\nu(\omega)\kappa - \frac{1}{2}4\pi N_c \sigma_{\delta_e}\kappa^2}}{\sqrt{R_x(\kappa,z)R_y(\kappa,z)R_\theta(\kappa,\omega)}}d\kappa\right]$$

$$\simeq \frac{1}{2\sqrt{1 + \frac{4\sigma_x^2}{w_0^2}}\sqrt{1 + \frac{4\sigma_y^2}{w_0^2}}\sqrt{1 + \mu_{\delta_e}^2 + \mu_{x'}^2 + \mu_{y'}^2 + \mu_{\theta_{\max}}^2}}\text{sinc}^2\left(\frac{\nu}{2\sqrt{1 + \mu_{\delta_e}^2 + \mu_{x'}^2 + \mu_{y'}^2 + \mu_{\theta_{\max}}^2}}\right) \quad (5.36)$$

where $\mu_{\theta_{\max}}^2 = \pi^2 N_c^2 (\Delta\omega/\omega)_\theta^2/3$ and we assumed a matched electron beam size $\sigma_e \lesssim w_0/2$. Inserting equation (5.36) into equation (5.35) and approximating for a narrow line (which is valid for relatively monochromatic CBS sources), it is possible to obtain:

$$\frac{dN}{d\omega} \simeq \frac{N_e \pi \alpha \omega a_0^2 N_c^2 \theta_{max}^2}{8\gamma_0^2 \omega_0^2} \frac{\delta(\omega - 4\gamma_0^2 \omega_0)}{\sqrt{1 + \frac{4\sigma_x^2}{w_0^2}}\sqrt{1 + \frac{4\sigma_y^2}{w_0^2}}\sqrt{1 + \mu_{\delta_e}^2 + \mu_{x'}^2 + \mu_{y'}^2 + \mu_{\theta_{max}}^2}} \frac{\pi \omega_0}{2\sqrt{3}} 4\gamma_0^2 \left(\frac{\Delta\omega}{\omega}\right)_s \quad (5.37)$$

and, after integrating equation (5.37), one eventually finds:

$$N \simeq \frac{\pi \alpha a_0^2 N_e N_c \gamma_0^2 \theta_{max}^2}{\sqrt{1 + \frac{4\sigma_x^2}{w_0^2}}\sqrt{1 + \frac{4\sigma_y^2}{w_0^2}}} = \frac{3\sigma_T N_0 N_e \gamma_0^2 \theta_{max}^2}{4\pi \sqrt{\frac{w_0^2}{4} + \sigma_x^2}\sqrt{\frac{w_0^2}{4} + \sigma_y^2}}. \quad (5.38)$$

Equation (5.38) coincides with equation (5.34) except for the number of electrons N_e at the numerator and the spatial overlap factors at the denominator, because inhomogeneous broadening does not change the total photon number, just the spectral density (see equation (5.36)). With cylindrical symmetry, the product of the square roots at the denominator would become $1 + 4\sigma_e^2/w_0^2$, recovering the scaling introduced in equation (5.6). In fact, $a_0^2 \propto N_0/w_0^2$. In particular, we have used the definition $a_0^2 = e^2 \varepsilon_0 E_0^2/\varepsilon_0 m_0^2 c^2 \omega_0^2 = 2e^2 N_0 \hbar/\pi \varepsilon_0 w_0^2 L m_0^2 c^2 \omega_0$. Multiplying N by f_{rep} and dividing by the solid angle of observation and by the effective size of the source, leads, by definition, to the source brightness:

$$B = \frac{f_{rep} N}{2\pi^2 \theta_{max}^2 \sqrt{\frac{w_0^2}{4} + \sigma_x^2}\sqrt{\frac{w_0^2}{4} + \sigma_y^2}} = \frac{3\dot{N}_X \gamma_0^2}{8\pi^3 \sqrt{\frac{w_0^2}{4} + \sigma_x^2}\sqrt{\frac{w_0^2}{4} + \sigma_y^2}} \simeq 0.012 \times \frac{\gamma_0^2 \dot{N}_X}{\Sigma_{e,0}^2}. \quad (5.39)$$

This is slightly larger than the value in equation (5.6), which considered the full solid angle of observation. However, the scaling of the brightness due to the relevant parameters is recovered. Moreover, we stress again the difference due to the absence of cylindrical symmetry in the calculations executed in this section compared to the simplified definition of brightness given in the introduction of this chapter. It is evident that equation (5.38) provides the final goal of our discussion. It indeed contains the specification of the Compton backscattered photons in terms of the parameters characterizing an actual experimental configuration. In the next section we discuss a few wise criteria to use when exploiting the previous conclusions in the design of a Thomson backscattering source.

5.4 A toy model of a real CBS radiation facility

For the toy model of a real CBS machine, we are going to consider the schematics in figure 5.9 and the reference values of table 5.1. The electron source is an electron gun that provides, for example, $\simeq 25$ MeV electrons ($\gamma_0 \simeq 50$). The electron gun will certainly consist of a cathode placed within a radio frequency (RF) cavity (generating an electron beam with an energy of a few MeV), focusing optics (for, example solenoids used to match the beam to a second RF cavity), and the second cavity, which is exploited to accelerate the beam to its final energy of 25 MeV. We

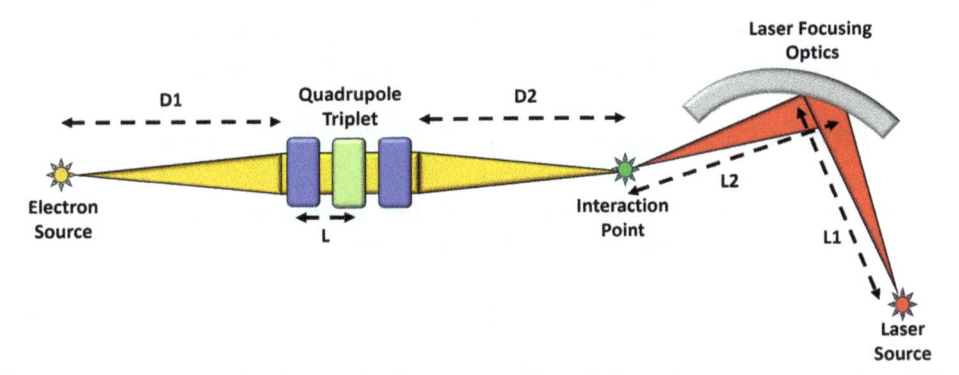

Figure 5.9. A toy model of a CBS source. The electron and laser sources have been drawn as point sources from which the two beams originate and diverge. The electron beam (yellow) is collected and focused in the two planes by means of a quadrupole triplet, where the first and last quad are focusing in one plane and the central is focusing in the the orthogonal plane. The laser beam (red) is focused by means of a curved mirror (e.g. parabolic). Here, the interaction occurs with a small angle between the two beams.

assume that the electron beam exits the RF accelerating cavity with a small divergence. The focusing optics for the electrons is then a quadrupole triplet. The central quadrupole is set to a higher current to provide a stronger focusing field compared to the other two. This compensates for the defocusing caused by the latter and achieves focusing in both planes. For example, here, we consider that the central quadrupole defocuses in x and has twice the focusing strength of the other two (its focal length is halved, i.e. $f_e/2$, compared to the focal lengths of the other two, i.e. $-f_e$). The Twiss parameters for the propagation of the electron beam are studied in the Comments and exercises section. However, here we directly report the result using the thin lens approximation, i.e. neglecting the terms proportional to $(L/f_e)^2$, and assuming a negligible electron divergence:

$$\beta_e \simeq \left(1 - \frac{4LD_2}{f_e^2}\right)\beta_e(0)$$

$$\alpha_e \simeq \frac{2L}{f_e^2}\beta_e(0) \tag{5.40}$$

$$\gamma_e \simeq 0.$$

The laser beam's Twiss parameters at the IP are provided by equations (4.98):

$$\beta_0 = \left(1 - \frac{2L_2}{f_0} + \frac{L_2^2}{f_0^2}\right)\beta_0(0) + \left[L_2^2 + L_1^2\left(1 - \frac{2L_2}{f_0} + \frac{L_2^2}{f_0^2}\right) - L_1\left(\frac{2L_2^2}{f_0} - 2L_2\right)\right]\gamma_0(0)$$

$$\alpha_0 = \left(\frac{1}{f_0} - \frac{L_2}{f_0^2}\right)\beta_0(0) + \left[L_1^2\left(\frac{1}{f_0} - \frac{L_2}{f_0^2}\right) - L_2 - L_1\left(1 - \frac{2L_2}{f_0}\right)\right]\gamma_0(0) \tag{5.41}$$

$$\gamma_0 = \frac{\beta_0(0)}{f_0^2} + \left(1 - \frac{2L_1}{f_0} + \frac{L_1^2}{f_0^2}\right)\gamma_0(0).$$

Based on the above, the laser beam's rms size at the IP can be calculated from the second momentum of the Wigner distribution $W(\eta, \eta', z)$, where $\eta = x, y$:

$$\sigma_0^2 = \int W(\eta, \eta', z = z_{int})\eta^2 d\eta d\eta' = \frac{2}{\pi \bar{\lambda}} \int e^{-\frac{1}{\lambda}[\gamma_0 \eta^2 + 4\beta_0 \eta'^2 + 4\alpha_0 \eta \eta']} \eta^2 d\eta d\eta' = \frac{\beta_0 \bar{\lambda}}{2}. \quad (5.42)$$

Similarly, the electron beam's rms size at the IP is calculated as the second momentum of the Liouville distribution $f(\eta, \eta', z)$:

$$\sigma_e^2 = \int f\eta, \eta', z = z_{int}\eta^2 d\eta d\eta' = \frac{1}{2\pi\varepsilon} \int e^{-\frac{1}{2\varepsilon}[\gamma_e \eta^2 + \beta_e \eta'^2 + 2\alpha_e \eta \eta']} \eta^2 d\eta d\eta' = \beta_e \varepsilon \quad (5.43)$$

where the IP $z = z_{int}$ is taken to equal zero, similarly to the examples of figures 5.7 and 5.8. For the reference parameters of table 5.1, $\beta_e = 1\, m$, thus $\sigma_e = \sqrt{\beta_e \epsilon_n/\gamma_0} \simeq 44$ μm. Assuming $\beta_e(0) = 10$ m, $4LD_2/f_e^2 = 0.9$. Reasonable values may be $L = 0.1$ m, $D_2 = 1$ m, and $f = 0.67$ m. We also assume that $\sigma_0 = 44$ μm and that the duration of the incident laser pulse is $\tau_0 = L/c = 1$ ps. This corresponds to an incident laser intensity $I = \widetilde{E}_0/2\pi\sigma_0^2\tau_0 \simeq 8 \times 10^{13}$ W cm^{-2}, i.e. to a relativistic parameter $a_0 \simeq 8 \times 10^{-3}$. The regime of interaction, therefore, is fully linear. The Rayleigh length is calculated to be $Z_R = 4\pi\sigma_0^2/\lambda_0 \simeq 2.4$ cm, which is significantly longer than the laser pulse duration. This already gives us the information that no diffractive effect is to be expected for the CBS line broadening. With a rep. rate of $f_{rep} = 1$ kHz, required to obtain $\dot{N}_X \simeq 10^8$ photons per second, we also consider a charge of 115 pC, corresponding to about $N_e = 7 \times 10^8$ electrons per beam. A laser pulse energy of 10 mJ, for a central photon wavelength of 1 μm, would correspond to $N_0 = 5 \times 10^{16}$. The brightness is calculated to be $B \simeq 0.01 \times \gamma_0^2 \dot{N}_X/\Sigma_{e,0}^2 \simeq 10^{12}$ s^{-1} mm^{-2} mrad^{-2}. In order to evaluate the spectral brightness, we have to define a value for the angular aperture of the detection line of the source. For instance, let the aperture be $1/2\gamma_0 = 10$ mrad. We should now take into account the electron energy spread. For instance, let us assume $\delta_e = 0.5\%$, and also, for simplicity, that $\sigma_\zeta = L/2$. Using equation (5.21) with the addition of the term at equation (5.30) and assuming that the dominant broadening terms are the angular μ_θ and the energy spread μ_{δ_e}, we obtain $B_s = 1.4 \times 10^{10}$ s^{-1}(0.1% bandwidth)$^{-1}$ mm^{-2} mrad^{-2}. The calculated relative bandwidth is, in fact, $\simeq 7\%$. The angle of interaction is $\phi_1 = 150°$, corresponding to a scattered photon energy of $\simeq 11$ keV. In the following we show the corresponding simulation of the dynamics of the beam interaction and of the radiation. Figure 5.10 confirms that diffraction is unimportant for the 'toy' source of this chapter, since the beam envelopes stay constant during the interaction. Moreover, the broadening is strongly affected by the electron energy spread, leading to a Gaussian core of the line shape that is reminiscent of the Gaussian energy spectrum. The long tail at low photon energies is instead related to the finite observation cone corresponding to a collection of Doppler redshifted photons.

In this chapter we have provided the practical tools necessary to deal with the design of a CBS source. The following section, Comments and exercises, is a useful integration of the matter treated in the course of the main body of the chapter.

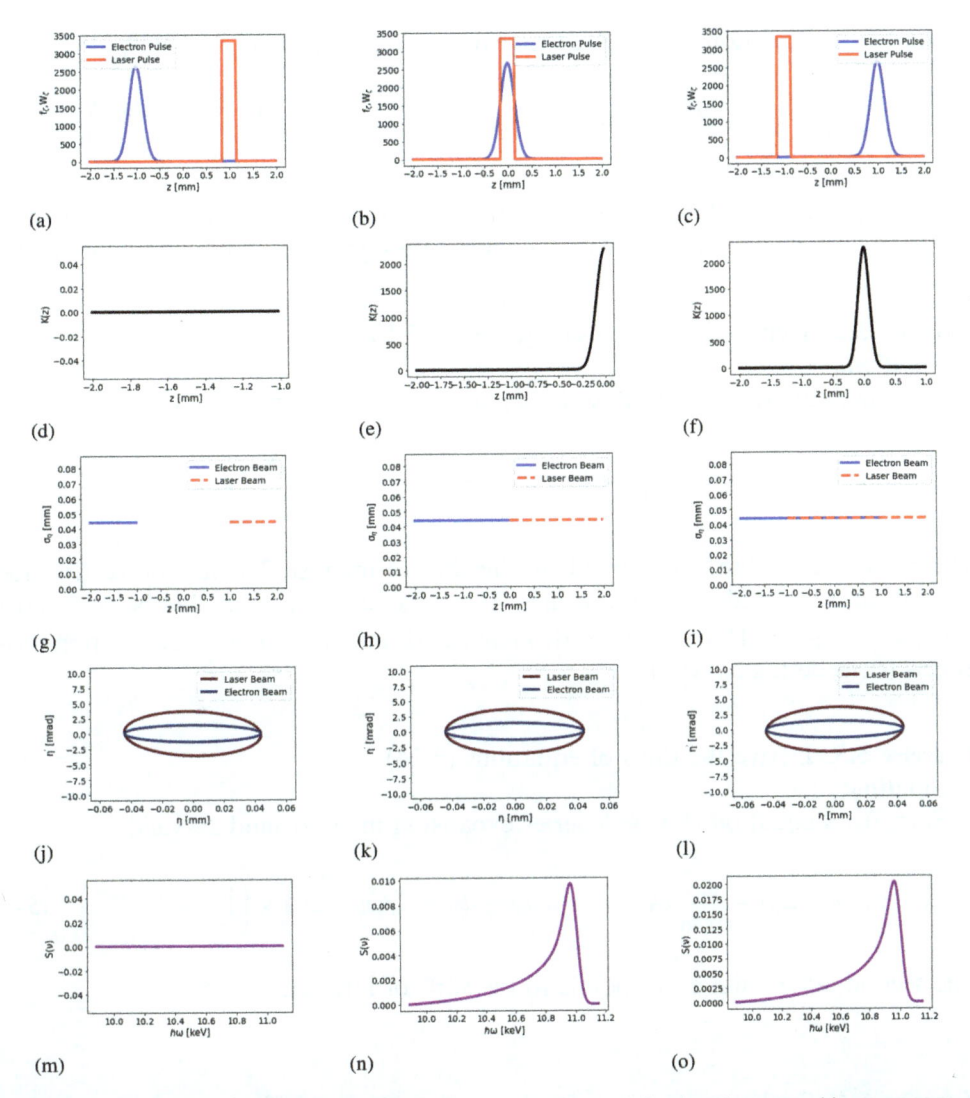

Figure 5.10. Beam–beam interactions at different time frames for the following electron and laser parameters: $\lambda_0 = 1$ μm, $N_c = 280$, $\gamma = 50$, $\sigma_{\delta_e} = 5 \times 10^{-3}$, $\sigma_{0x} = \sigma_{0y} = \sigma_{ex} = \sigma_{ey} = 44$ μm, $L = 300$ μm, $\sigma_\zeta = 150$ μm, and $\gamma_0 \varepsilon_\eta = 1.5$ μm. **Figures (a)–(c)**: the collision of the two incident pulses. **Figures (d)–(f)**: the function $K(z)$, indicating the temporal overlap. **Figures (g)–(i)**: the transverse beam sizes of the two colliding beams. **Figures (j)–(l)**: the transverse phase-space ellipses of the two beams. **Figures (m)–(o)**: the broadened and shifted CBS spectral line.

5.5 Comments and exercises

5.5.1 Exercises

Exercise 5.1. Derive the first of equations (5.14).
Solution:

Start from equation (5.12). A series expansion in γ around γ_0 yields:

$$\omega_s \simeq \omega_s(\gamma = \gamma_0) + \frac{d\omega_s}{d\gamma}(\gamma = \gamma_0) \times \gamma - \gamma_0 \simeq \omega_s(\gamma = \gamma_0) \times \left(1 + 2\frac{\gamma - \gamma_0}{\gamma_0}\right). \quad (5.44)$$

The second addendum corresponds to the shift that is of interest here. In fact, $d\nu/\nu\big|_{\nu=0} = -d\omega_s/\omega_s$. This explains the opposite sign of the shift in equation (5.14).

Exercise 5.2. Derive the second of equations (5.14).
Solution:
Consider equation (5.12). A series expansion in θ_e around zero yields:

$$\omega_s \simeq \omega_s(\theta_e = 0) + \frac{1}{2}\frac{d^2\omega_s}{d\theta_e^2}(\theta_e = 0) \times \theta_e^2 \simeq \omega_s(\theta_e = 0) \times \left(1 - \gamma_0^2\theta_e^2\right). \quad (5.45)$$

The second addendum corresponds to the shift of interest. The first derivative does not appear, since it identically vanishes. The parameter θ_e is related to both x' and y' via $\theta_e^2 = x'^2 + y'^2$. Here we have also neglected the contributions due to nonlinear effects (considering $a_0 \ll 1$).

Exercise 5.3. Derive the third of equations (5.14).
Solution:
Start from equation (5.12). A series expansion in ω_0 around $\bar{\omega}_0$ yields:

$$\omega_s \simeq \omega_s(\omega_0 = \bar{\omega}_0) + \frac{d\omega_s}{d\omega_0}(\omega_0 = \bar{\omega}_0) \times \omega_0 - \bar{\omega}_0 \simeq \omega_s(\omega_0 = \bar{\omega}_0) \times \left(1 + \frac{\omega_0 - \bar{\omega}_0}{\bar{\omega}_0}\right). \quad (5.46)$$

The second addendum corresponds to the shift of interest.

Exercise 5.4. Evaluate equation (5.18) to get equation (5.19).
Solution:
First solve the following integral:

$$\int d\eta S(\nu + \delta\nu_\eta) f_\eta(\eta, \eta'; z) \propto \int d\eta e^{-i\delta\nu_\eta \kappa} \frac{N_e}{2\pi\varepsilon_\eta} e^{-\frac{\gamma_{e\eta}\eta^2 + 2\alpha_{e\eta}\eta\eta' + \beta_{e\eta}\eta'^2}{2\varepsilon_\eta}} a_0^2 e^{-\frac{2\eta^2}{w^2(z)}} \quad (5.47)$$

where, as usual, $\eta = x, y$ and the shift $\delta\nu_\eta$ is given by the fourth equation of (5.14), i.e.:

$$\delta\nu_\eta = -2\pi N_c \frac{a_0^2 \eta^2}{\left(1 + \frac{a_0^2}{2}\right)w_0^2}. \quad (5.48)$$

Exercise 5.5. Demonstrate that the broadening related to the shift $(\delta\omega/\omega)_{\delta_0}$ is null for a rectangular laser pulse.
   Solution:
   The spectrum of a rectangular laser pulse is given by equation (5.17):

$$W_{\delta_0}(\delta_0) = \text{sinc}^2 \pi \delta_0 N_c. \tag{5.49}$$

The integral at equation (5.18) eventually corresponds to a convolution between two sinc functions: the first is related to the natural spectrum of CBS radiation and the second to the laser pulse spectrum. The width of the first sinc function is proportional to ω_s, while the second is proportional to ω_0. Thus, since $\omega_s = 4\gamma_0^2 \omega_0 \gg \omega_0$, the first sinc is much wider. The convolution between two sinc functions is always equal to the largest of the two. Therefore, we have shown that the laser spectral bandwidth does not contribute to the spectral broadening of the CBS source. It is important to stress that this is true only for rectangular laser pulses; otherwise, there would be an effect. Therefore, to evaluate equation (5.18) and thus obtain equation (5.19), the following integral must be considered:

$$\int d\delta_0 S\nu - 2\pi N_c \delta_0 W_{\delta_0}(\delta_0) = N_c \int d\delta_0 \text{sinc}^2\left(\frac{\nu - 2\pi N_c \delta_0}{2}\right) \text{sinc}^2\left(\pi \delta_0 N_c\right) = S\nu. \tag{5.50}$$

Exercise 5.6. Calculate the CBS line broadening for a Gaussian laser pulse.
   Solution:
   The spectrum of a Gaussian laser pulse is given by:

$$W_{\delta_0}(\delta_0) = \frac{e^{-\frac{\delta_0^2}{2\sigma_{\delta_0}^2}}}{\sqrt{2\pi}\,\sigma_{\delta_0}}. \tag{5.51}$$

According to equation (5.18), the integral to be solved is:

$$\int d\delta_0 e^{-i[\nu - 2\pi N_c \delta_0]\kappa} W_{\delta_0}(\delta_0) = e^{-i\nu\kappa - \frac{1}{2} 2\pi N_c \sigma_{\delta_0} \kappa^2} \tag{5.52}$$

which corresponds to a broadened line given by:

$$\langle S(\nu) \rangle = 2 \int dz K(z) Re\left[\int_0^1 (1-\kappa) \frac{e^{-i\nu\kappa - \frac{1}{2} 4\pi N_c \sigma_e \kappa^2 - \frac{1}{2} 2\pi N_c \sigma_{\delta_0} \kappa^2}}{\sqrt{R_x(\kappa, z) R_y(\kappa, z)}} d\kappa\right]. \tag{5.53}$$

Therefore, the spectral bandwidth of the laser can have an impact on line broadening in any case except that of rectangular pulses.

Exercise 5.7. Calculate the Twiss parameters of an electron beam propagating through the lattice given by a drift, a triplet of quadrupoles (such as the one considered in the toy model of this chapter), and another drift.

Hint:
Use the matrix approach studied in chapter 3, exploiting the definition of the drift matrix of equation (3.30) and the quadrupole matrix definition of equation (3.35). For a beam diverging from a point source, according to equations (3.36), use $\beta_e(0) = \sqrt{\sigma_\eta^2/\varepsilon}$, $\alpha_e(0) = 0$, and $\gamma_e(0) = \sqrt{\sigma_\eta'^2/\varepsilon}$.

Exercise 5.8. Based on the previous exercise, derive equations (5.40).
Hint:
Expand the result of the previous exercise to the first order in L/f_e.

Exercise 5.9. Use equation (5.34) to estimate the photon number of the toy model source of this chapter.
Solution:
Given the parameters considered for the toy model source, we obtain $N \simeq 3.5 \times 10^4$. Finally multiplying by $f_{\text{rep}} = 1000 \text{ s}^{-1}$, we obtain $\dot{N}_X \simeq 3.5 \times 10^7 \text{ s}^{-1}$, which is comparable to the total rate $\dot{N}_X = 10^8 \text{ s}^{-1}$ that we targeted for the source, obtained by collecting the CBS radiation over the full solid angle.

Exercise 5.10. Calculate the total bandwidth $\Delta\omega$ of a linear CBS source.
Hint:
Calculate the standard deviation of ω over the spectral distribution given by equation (2.101).

Exercise 5.11. In the linear regime of interaction, calculate the spectral shape of a Gaussian field $A(\zeta)$.
Solution:
Use the expression $A(\zeta) = A_0 \mathcal{P}(\zeta)$, where A_0 is the peak of the vector potential and ζ the longitudinal profile of the field. After recalling equation (2.76), we set $\Psi_{3x} \to 0$ due to the linear regime of interaction:

$$\mathcal{H}_{x,y,z} = \left[\hat{n} \times \sum_{h=-\infty}^{h=+\infty} \int d\zeta' \frac{d\vec{r}_e}{d\zeta'} e^{i(\Psi_1 - h\frac{\omega_0}{c})\zeta'} e^{ih\phi_x} J_h - \Psi_{2x}, 0; 0, 0 \bigg| e^{il(\phi_y - \phi_x)} \right]_{x,y,z} \quad (5.54)$$

where we have also assumed a linearly polarized incident photon beam traveling along x, thus $\Psi_{2y} = 0$. Using the limit cases found in the comments at the end of chapter 2 and equation (2.72), we can recast equation (5.54) into:

$$\mathcal{H}_{x,y,z} = \left[\hat{n} \times \sum_{h=-\infty}^{h=+\infty} \int d\zeta' \frac{d\vec{r}_e}{d\zeta'} e^{i(\Psi_1 - h\frac{\omega_0}{c})\zeta'} e^{i\phi_x} J_h - \Psi_{2x} \right]_{x,y,z} . \quad (5.55)$$

For the sake of simplicity, let us consider the $\mathcal{H}_y$ component only, which is the most relevant for small observation angles ($\theta \sim 0$), and also assume that $\phi_x = 0$. This allows us to consider only $dx/d\zeta$ in equation (5.55). For a particle without initial x-momentum, equation (5.55) is expressed as:

$$\mathcal{H}_y = \frac{a_0}{2\gamma_0} \sum_{h=-\infty}^{h=+\infty} \int d\zeta' \mathcal{P}(\zeta') e^{i(\Psi_1 - h\frac{\omega_0}{c})\zeta'} J_h - \Psi_{2x}. \quad (5.56)$$

Let us now express $\mathcal{P}(\zeta) = e^{-\frac{\zeta^2}{4\sigma_\zeta^2}} \cos \frac{\omega_0 \zeta}{c}$ inside equation (5.56):

$$\mathcal{H}_y = \frac{a_0}{4\gamma_0} \sum_{h=-\infty}^{h=+\infty} \int d\zeta' e^{-\frac{\zeta'^2}{4\sigma_\zeta^2}} \left[e^{i\frac{\omega_0\zeta'}{c}} + e^{-i\frac{\omega_0\zeta'}{c}} \right] e^{i(\Psi_1 - h\frac{\omega_0}{c})\zeta'} J_h - \Psi_{2x} \quad (5.57)$$

where we have used Euler's formula for the cosine. Following suitable reindexing, equation (5.57) can easily be transformed into:

$$\mathcal{H}_y = \frac{a_0}{4\gamma_0} \sum_{h=-\infty}^{h=+\infty} \int d\zeta' e^{-\frac{\zeta'^2}{4\sigma_\zeta^2}} [J_{h+1}(-\Psi_{2x}) + J_{h-1}(-\Psi_{2x})] e^{i(\Psi_1 - h\frac{\omega_0}{c})\zeta'}. \quad (5.58)$$

Due to the linear regime of interaction, we set $h = 1$ and $\Psi_{2x} \to 0$; therefore:

$$\mathcal{H}_y = \frac{a_0}{4\gamma_0} \int d\zeta' e^{-\frac{\zeta'^2}{4\sigma_\zeta^2}} e^{i(\Psi_1 - h\frac{\omega_0}{c})\zeta'} = \frac{a_0 \sqrt{\pi} \sigma_\zeta}{2\gamma_0} e^{-(\Psi_1 - \frac{\omega_0}{c})^2 \sigma_\zeta^2}. \quad (5.59)$$

Consequently, the line shape is now Gaussian instead of a sinc function as expressed in equation (2.83). Finally, the spectral–angular distribution is simply found to be:

$$\frac{d^2 E}{d\omega d\Omega} \simeq \frac{\mu_0 e^2 \omega^2}{4\pi^2 c} \frac{a_0^2}{16\gamma_0^2} \left(1 + \frac{\theta^4 \omega^2}{4\omega_0^2} \cos^2 \phi - \frac{\omega}{\omega_0} \theta^2 \cos^2 \phi \right) \sigma_\zeta^2 e^{-2(\Psi_1 - \frac{\omega_0}{c})^2 \sigma_\zeta^2}. \quad (5.60)$$

It can easily be checked that no significant broadening is expected from a Gaussian pulse compared to a flat counterpart. In this respect, it can be argued that the pulse shape is not particularly significant, at least for long interaction times (in the Gaussian case, this means large σ_ζ).

Exercise 5.12. Calculate the limit of equation (5.60) for $\sigma_\zeta \to \infty$ in order to get an expression in terms of a Dirac delta.
Solution:
We use the definition:

$$\lim_{\sigma_\zeta \to \infty} \sigma_\zeta e^{-2(\Psi_1 - \frac{\omega_0}{c})^2 \sigma_\zeta^2} = \sqrt{\frac{\pi}{2}} \delta\left(\Psi_1 - \frac{\omega_0}{c} \right). \quad (5.61)$$

Equation (5.60) becomes:

$$\frac{d^2 E}{d\omega d\Omega} \simeq \frac{\mu_0 e^2 \omega^2}{4\pi^2 c} \frac{a_0^2}{16\gamma_0^2} \left(1 + \frac{\theta^4 \omega^2}{4\omega_0^2} \cos^2 \phi - \frac{\omega}{\omega_0} \theta^2 \cos^2 \phi \right) \sigma_\zeta \sqrt{\frac{\pi}{2}} \delta\left(\Psi_1 - \frac{\omega_0}{c} \right). \quad (5.62)$$

Exercise 5.13. Express the Dirac delta equation (5.62) in terms of θ.
Solution:
From the properties of the Dirac delta function:

$$\delta\left(\Psi_1 - \frac{\omega_0}{c}\right) = \delta\left[\frac{\omega}{4\gamma_0^2 c}(1+\gamma_0^2\theta^2) - \frac{\omega_0}{c}\right] = \frac{2c\delta\theta - \theta^*}{\omega\theta^*}. \qquad (5.63)$$

Exercise 5.14. Starting from equation (5.62), calculate the spectral–angular distribution of the scattered photon number of a Gaussian incident pulse.
Solution:
Dividing equation (5.62) by $\hbar\omega$ and using the result from the previous exercise yields:

$$\frac{d^2 N}{d\omega d\Omega} \simeq \frac{\alpha\sigma_\zeta}{16\pi c}\frac{a_0^2}{\gamma_0^2}\left(1 + \frac{\theta^4 \omega^2}{4\omega_0^2}\cos^2\phi - \frac{\omega}{\omega_0}\theta^2\cos^2\phi\right)\sqrt{2\pi}\,\frac{\delta\,\theta-\theta^*}{\theta^*}. \qquad (5.64)$$

We notice that equation (5.64) is equivalent to equation (2.99) with the replacement $\sqrt{2\pi}\,\sigma_\zeta \to \zeta$.

Exercise 5.15. Calculate the ponderomotive force correction to the Thomson backscattered radiation in the linear regime of interaction.
Solution:
We use the results (equations (2.205)) given in the Comments and exercises section of chapter 2, for which:

$$\begin{aligned} x &\simeq x_0 \cosh\left(\frac{a_0\zeta}{2\gamma_0 w_0}\right) + \frac{a_0 c}{2\gamma_0\omega_0}\sin\left(\frac{\omega_0\zeta}{c}\right) \\ y &= y_0 \cosh\left(\frac{a_0\zeta}{2\gamma_0 w_0}\right). \end{aligned} \qquad (5.65)$$

The spectral–angular distribution of the radiation is found via equation (2.62):

$$\frac{d^2 E}{d\omega d\Omega} = \frac{\mu_0 e^2\omega^2}{16\pi^3 c}\left|\hat{n}\times\int_0^\zeta d\zeta'\frac{d\vec{r}_e}{d\zeta'}e^{i\left[\Psi_1\zeta'-\Psi_{2x}\sin\left(\frac{\omega_0\zeta'}{c}\right)\right]-i\frac{\omega}{c}\cosh\left(\frac{a_0\zeta'}{2\gamma_0 w_0}\right)x_0\cos\phi\sin\theta + y_0\sin\phi\sin\theta}\right|^2. \quad (5.66)$$

We work out equation (5.66) using the Jacobi–Anger identity:

$$\frac{d^2 E}{d\omega d\Omega} = \frac{\mu_0 e^2\omega^2}{16\pi^3 c}\left|\hat{n}\times\sum_n J_n(\Psi_{2x})\int_0^\zeta d\zeta'\frac{d\vec{r}_e}{d\zeta'}e^{i(\Psi_1 - n\frac{\omega_0}{c})\zeta' - i\frac{\omega r_0}{c}\cosh\left(\frac{a_0\zeta'}{2\gamma_0 w_0}\right)}\right|^2 \qquad (5.67)$$

where $r_0 = x_0\cos\phi\sin\theta + y_0\sin\phi\sin\theta$. We also recognize that $\cosh(ix) = \cos x$, therefore:

$$\frac{d^2 E}{d\omega d\Omega} = \frac{\mu_0 e^2\omega^2}{16\pi^3 c}\left|\hat{n}\times\sum_{n,l} J_n(\Psi_{2x})(-i)^l J_l\left(\frac{\omega r_0}{c}\right)\int_0^\zeta d\zeta'\frac{d\vec{r}_e}{d\zeta'}e^{i(\Psi_1 - h\frac{\omega_0}{c})\zeta'}e^{-\frac{la_0}{2\gamma_0 w_0}\zeta'}\right|^2. \qquad (5.68)$$

5.5.2 Comments on the definition and practical aspects of the brightness of light sources

The units used for the brightness of synchrotron radiation sources are known to be:

$$N_f[s^{-1}]\text{mm} \cdot \text{mrad}^{-2}\, (0.1\%\ \text{bandwidth})^{-1}$$

It has been argued that this choice is a departure from SI units. The reason for this choice is presumably due to the use of micrometers or microradians, which are used to characterize the transverse size and divergence of the emission source.

The definition of brightness traces back to that of luminous flux, identified by the unit *lumen* (symbol lm), which is linked to the unit known as *candela* (cd) by the identity:

$$1\ \text{lm} = 1\ \text{cd} \times 1\ \text{sr}$$
$$1\ \text{cd} = \frac{1}{683}\frac{\text{W}}{\text{sr}}. \tag{5.69}$$

In more quantitative terms, we note that the candela is the unit of luminous intensity in a given direction produced by a source emitting radiation at a frequency of 540×10^{12} Hz. The quantity $1/683$ W sr^{-1} specifies the luminous efficiency at this frequency.

In figure 5.11 we give a practical idea of luminosity using an ordinary lamp and also illustrate the meaning of a unit of solid angle. The *steradian* (sr) is a unit that measures solid angles (Ω) defined by the solid angle projected on the surface of a sphere. For a radius r and an area of $A = r^2$, the solid angle $\Omega = \frac{A}{r^2} = \frac{r^2}{r^2} = 1[\text{sr}]$. Given the definition of the candela, we can easily express it in terms of photon flux, namely:

$$1\ \text{cd} = \frac{h\nu}{683}\frac{\dot{n}}{\text{sr}},\quad \nu = 540 \times 10^{12}\ \text{Hz}$$
$$\frac{\dot{n}}{\text{sr}} = 1.909 \times 10^{21}\ \text{cd}. \tag{5.70}$$

Accordingly, the photon flux per unit solid angle can be expressed in terms of candela.

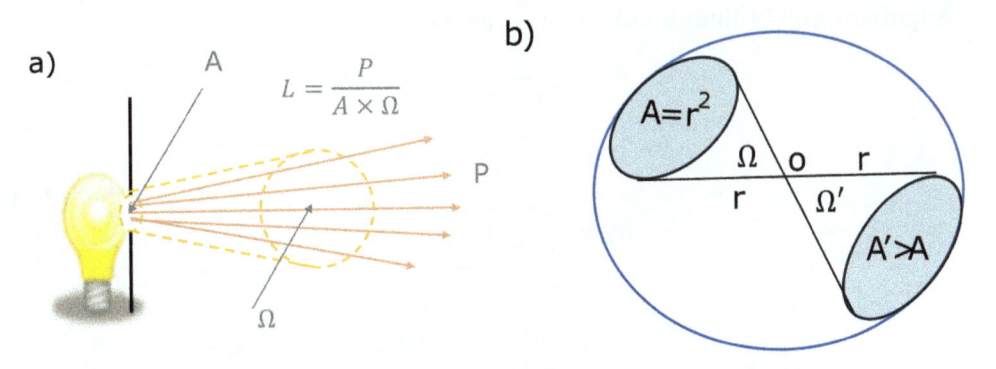

Figure 5.11. Illustrations of: (a) luminosity and (b) solid angle.

We have dwelt on these units, historically adopted in photometry, to give an idea of the conceptual framework in which the definition of certain units is grounded.

In this section, we give a systematic description of the notions we have 'wildly' exploited in this chapter and we also clarify the associated concepts to avoid the confusion that often arises between the definition of brightness and that of brilliance.

Although x-ray source brightness was initially defined for devices (synchrotron radiation, FELs...) based on undulators, we have seen that it can be extended to CBS without any significant change.

Roughly speaking, the quality of an x-ray beam is characterized by three different types of properties:

a) **Single-electron properties**

The photon flux F belongs to this category. It is a quantity that depends on the electron energy and current, photon energy, interaction length, and undulator K or laser a_0 parameters.

b) **Emittance-based properties**

The most representative parameters within this class are the brilliance B and brightness B_r.

The first is defined as the photon flux per unit transverse phase-space area (namely the source emittance). A physical quantity associated with B is the transversally coherent photon flux F_c.

The brightness measures the the flux per unit area (intensity) or flux per solid angle. The evaluation of these quantities requires a knowledge of the electron beam size and the divergence in both transverse directions, i.e x and y.

c) **Properties that include longitudinal phase space**

The most important quantity within this group is the peak brightness $\hat{B}$ discussed below in this section. The introduction of this quantity may cause confusion. It is indeed appropriate in the study of synchrotron radiation sources, where a distinction between average (B) and peak ($\hat{B}$) brightness occurs. Within this framework the definition of B requires the use of the electron beam average current, whereas the CBS case uses the peak current.

We list the various definitions below, even though some of them have already been introduced in the text.

Brightness and brilliance are specified below:

$$\begin{aligned} B_r &= \frac{F}{2\pi \bar{\sigma}_{x'} \bar{\sigma}_{y'}}, \\ B &= \frac{F}{2\pi^2 \bar{\varepsilon}_x \bar{\varepsilon}_y} \\ \bar{\sigma}_{\eta'} &= \sqrt{\sigma_{\eta'}^2 + \sigma_{r',\eta}^2} \\ \bar{\sigma}_{\eta} &= \sqrt{\sigma_{\eta}^2 + \sigma_{r,\eta}^2} \\ \bar{\varepsilon}_{\eta} &= \bar{\sigma}_{\eta'} \bar{\sigma}_{\eta}. \end{aligned} \qquad (5.71)$$

We have loosely mentioned the (transverse and temporal) coherent properties of an x-ray CBS source. In applications, it is important to adequately specify the fractional components of the source.

The conditions determining the spatial coherence can be understood as follows. For an emitting source with transverse lengths $d_{x,y}$, we can determine that the optical path difference between the waves emitted by the two ends is $\delta l = d\,\delta\vartheta$, where $\delta\vartheta$ is the angle formed with the emission axis (see figure 5.12). The use of the 'uncertainty principle' we reported in the description of light propagation allows us to relate $\delta\vartheta$ to the reduced wavelength of the emitted radiation, namely:

$$2\delta\vartheta_\eta \delta\eta \cong \bar{\lambda} \qquad (5.72)$$
$$d = 2\delta\eta.$$

The radiation is transversely coherent if:

$$\delta\vartheta_\eta \cong \frac{\bar{\lambda}}{2\delta\eta}. \qquad (5.73)$$

The portion of transversely coherent radiation is accordingly given by:

$$F_c = \frac{\delta\vartheta_x \delta\vartheta_y}{2\pi\,\bar{\sigma}_{x'}\bar{\sigma}_{y'}} F \cong \left(\frac{\bar{\lambda}}{2}\right)^2 \frac{F}{2\pi\,\bar{\sigma}_{x'}\bar{\sigma}_{y'}\delta x \delta y}. \qquad (5.74)$$

Replacing δx, δy with the rms transverse source dimensions $\bar{\sigma}x$, $\bar{\sigma}y = \frac{\bar{\sigma}_{x,y}}{\sqrt{2\pi}}$, we find:

$$F_c \cong \left(\frac{\lambda}{2}\right)^2 \frac{F}{2\pi^2\,\bar{\sigma}_{x'}\bar{\sigma}_{y'}\bar{\sigma}_x\bar{\sigma}_y} = \left(\frac{\lambda}{2}\right)^2 \frac{F}{2\pi^2\,\bar{\varepsilon}_x\bar{\varepsilon}_y} = \left(\frac{\lambda}{2}\right)^2 B. \qquad (5.75)$$

The ratio of coherent flux to total flux is therefore given by:

$$\frac{F_c}{F} = \left(\frac{\lambda}{2}\right)^2 \frac{1}{2\pi^2\,\bar{\varepsilon}_x\bar{\varepsilon}_y}. \qquad (5.76)$$

If $\lambda/4\pi \cong \bar{\varepsilon}_{x,y}$, then $F_c = F$.

Before concluding, we mention the peak brightness, which is a quantity that brings the longitudinal phase space into the game. In this context, we recall that the longitudinal emittance is defined as:

$$\varepsilon_\varepsilon = \sigma_\varepsilon \sigma_\tau. \qquad (5.77)$$

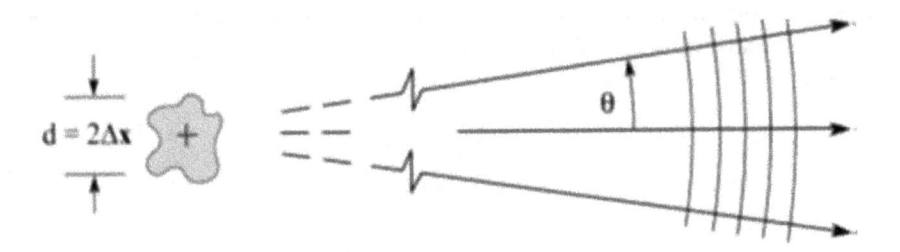

Figure 5.12. An emitting source, showing its transverse dimensions and associated emission angles.

The associated brightness is defined as:

$$\hat{B} = \frac{F_p}{(2\pi)^3 \bar{\varepsilon}_x \bar{\varepsilon}_y \varepsilon_E} = \frac{1}{2\pi \, f \varepsilon_E} B \qquad (5.78)$$

where $F_p = F/f$ gives the photons per pulse and f is the number of electron bunches per second.

Further comments can be found in the literature listed at the end of the chapter.

5.5.3 A phenomenological perspective on CBS brightness

It is sometimes convenient to use approximations which allow us to understand, in simpler terms, the physical content of results obtained within a rigorous context. We discuss below approximate relationships of the CBS spectral lines involving the so-called Gaussian approximation, which played a notable role in the 'design' theory of FEL devices. For this purpose we note that

$$S(\nu) = \left[\frac{\sin\left(\frac{\nu}{2}\right)}{\left(\frac{\nu}{2}\right)} \right]^2 \simeq e^{-\frac{\nu^2}{2\sigma^2}}, \quad \sigma \simeq 2.489$$

$$\nu = 2\pi N_c \left(1 - \frac{\omega}{\omega_s}\right). \qquad (5.79)$$

We use the procedure discussed in this chapter to evaluate the effect of the inhomogeneous broadening and limit ourselves, for the moment, to the broadening induced by the energy spread and to the linear regime ($a_0^2 \ll 1$). We can now write:

$$S(\nu, \mu_{\delta_e}) \cong \frac{1}{\sqrt{2\pi} \, \sigma_\varepsilon} \int_{-\infty}^{+\infty} e^{-\frac{(\nu + 4\pi N_c \delta_e)^2}{2\sigma^2}} e^{-\frac{\delta_e^2}{2\sigma_{\delta_e}^2}} d\varepsilon = \sqrt{2\pi} \, \sigma \, G\nu, \sigma \, \mu_{\delta_e}$$

$$G \, \nu, \sigma(\mu_{\delta_e}) = \frac{e^{-\frac{\nu^2}{2\sigma\mu_{\delta_e}^2}}}{\sqrt{2\pi} \, \sigma\mu_{\delta_e}}, \quad \sigma(\mu_{\delta_e}) = \sigma\sqrt{1 + \mu_{\delta_e}^2} \qquad (5.80)$$

which is expressed in the form of a Gaussian function with an rms that depends on the energy-spread-induced inhomogeneous broadening parameter, namely a line broadening of

$$\left(\frac{\Delta\omega}{\omega}\right)_s = \left(\frac{\Delta\omega}{\omega}\right)_0 \sqrt{1 + \mu_{\delta_e}^2}. \qquad (5.81)$$

The validity of equation (5.80) has been checked by comparing it with the numerical handling of the integral convolutions reported in this chapter. It is important to understand the importance of the inhomogeneous broadening from the perspective of the measured spectrum per unit angle. For this reason, we go back to equations (5.23) and (2.96):

$$\frac{d^2N}{d\omega\, d\theta} \cong N(\theta)\, \text{sinc}^2\!\left[\frac{\nu\theta}{2}\right]$$

$$N(\theta) = \frac{\alpha\omega a_0^2 N_c^2 \theta}{128\pi^3 \omega_0^2 \gamma_0^2}\left(1 + 4\left(\frac{\omega}{4\gamma_0^2\omega_0}\right)\gamma_0\theta^2\left(\left(\frac{\omega}{4\gamma_0^2\omega_0}\right)\gamma_0\theta^2 - 1\right)\right) \quad (5.82)$$

$$\nu\,\theta = 2\pi N_c\left(\left(\frac{\omega}{4\gamma_0^2\omega_0}\right)1 + \gamma_0\theta^2 - 1\right).$$

The inclusion of the inhomogeneous broadening associated with the energy spread yields:

$$\frac{d^2N}{d\omega d\theta} \cong \sqrt{2\pi}\,\sigma N(\theta) G(\nu,\, \sigma\mu_\varepsilon)$$

$$G\nu,\, \sigma\mu_{\delta_e} = \frac{e^{-\frac{\nu(\theta)^2}{2\sigma^2\mu_{\delta_e}}}}{\sqrt{2\pi}\,\sigma\mu_{\delta_e}}. \quad (5.83)$$

The inclusion of the beam angular divergence is accomplished similarly (the assumption of the linear regime excludes sizeable contributions from the spatial coordinates):

$$\frac{d^2N}{d\omega d\theta} \cong \frac{N(\theta)}{\Sigma_{x,y}} \frac{e^{-\frac{\nu(\theta)^2}{2\sigma^2\mu_{\delta_e}}}}{\sqrt{\left(1+\mu_{\delta_e}^2\right)\left(1+\frac{\nu(\theta)\pi\mu_{x'}}{\sigma^2}\right)\left(1+\frac{\nu(\theta)\pi\mu_{y'}}{\sigma^2}\right)}} \quad (5.84)$$

$$\mu_{\eta'} = \frac{4N_l\gamma_0\,\varepsilon_\eta}{\beta_\eta}, \quad \Sigma_{x,y} = 2\pi\sqrt{\left(\sigma_x^2 + \frac{w_0^2}{4}\right)\left(\sigma_y^2 + \frac{w_0^2}{4}\right)}$$

which holds for small values of $\mu_{\eta'}(<0.5)$ and $\eta = x, y$. The previous relation yields a more definite idea of the meaning of inhomogeneous broadening, which reduces the spectral line shape to a Gaussian/Lorentzian distribution, along with the relevant statistical attributes. Before going further, we write equation (5.82) in a more convenient form for our purposes:

$$\frac{d^2\tilde{N}}{dr d\bar{\theta}} \cong \tilde{N}(\bar{\theta})\, \text{sinc}^2\!\left[\frac{\nu\bar{\theta}}{2}\right]$$

$$\tilde{N} = \frac{N}{N^*},\quad N^* = \frac{3\pi}{(2\pi)^4}N_c,\quad N = \frac{2}{3}\alpha\pi\, N_c a_0^2,\quad \bar{\theta} = \gamma_0\theta \quad (5.85)$$

$$r = \frac{\omega}{2\gamma^2\omega_0},\quad \tilde{N}(\bar{\theta}) = r\bar{\theta}\,[1 + 2r\bar{\theta}^2 r\bar{\theta}^2 - 1]$$

$$\nu\,\bar{\theta} = 2\pi N_c r 1 + \bar{\theta}^2 - 1.$$

As already underscored, the inclusion of the broadening effect allows us to write the first equation of (5.85) as follows:

$$\frac{d^2\bar{N}}{dr d\bar{\theta}} \cong \bar{N}\bar{\theta}\,\langle S(\nu(\bar{\theta}))\rangle. \qquad (5.86)$$

In figure 5.13 we have shown the behavior of equation (5.86) versus $\bar{\theta}$ for different values of the e-beam energy spread. The spectral number of photons is obtained by integrating equation (5.86) with respect to $\bar{\theta}$:

$$\frac{d\bar{N}}{dr} \cong \int_0^{\bar{\theta}_{max}} \bar{N}\,\bar{\theta}\,\langle S(\nu(\bar{\theta}))\rangle d\bar{\theta} \qquad (5.87)$$

The above quantity is shown in figure 5.14, where it has been plotted versus r for different values of $\bar{\theta}_{max}$. Each figure contain a comparison between homogeneously and inhomogeneously broadened cases. It is important to emphasize (as underscored in the previous chapter) that with an increase in $\bar{\theta}_{max}$, the effect of beam qualities becomes less significant. Within this framework, the integration of equation (5.86) on the variables r, $\bar{\theta}$ eventually yields the 'number' of photons:

$$\Delta\bar{N} = \int_{r_{min}}^{r_{max}} dr \int_0^{\bar{\theta}_{max}} \frac{d^2\bar{N}}{dr d\bar{\theta}}\,d\theta = \bar{N}(\Delta r, \bar{\theta}_{max}, \text{inh. broadening param}) \qquad (5.88)$$

$$\Delta r = r_{max} - r_{min}.$$

The plot in figure 5.15 shows the weak dependence of $\Delta\bar{N}$ on the inhomogeneous broadening parameters. The previous comments are a rephrasing of the results

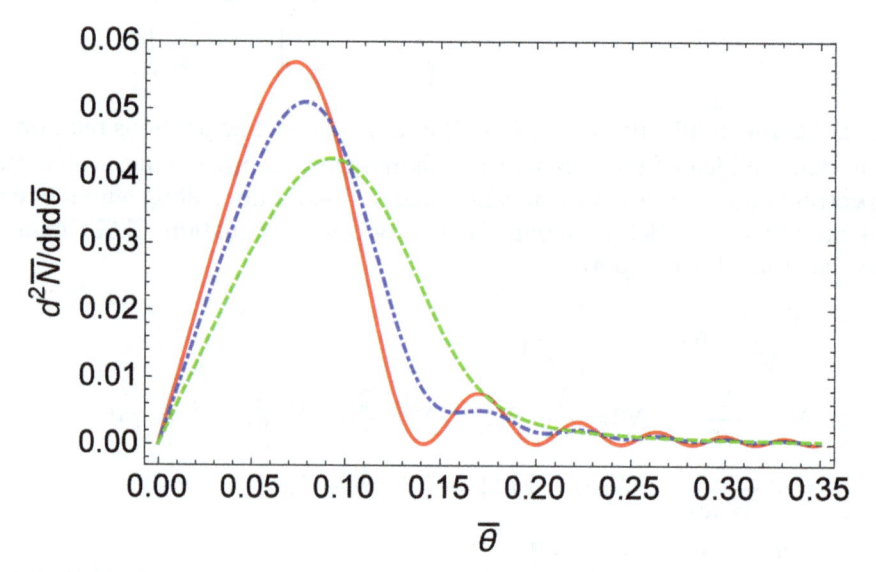

Figure 5.13. Photon flux per unit volume for $r = 1$ and different values of the e-beam energy spread μ_{δ_e} (continuous line $\mu_{\delta_e} = 0$, dot-dashed $\mu_{\delta_e} = 0.5$, dashed $\mu_{\delta_e} = 1$.)

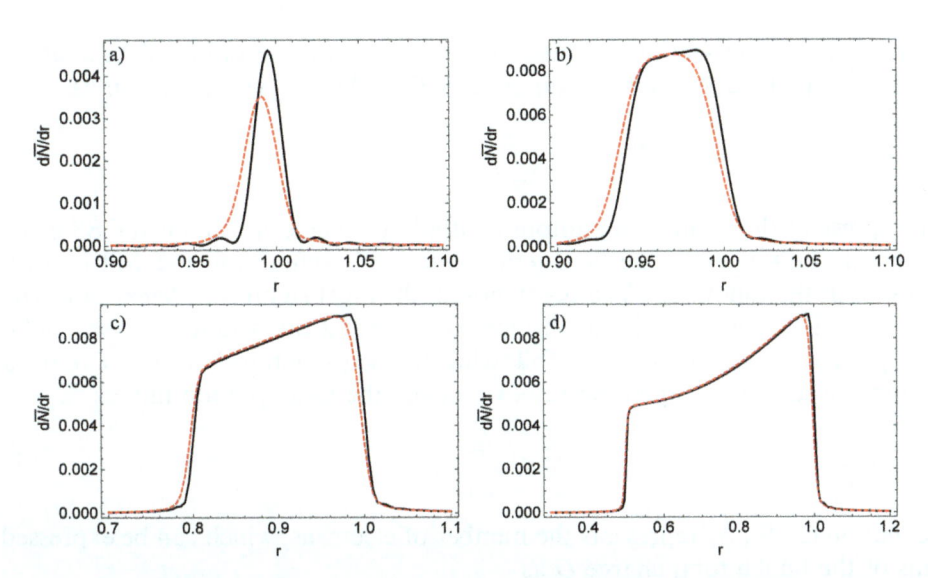

Figure 5.14. A comparison of $d\bar{N}/dr$ versus r with and without spread ($\mu_{\eta'} = 0.5$, $\mu_{\delta e} = 0.5$) for different values of θ_{max}: (a) $\theta_{max} = 0.1$, (b) $\theta_{max} = 0.25$, (c) $\theta_{max} = 0.5$, (d) $\theta_{max} = 1$.

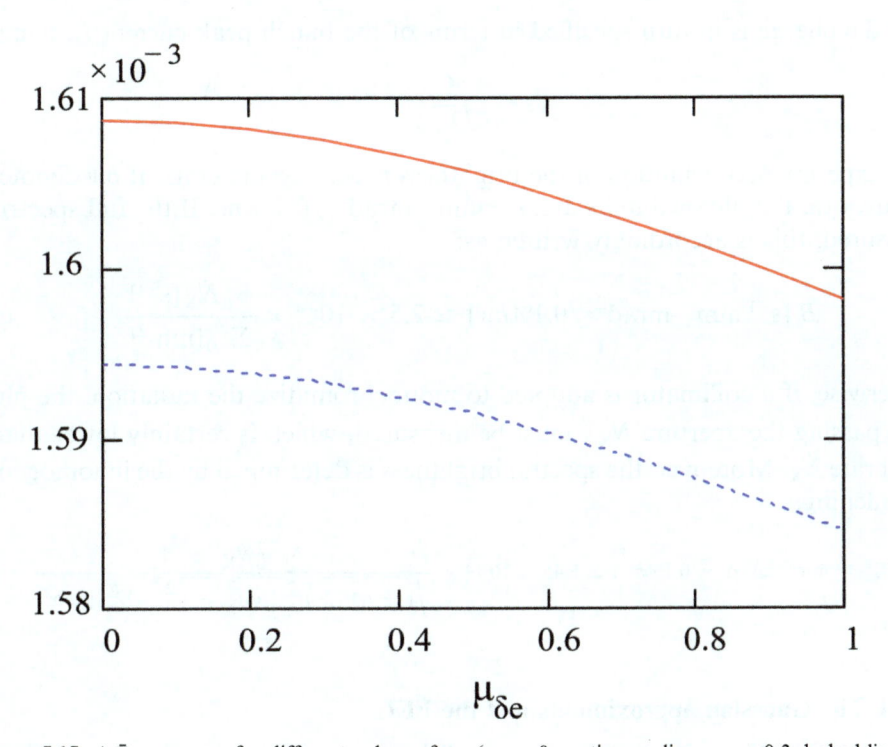

Figure 5.15. $\Delta\bar{N}$ versus μ_{δ_e} for different values of $\mu_{\eta'}$ ($\mu_{\eta'} = 0$ continuous line, $\mu_{\eta'} = 0.3$ dashed line).

obtained in the later sections of this chapter. A significant product of the analysis of chapter 5 was the derivation of equation (5.38), which can be rewritten as:

$$N_x \simeq \frac{3}{4\pi} \frac{\sigma_T}{\Sigma_{e,0}} N_0 N_e \gamma_0 \theta_{max}^2. \tag{5.89}$$

This represents the number of photons emitted by a single electron in a CBS process. As already underscored, the final term $(\gamma_0 \theta_{max})^2$ is significant and will be further discussed in the following. As already noted, the most significant *figure of merit* for any x-ray source is the relevant brightness. A practical definition based on *wise* assumptions can be obtained by following the steps outlined below. Taking into account the scattering repetition rate, we define the x-ray photon flux as

$$\dot{N}_x \cong \frac{f_{rep}}{2\pi} N_0. \tag{5.90}$$

In equation (5.89), N_e represents the number of electrons, which can be expressed in terms of the bunch total charge Q as

$$N_e = \frac{Q}{|e|} \tag{5.91}$$

and the charge is in turn specified in terms of the bunch peak current $\hat{I}_e$, namely

$$I_e = \frac{Q}{\sqrt{2\pi}\,\sigma_\tau}. \tag{5.92}$$

The experimental definition of the brightness needs specific units. If bw denotes the bandwidth, the physical units are: s^{-1} mm · mrad^{-2}/0.1%bw. If the full spectrum is measured, this is accordingly written as:

$$B_s [s^{-1}\,\text{mm}\cdot\text{mrad}^{-2}/0.1\%bw] \simeq 2.5 \times 10^{-4} \times \frac{\gamma_0^2 \dot{N}_X [s^{-1}]}{\Sigma_{e,0}^2 [\text{mm}^2]}. \tag{5.93}$$

Otherwise, if a collimator is adopted to monochromatize the radiation, the photon rate passing the aperture $\dot{N}_X^{(A)}$ must be measured, which is certainly lower than the total rate $\dot{N}_X$. Moreover, the spectral brightness is determined by the inhomogeneous broadening:

$$B_s [s^{-1}\,\text{mm}\cdot\text{mrad}^{-2}/0.1\%bw] \simeq 1.45 \times 10^{-5} \times \frac{N_c \gamma_0^2 \dot{N}_X^{(A)}[s^{-1}]}{\sqrt{1 + \mu_{\delta_e}^2 + \mu_{x'}^2 + \mu_{y'}^2 + \mu_{\theta_{max}}^2}\,\Sigma_{e,0}^2 [\text{mm}^2]}. \tag{5.94}$$

5.5.4 The Gaussian approximants and the FEL

The use of the Gaussian approximation of the spectral line might be viewed as a low-level trick, but we should like to underline that this is not the case. It has played an important role in the theory of FELs (see the bibliography at the end of the chapter).

The following exercises should give the reader further confidence in the use of the method and offer an idea of how it should be exploited in a different context.

Exercise 5.16. Use the Gaussian method to approximate the first-order FEL gain.
Solution:
It is well known that ***Madey's theorem*** (see the bibliography) states that the FEL gain is linked to the derivative of the spontaneous emission line shape, namely

$$G_1(\nu) \propto -\pi \frac{\partial}{\partial \nu}\left[\operatorname{sinc}\left(\frac{\nu}{2}\right)^2\right]. \quad (5.95)$$

It is evident that the use of the approximation leads to the gain function

$$G_1(\nu) = -\pi \frac{\nu}{\sigma^2} e^{-\frac{\nu^2}{2\sigma^2}}. \quad (5.96)$$

The comparison is shown in figure 5.16, and the conclusion is that it appears only reasonable. The reason for this is that the maximum gain is shifted towards larger ν, where the Gaussian sinc-square approximation becomes less accurate

Notwithstanding the above, the use of approximations based on Gaussians is particularly useful to draw analytical results that can be used to benchmark more accurate numerical computations.

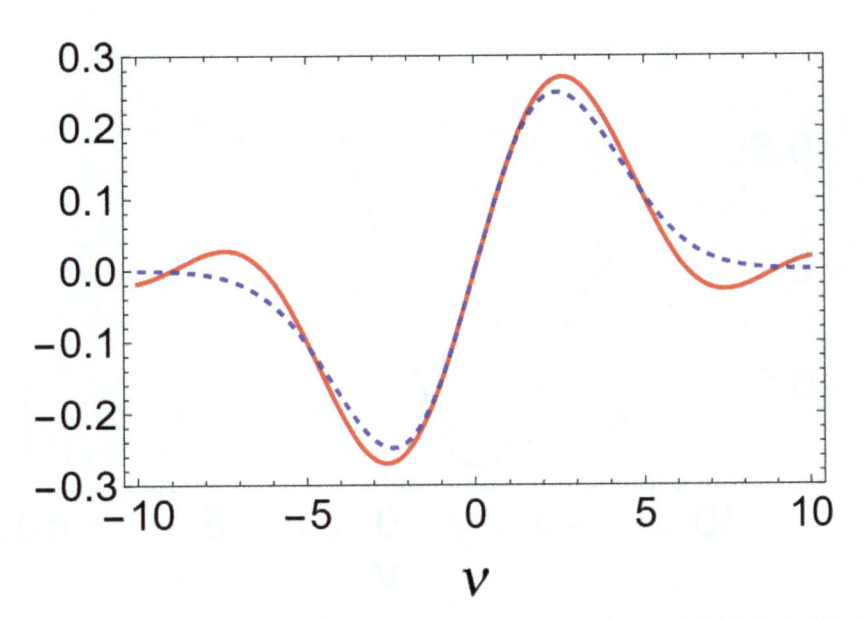

Figure 5.16. A comparison between equations (5.95) (continuous line) and (5.96) (dashed line).

Exercise 5.17. We invite the reader to prove an alternative approximation of equation (5.95) that represents a better agreement with the exact form.
Hint: Try an approximation of the type

$$G_1(\nu) \cong \frac{\pi}{a_1} \exp\left(-\frac{\nu^2}{b_1}\right) \sin\left(\frac{\nu}{c_1}\right) \cos\left(\frac{\nu}{d_1}\right) \cos\left(\frac{\nu}{e_1}\right) \quad (5.97)$$
$$a_1 = 3.0 \quad b_1 = 209.11 \quad c_1 = 2.0$$
$$d_1 = 5.7212 \quad e_1 = 9.8361$$

and show that the maximum deviation is less than 2% for $|\nu| < 10$ (see figure 5.17)

Exercise 5.18. Explain why it is convenient to approximate equation (5.95) with (5.97).
Hint:
Show that the convolution

$$\widetilde{G}_1(\nu, \mu_\varepsilon) = \int_{-\infty}^{+\infty} f(\varepsilon) G_1(\nu + 4\pi N \varepsilon) d\varepsilon \quad (5.98)$$

can be obtained using (5.97) but not with (5.95). It is indeed easily confirmed that the integral reduces to a form of the type

$$I = \int_{-\infty}^{+\infty} e^{-A\varepsilon^2 + B\varepsilon} d\varepsilon. \quad (5.99)$$

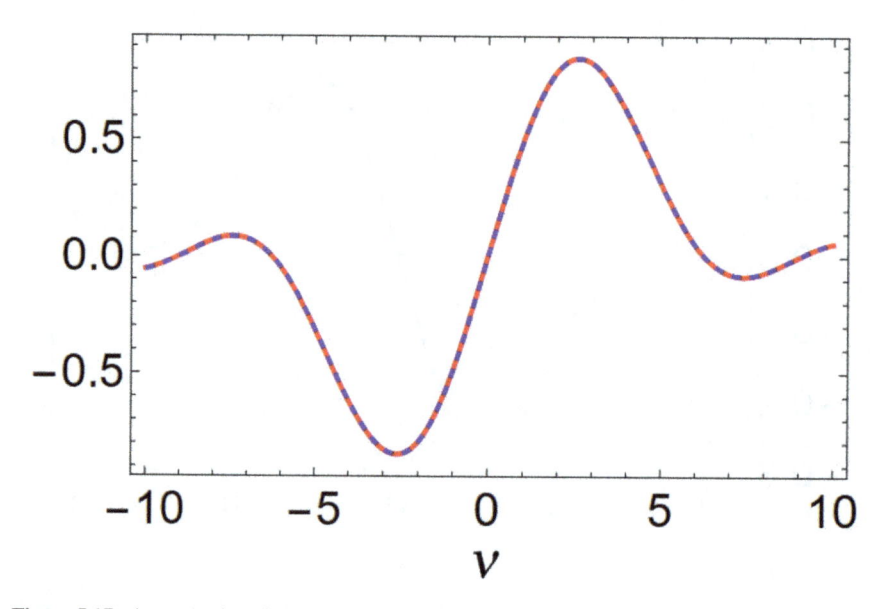

Figure 5.17. A comparison between equations (5.95) (continuous line) and (5.97) (dotted line).

Exercise 5.19. Show that the same expression (with different values of the *a, b, c, d,* and *e* coefficients) allows us to approximate the functions

$$G_2(\nu) = \frac{\pi^2}{3\nu^6}[84(1 - \cos(\nu)) - 60\nu \sin(\nu) + 3\nu^2 + 15\nu^2 \cos(\nu) + \nu^3 \sin(\nu)]$$

$$G_3(\nu) = \frac{\pi^3}{60\nu^9}[11520(1 - \cos(\nu)) - 9000\nu \sin(\nu) + 360\nu^2 + 2880\nu^2 \cos(\nu) \\ + 480\nu^3 \sin(\nu) - 20\nu^4(1 + 2\cos(\nu)) - \nu^5 \sin(\nu)]$$

(5.100)

and get the specific values of the fitting coefficients.

Hint:
Use a minimization procedure and show that the parameters

$$a_2 = 40.0 \quad b_2 = 58.01 \quad c_2 = 5.2427 \quad d_2 = 78.9575 \quad e_2 = 142.589$$
$$a_3 = 11\,461 \quad b_3 = 128.60 \quad c_3 = 2.9317 \quad d_3 = 9.0363 \quad e_3 = 13.369$$

guarantee the same approximation level for the case $G_1(\nu)$.

The reason that we have mentioned the functions $G_{2,3}(\nu)$ is because they are higher-order corrections to the FEL gain and measure the deviation from the low-gain regime.

We have mentioned this type of approximation for two reasons: because obtaining it is not straightforward and because the FEL is a stimulated CBS device.

Part of the second section of this chapter was taken from Dattoli G, Palma E D and Petrillo V 2023 A collection of formulae for the design of Compton backscattering x-ray sources *Appl. Sci.* **13** 2645, CC BY 4.0.

References and further reading

For the definition and use of brightness, see the following:

Zhu H and Blackborow P 2018 Understanding radiance (brightness), irradiance and radiant flux *Tech. Rep.* available on line at: https://www.energetiq.com/technote-understanding-radiance-brightness-irradiance-radiant-flux. Energetiq Technology

Mills D M *et al* 2005 Report of the Working Group on Synchrotron Radiation Nomenclature—brightness, spectral brightness or brilliance? *J. Synchrotron Radiat.* **12** 385

Shen Q 2001 *X-ray Flux, Brilliance and Coherence of the Proposed Cornellenergy-recovery Synchrotron Source CHESS 01-002* Cornell University https://www.classe.cornell.edu/rsrc/Home/Research/ERL/ErlPubs2001/ERLPub01_4.pdf

Krafft G A and Priebe G 2010 Compton sources of electromagnetic radiation *Rev. Accel. Sci. Technol.* **3** 147–63

Brown W J and Hartemann F V 2004 Three-dimensional time and frequency-domain theory of femtosecond x-ray pulse generation through Thomson scattering *Phys. Rev. ST-Accel. Beams* **7** 060703

Dattoli G, Palma E D and Petrillo V 2023 A collection of formulae for the design of Compton back-scattering x-ray sources *Appl. Sci.* **13** 2645

For the theory of inhomogeneous broadening for magnetic undulator radiation and free electron lasers, see the following:

Dattoli G *et al* 1984 Lawson–Penner limit and single passage free electron lasers performances *IEEE J. Quant. Electron.* **20** 637–46

Kim K J 1986 A new formulation of synchrotron radiation optics using the Wigner distribution *Insertion Devices for Synchrotron Sources* vol 0582 ed Lindau I E and Tatchyn R O (Bellingham, WA: SPIE) pp 2–9 doi: https://dx.doi.org/10.1117/12.950906

Dattoli G and Renieri A 1985 Experimental and theoretical aspects of the free-electron laser *Laser Handbook* (Amsterdam: Elsevier) pp 1–141 doi: https://dx.doi.org/10.1016/B978-0-444-86927-2.50005-X

Colson W B, Gallardo J C and Bosco P M 1986 Free-electron-laser gain degradation and electron-beam quality *Phys. Rev. A* **34** 4875–81

Ciocci F, Dattoli G and Torre A 1994 The effect of emittance inhomogeneous broadening on the brightness of a magnetic undulator *IEEE J. Quant. Electron.* **30** 793–9

Xie M 1995 Design optimization for an x-ray free electron laser driven by SLAC linac *Proc. Particle Accelerator Conf.* (Piscataway, NJ: IEEE) pp 183–5 doi: https://dx.doi.org/10.1109/PAC.1995.504603

Dattoli G and Ottaviani P L and Pagnutti S 2007 *Booklet of FEL Design* URL: http://fel.enea.it/booklet/pdf/Booklet_for_FEL_design.pdf

Ghaith A *et al* 2021 Undulator design for a laser-plasma-based free-electron-laser *Phys. Rep.* **937** 1–73 https://www.sciencedirect.com/science/article/pii/S0370157321003434

For Inhomogeneous Broadening in Compton Sources, see the following:

Deitrick K *et al* 2013 *The ODU CAS Inverse Compton Source Design* http://toddsatogata.net/Papers/2013-09-03-ComptonSource-2up.pdf

Hartemann F V *et al* 2005 High-energy scaling of Compton scattering light sources *ST Accel. Beams* **8** 100702

Petrillo V *et al* 2012 Photon flux and spectrum of γ-rays Compton sources *Nucl. Instrum. Methods Phys. Res. A* **693** 109–16

Hajima R 2016 Status and perspectives of Compton sources *Phys. Procedia* **84** 35–9 https://www.sciencedirect.com/science/article/pii/S1875389216303029

Deitrick K E 2017 Inverse Compton light source: a compact design proposal. *PhD Dissertation* Old Dominion University Available on line at: https://digitalcommons.odu.edu/physics_etds/7

Krämer J *et al* 2018 Making spectral shape measurements in inverse Compton scattering a tool for advanced diagnostic applications *Sci. Rep.* **8** 1398

Hornberger B *et al* 2019 A compact light source providing high-flux, quasi-monochromatic, tunable x-rays in the laboratory *Proc. SPIE* **11110** 1111003

Petrillo V *et al* 2022 Synchronised teraHertz radiation and soft x-rays produced in a FEL oscillator *Appl. Sci.* **12** 8341

Chapter 6

CBS sources

6.1 Introduction

In the previous chapters we have given a broad view of the physics underlying the Compton backscattering (CBS) processes and the ancillary systems and beam and radiation transport devices necessary for the relevant design and construction. Figure 6.1 classifies the CBS sources spread around the world, including the sources under design, those in the commissioning process, and those already/still operational. Figure 6.2 shows the layouts of existing CBS devices, displaying different design

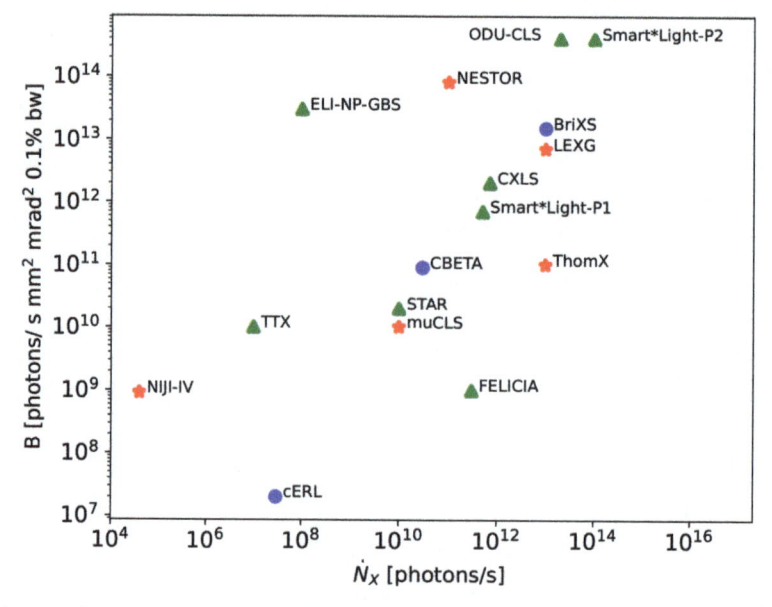

Figure 6.1. An overview of CBS sources including sources under design, those in the commissioning process, and those already/still operational. The sources are arranged by emission rate and brightness. Key: **MuCLS** – Munich Compact Light Source, TTX – Tsinghua Thomson-Scattering X-ray Source. Red: based on electron recovery Linacs; green: based on Linacs; blue: based on storage rings.

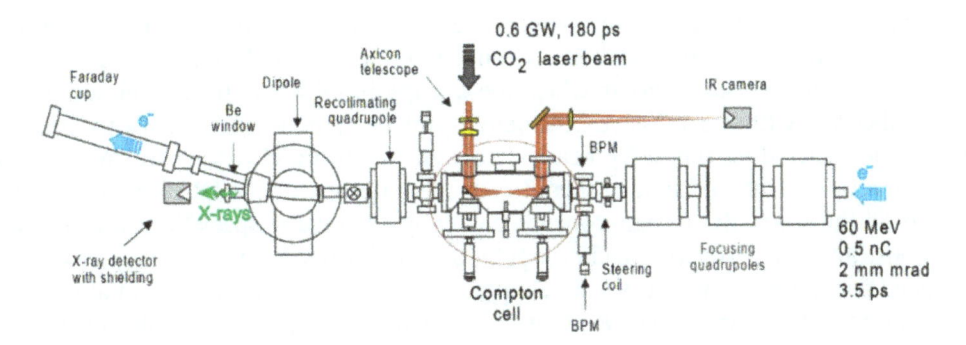

(a) Brookhaven National Laboratory high flux Thomson scattering experiment (2000).

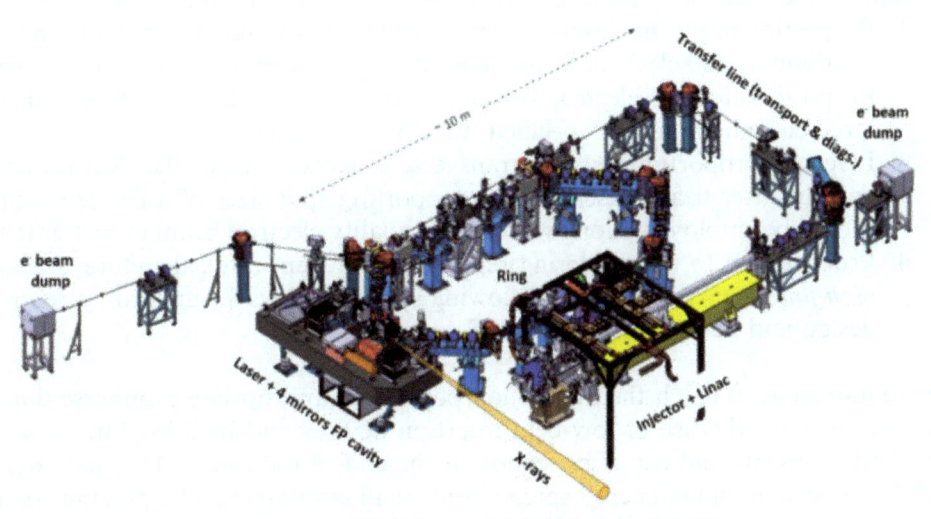

(b) Lyncean CBS Compact Light Source (2019).

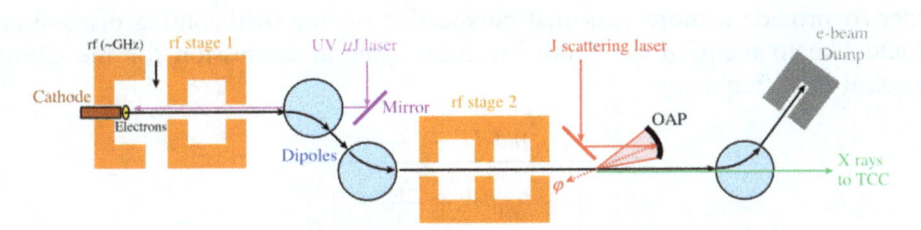

(c) Schematic of an e-beam Compton scattering x-ray source for OMEGA (2023).

Figure 6.2. Different concepts for CBS sources (for the specific details, see the references at the end of the chapter). (a) Reprinted from Pogorelsky *et al* 2000 Demonstration of 8×10^{18} photons/second peaked at 1.8 Å in a relativistic Thomson scattering experiment *Phys. Rev. ST Accel. Beams* **3** 090702, CC BY 3.0. (b) Reprinted from Dupraz *et al* 2020 The ThomX ICS source *Phys. Open* **5** 100051, CC BY_NC_ND. (c) Adapted figure with permission from American Physical Society, CC BY 4.0. Reprinted from Rindernecht et al 2024 Electron-beam-based Compton scattering x-ray source for probing high-energy-density physics *Phys. Rev. Accel. Beams* **27** 034701.

concepts for the generation of quasi coherent x-ray radiation using an e-beam provided by an accelerator and an external laser. Figure 6.2(a) shows the concept used almost three decades ago, in which the Compton cell where the interaction occurs was embedded within a massive outer environment. This was necessary to host the diagnostics (including the x-ray beam detection), to bring electrons and photons to the interaction point, and to take them to the beam dump and the IR camera. Over the course of the years the design concept has evolved toward specialized and modest solutions, and the associated technology is indeed evolving toward compact devices designed for specific applications. In chapter 1 we developed a few criteria aimed at a quick design of a CBS device. Here, we discuss the same problem while considering what we outlined in the following chapters. Going back to the 'fundamental' recommendations at the beginning of this book, we recall that the x-ray flux is:

1. Proportional to the laser energy and hence to the number of photons overlapping the electron beam; accordingly, powerful lasers are necessary.
2. Proportional to the electron bunch charge; therefore, beams with sufficiently large current should be specified for CBS operations.
3. Inversely proportional to the transverse cross sections of the electron beam and the laser beam; laser cavities supporting spot sizes of a few tens of μm should be employed, along with a high-quality electron beam to scatter from.
4. Proportional to the scattering repetition rate (rep. rate); therefore, stacking **high-finesse** cavities (see the following sections) plays a significant role in the design and development of a CBS source.

The requirement to reach the interaction point with appropriate transverse dimensions necessitates the use of low-beta insertion devices and high-brightness beams (see the Comments and exercises section at the end of the chapter), namely beams with large current, small energy spread, and small emittances. The previous points fix the technological steps that characterize the design and construction of high-brightness x-ray sources using modestly sized and relatively low-cost devices. In order to provide a more educated perspective on the forthcoming discussion, we should like to mention the following more general expression for the Compton backscattered frequency:

$$\omega_s \simeq \frac{2\gamma_0^2 n \omega_0 (1 - \beta_0 \cos \phi_1)}{1 + (\gamma_0 \theta)^2 + \frac{a_0^2}{2} + 2\frac{\chi n}{a_0}} \tag{6.1}$$

where ϕ is the angle shown in figure 6.3 (whose role will be discussed in the forthcoming sections that discuss other Thomson scattering geometry), the parameter χ is the ratio between the laser's electric field strength (in the electron rest frame) and the strength associated with the Sauter–Schwinger field, namely:

$$\chi = \gamma_0 \frac{a_0}{a_{ss}} \tag{6.2}$$

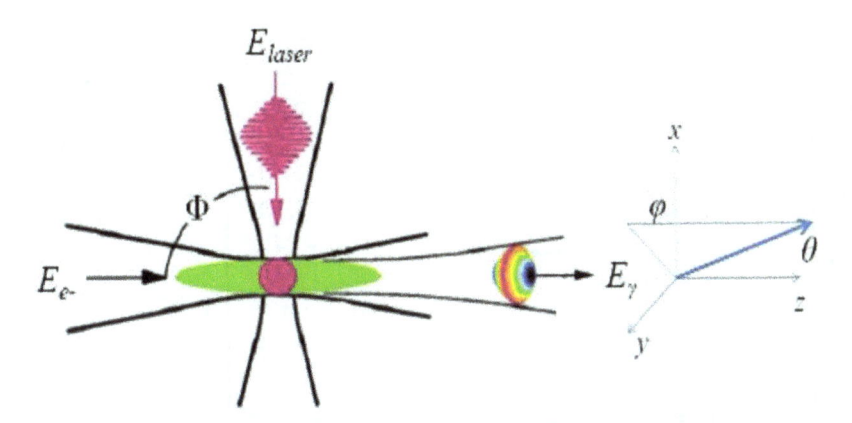

Figure 6.3. A CBS configuration that does not require a head-on geometry.

and n represents the number of photons scattered per event by a single electron (this characterizes the multiphoton CBS discussed in the second volume). In the discussion developed so far, we have limited ourselves to single-photon processes: keeping $\phi = 0$, considering that $n = 1$, and noting that $\frac{\chi}{a_0} = \gamma \frac{\lambda_e}{\lambda_l} \ll 1$, equation (6.1) reduces to the cases discussed so far. Here we consider the weak-field case and ignore the nonlinear corrections that are unessential within the context of the present worldwide effort aimed at providing CBS x-ray sources. We have written and commented on the CBS backscattering frequency in the form of equation (6.1) to emphasize the wealth of possibilities it contains, which have been explored so far and will be further developed in the second volume.

6.2 Further comments on the bandwidth of CBS sources

The points we have raised in the previous sections have largely been discussed in many research papers describing the design analysis of CBS sources and the development of numerical software devoted to benchmarking the experimental data. Taking into account that the design concept of a CBS source depends on the expected application of the delivered photon beam, we underscore that the feasibility of CBS-based experiments and their reliability in terms of statistical precision are strongly dependent on the photon flux and bandwidth of the beam driving the process. We have argued that the effect of bandwidth broadening in Compton radiation sources can be controlled by a collimator to restrict the scattering angle of photons transported to the experimental station; the collimator thus acts as a monochromator. We have also mentioned that the achievable bandwidth is limited by the inhomogeneous properties of the electron and laser beams used to produce the Compton scattering; therefore, the collimation angle should be chosen wisely. Before addressing this specific aspect of the problem, we comment on the plot shown in figure 6.4 showing various radiation spectra versus the photon energy for the NewSUBARU CBS source (see the bibliography at the end of the chapter) calculated for an emittance of 3.8 nm. In figure 6.5 shows the same quantities

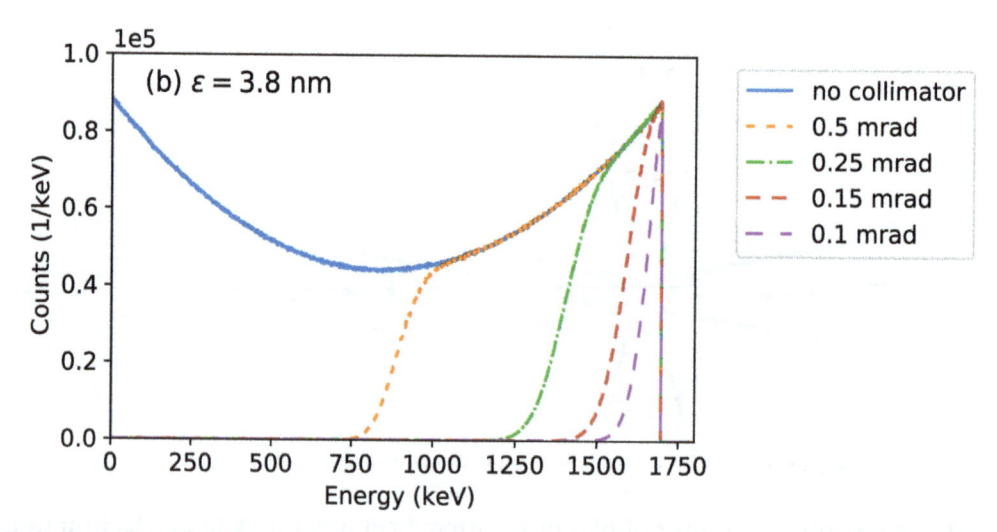

Figure 6.4. The CBS spectra of the NewSUBARU project. The lines represent counts obtained without a collimator and using collimators with opening angles of 0.5 mrad, 0.25 mrad, 0.15 mrad, and 0.1 mrad.

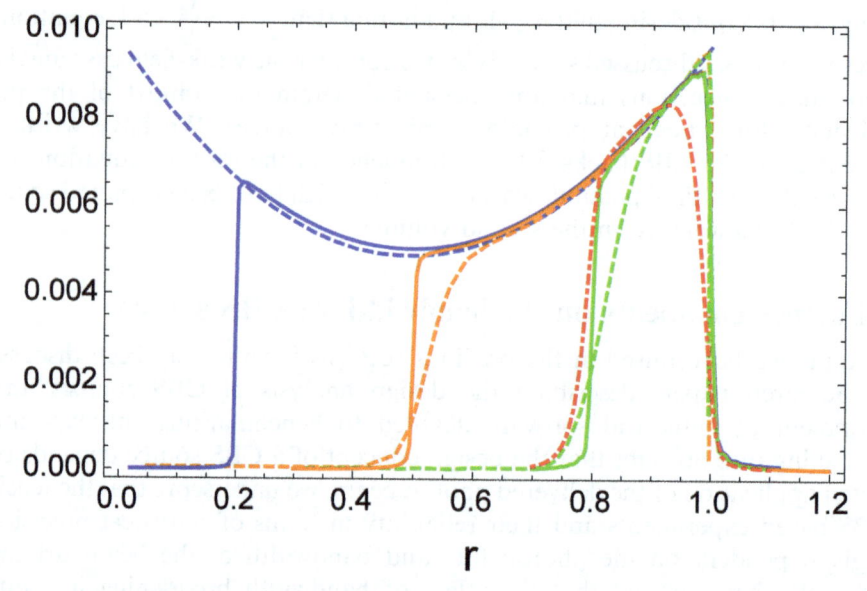

Figure 6.5. Same as figure 6.4, evaluated with the spectral line integral representation reported in chapter 5 (see equation (5.25)). The figure includes also the effect of the inhomogeneous broading (dot-dashed line) compared to the homogenous case at $\theta = 0.5$ mrad.

evaluated using the integral representations reported in chapter 5. A comparison of figures 6.4 and 6.5 leads to the conclusion that the plots, even though calculated using different procedures, are qualitatively similar. The differences, presumably due to the fact that the results in figure 6.4 are obtained for a strongly symmetric electron beam, are not further discussed here. Furthermore, we underline that the photon energy spectrum in figure 6.4 is indeed evaluated using a full quantum calculation, which, for the range of parameters discussed so far, does not create a significant deviation from the classical treatment. The theoretical analyses underlying the design and the analysis of the different worldwide sources are essentially equivalent, and selected examples are reported and discussed in the Comments and exercises section. In the next section we provide a paradigmatic example of design procedure.

In order to illustrate the steps to be followed in the design of a CBS source, we adopt the point of view developed in Rinderknecht et al (2024) (for further details, see the bibliography at the end of the chapter), which, in our opinion, represents a highly professional way of addressing this type of problem. Accordingly, we express the field amplitude in terms of the practical units reported below:

$$a_0 \simeq 0.857 \lambda_0 [\mu m] \sqrt{I_{18}} \tag{6.3}$$

where λ_0 and I_{18} represent the laser wavelength in μm and its power density (in units of $I_{18} = 10^{18}$ W cm^{-2}), respectively. We underscore again that the scattered photons are monoenergetic at any collection angle, but their collection at different angles plays a part in reducing the source monochromaticity. The bandwidth due to the collecting angles scales according to (see Comments and exercises section):

$$\left(\frac{\Delta \omega}{\omega}\right)_\theta \simeq 0.29 \gamma_0 \theta^2 - 0.13 \gamma_0 \theta^4 \tag{6.4}$$
$$\gamma_0 \theta < 0.7.$$

Inverting the previous relation, we obtain the collector angle in terms of the bandwidth:

$$\theta(\Delta) = \frac{1}{\gamma_0} \sqrt{\frac{1 - \sqrt{1 - 6.183\Delta}}{0.897}} \tag{6.5}$$

which helps us to fix an acceptable bandwidth Δ that is compatible with the required degree of monochromaticity, thus defining the corresponding angular aperture of the monochromator. Another important quantity to be underscored is the fractional number of photons collected within a solid angle. We note that this scales with the collection angle as follows (see the Comments and exercises section):

$$f_C \simeq 1.49 (\gamma_0 \theta)^2 - 2.18 (\gamma_0 \theta)^4 \tag{6.6}$$
$$\gamma_0 \theta < 0.4.$$

Figure 6.6 is a plot of f_C versus Δ. It is worth stressing that a bandwidth of around 2% (corresponding to an angular aperture of 2.67 mrad for $\gamma_0 \simeq 100$) allows the collection of just 10% of the scattered photons. The line broadening due to the photon collimator adds (quadratically) to that induced by the electron and photon beam qualities. The problem has been widely debated in the current literature and, for example, various substantially equivalent expressions (see the Comments and exercises section) have been discussed in the past.

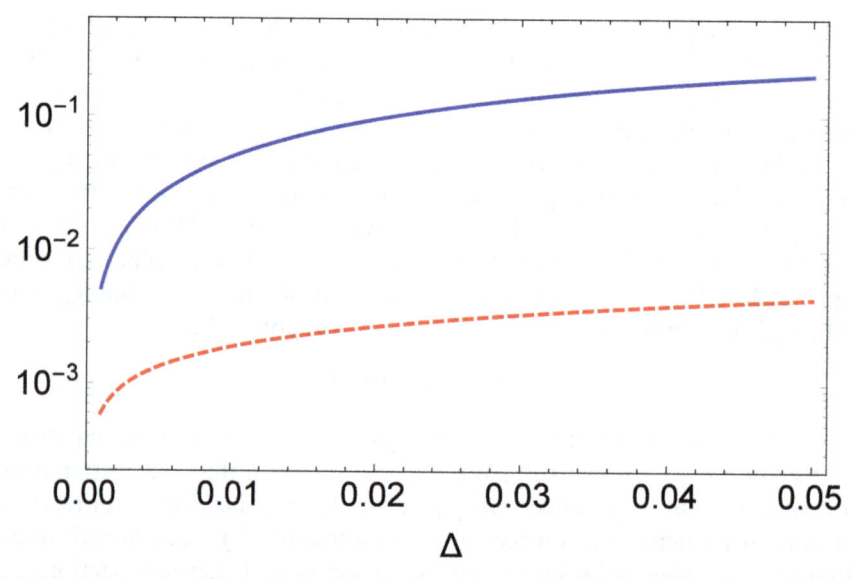

Figure 6.6. The fractional numbers of collected photons (f_C) (continuous line) and the aperture angle (6.5) for $\gamma_0 = 100$ (dashed line). f_c is a pure number and the aperture is measured in rad.

Table 6.1. Laser beam parameters of selected operational CBS sources.

Source Name	Laser Power	Rep. rate (f_{rep})	Focal spot (σ_0)
MuCLS	30 W (average)	64.91 MHz	$\simeq$50 μm
TTX	16 TW (peak)	10 Hz	$\simeq$30 μm
NIJI-IV	2 W (average)	10.1 MHz	$\simeq$500 μm
ThomX	100 W (average)	33.3 MHz	$\simeq$40 μm

6.3 Laser systems in operational CBS sources

The precise definition of the characteristics of the incident photon pulse in CBS sources is a key element for the design of such radiation machines. In this section, we analyze and discuss the parameters of real laser systems used to provide the incident photons in CBS sources that are currently operational worldwide. Specifically, we focus on four sources: the Munich Compact Light Source (**MuCLS**), the Tsinghua Thomson-Scattering x-ray Source (**TTX**), the source based on the free-electron laser (FEL) **NIJI-IV**, and the **ThomX** Inverse Compton Scattering (ICS) source (currently in commissioning, 2023). The laser parameters characterizing the abovementioned sources are reported in table 6.1. The laser system of the **MuCLS** source is based on neodymium-doped yttrium aluminum garnet (Nd:YAG) technology operating at a wavelength of $\lambda_0 = 1064$ nm. Such a system delivers up to 30 W of average power, corresponding to a train of laser pulses carrying $\simeq$0.5 μJ of energy each, delivered at a rep. rate $f_{rep} = 64.91$ MHz. The single-pulse duration is $L/c = 26$ ps, corresponding to a peak power $\lesssim 18$ kW. The relativistic parameter corresponding to 18 kW is about $a_0 \simeq 10^{-5}$, revealing the linearity of the Thomson backscattering source. The **MuCLS** source reaches high photon emission rates $\dot{N}_X \gtrsim 10^{10}$ photons s^{-1} due to its high rep. rate and to the fact the the laser energy is stored within an optical cavity that increases the average laser power to 350 kW.

The laser system of the **TTX** source is based on Ti:Sa technology operating at the wavelength $\lambda_0 = 800$ nm. This system delivers up to 16 TW = 16×10^{12} W of peak power, i.e. single pulses carrying 500 mJ in a period of 30 fs. This system is very different compared to the system used in the **MuCLS** source. Indeed, the latter employs a high peak power laser, while the second source takes advantage of a high average power laser system achieved via a high rep. rate. The rep. rate of the **TTX** source is relatively low ($f_{\rm rep} \leqslant 10$ Hz); however, high average fluxes can be reached since the number of photons per single pulse is rather high (up to 10^7). The relativistic parameter at the full laser intensity is fairly large, $a_0 \lesssim 0.5$, creating the possibility of delivering sizeable higher-order harmonics.

The laser system of the **NIJI-IV** is very different from the previous two, since it is based on an FEL. The FEL is used as an insertion device of a compact storage ring with a revolution frequency of 10.1 MHz. Moreover, the undulator is embedded within an optical cavity with an intracavity power in the 2 W range in the infrared (IR). The beam size at the interaction point is 0.8 mm in x and 0.3 mm in y, giving a geometric average of about 0.5 mm. The electron beam energy is rather high compared to the abovementioned sources ($\gtrsim 300$ MeV compared to <100 MeV), therefore more γ-rays are produced than hard x-rays. The interaction regime is linear.

Finally, the laser system of the **ThomX** source is, as for the **MuCLS** source, based on Nd:YAG technology operating at the wavelength $\lambda_0 = 1064$ nm. Here, the optical cavity is designed to reach impressive levels of stored average power as high as 1 MW. This power level, together with the rather high rep. rate of $f_{\rm rep} = 33.3$ MHz, can ensure high photon emission rates $\dot{N}_X > 10^{12}$ photons s^{-1} while keeping the interaction in the linear regime (low peak laser power).

6.4 Electron accelerators in operational CBS sources

To parallel the previous section, we briefly analyze the characteristics of operational CBS sources in terms of their electron accelerators. In fact, the proper definition of the electron beam parameters is as important as the definition of the laser beam parameters. The two must be matched to optimize the yield of the source and/or to tune it to the working range of interest. In table 6.2 we list the parameters of the accelerator machines exploited for the operational CBS sources considered as practical examples for this chapter.

As a general consideration, we should mention that the large range of possible electron energies allows CBS sources to be tuned, paving the way for more applications and users. Indeed, as we know, the scattered photon energy scales with the square of the electron energy. Furthermore, some machines are characterized by their average current and some by the charge per electron bunch. This mirrors the fact

Table 6.2. Electron beam parameters in selected operational CBS sources.

Source Name	Electron Energy	Charge/Current	Pulse Duration	Focal spot (σ_e)
MuCLS	29–45 MeV	16 mA	50 ps	$\simeq$50 μm
TTX	50–350 MeV	>200 pC	<2 ps	$\simeq$30 μm
NIJI-IV	$\gtrsim$300 MeV	$\gtrsim$5 mA	$\lesssim$10 ps	$\simeq$500 μm
ThomX	50–70 MeV	50–1000 pC	10–20 ps	45–100 μm

that when electron bunches are injected into a storage ring, they fill different RF buckets, producing a temporal pattern with the same frequency as the rep. rate of the CBS source and corresponding to an average current determining the average flux of the x-ray photon source. The **MuCLS** source also consists of an electron ring; however, it is described using the charge per bunch to stress the fact that in addition to a relatively large average power (proportional to the number of scattered photons per second), the source can also provide considerable peak power. Finally, the **TTX** includes a high-frequency (X-band), i.e. compact, linear accelerator (linac), that provides single bunches of the charge reported in table 6.2 and therefore allows the realization of single scattering events at the rate reported in table 6.1.

6.5 A comparison between theory and experiment

In this section we would like to make use of the results obtained in the previous chapters to explain the experimental observations made at operational CBS sources. For example, in the case of the **MuCLS** source, the stated brightness that has been measured at the facility is $B_s = 1.2 \times 10^{10}$ photons s^{-1} (0.1% bandwidth)$^{-1}$ mm^2 mrad^2. Revisiting equations (5.6) and (5.7) and considering that $\gamma \simeq 90$ (45 MeV of maximal electron energy is available at the facility), the declared x-ray beam size $\simeq$50 μm, and the declared emission rate of 4.5×10^{10} photons s^{-1}, we obtain $B_s \simeq 1.5 \times 10^8$ photons s^{-1} (0.1% bandwidth)$^{-1}$ mm^2 mrad^2. It is evident that the two values are not in agreement, but there is an explanation for this. First of all, the declared x-ray divergence of the source is 4 mrad, while equation (2.108) would give 20 mrad. Thus, it is clear that the declared rate for $\dot{N}_X$ does not refer to the full rate of the source, but, as it should for a correct evaluation of the brightness, to the number of photons emitted per second in the observation cone fixed by the aperture of $\theta_{max} \simeq 4$ mrad. Therefore, revisiting equation (5.6) and considering an x-ray divergence of 4 mrad, we obtain $B = 1.1 \times 10^{12}$ photons s^{-1} mm^2 mrad^2. Let us now take a closer look the bandwidth. The declared bandwidth is 5%. Thus, we have to consider equation (5.21) with the addition of the term in equation (5.30) and, keeping in mind that for **MuCLS** $N_c \simeq 7300$, we obtain:

$$\sqrt{1 + \mu_\theta^2} \simeq 3.7\% \qquad (6.7)$$

The above value is rather close to 5%, showing that the dominant term for the broadening is derived from the spectral–angular correlation of the CBS source. However, the rest of the bandwidth is determined by the other terms reported in equation (5.21). The rep. rate of the machine is $f_{\text{rep}} = 65$ MHz, corresponding to an RF bucket for the electron beam in the storage ring of the same size. The declared electron bunch length is $L_e/c = 50$ ps. Multiplying the bunch length by the rep. rate, we guess at a relative energy spread $\delta_e \simeq 0.003$. We can, accordingly, verify that:

$$\sqrt{1 + \mu_\theta^2 + \mu_{\delta_e}^2} \simeq 3.8\% \tag{6.8}$$

i.e. the contribution due to the electron energy spread is negligible. For the reasons explained in chapter 5, the diffractive effects due to $\mu_{x'}$ and $\mu_{y'}$ are in many cases, as in this one (where no tight focusing is exploited), negligible. Therefore, the discrepancy between the declared value of 5% and the estimated value of 3.8% may be due to a slightly different definition of $\theta_{\max}$. A small error in the measurement of $\theta_{\max}$ would also make our estimation agree with the declared value. In fact, taking $\theta_{\max} \simeq 4.6$ mrad yields the desired value of 5%. The spectral brightness estimated by equation (5.7) is now $B/(0.1\% \text{ bandwidth}) \simeq 2.3 \times 10^{10}$, which is now much closer to the declared value than the value that we estimated at the beginning of this section. All the above reasoning is meant to train the reader's mind on the comparison between experiment and theory. However, we are quite happy that the formulas derived in this book reproduce, with a high degree of reliability, actual experimental results. Continuing the comparison, we check whether equation (5.22) satisfactorily accounts for the measurements made at **MuCLS**. For $\gamma \simeq 90$ and $\hbar\omega_0 \simeq 1.1$ eV, the formula yields $E_s = \hbar\omega_s \simeq 35.6$ keV, whereas the stated value is 35 keV.

6.6 Applications and the costs of CBS sources

CBS sources have the advantage of being 'fairly' monochromatic. Monochromatic sources of x-rays are not very popular, since they are technologically demanding. X-ray FELs can provide very high fluxes of monochromatic hard photons (in this respect, they are the best-performing sources), but such facilities are not numerous (and access is limited by the large number of user requests), and they require huge investments (of the order of over 1 billion Euros). In general, x-ray sources are useful in many areas and fields of science and technology. In table 6.3 we list several well-known applications of x-ray sources and the ranges that they require.

It should be mentioned that applications such as x-ray crystallography and medical diagnostics and therapy really benefit from monochromatic sources, since for the former only specific lattice vectors can be excited in the study of crystal structures, and for the latter a well-defined photon energy may correspond to better-defined images and precisely-set penetration depths. For baggage screening, there may be no need for highly monochromatic sources, but for similar applications, related, for example, to safety (such as the identification of explosive materials), a narrow photon spectrum may be useful. In figure 6.7 we show the color map of the scattered photon energy of a CBS source vs the energy of the incident photons (in terms of their wavelength) and electrons. It should be noted that for electron energies of the order of many hundreds of MeV up to 1 GeV, the γ-ray range corresponding to photon energies of the order of several hundreds of keV to MeVs can be explored. Such a possibility paves the way for another vast field of applications based on the

Table 6.3. Notable applications of x-rays and the corresponding ranges of interest.

Applications	X-ray range
X-ray crystallography	1–20 keV
Medical diagnostics	20–60 keV
Baggage screening	80–9000 keV
Medical therapy	3–10 MeV

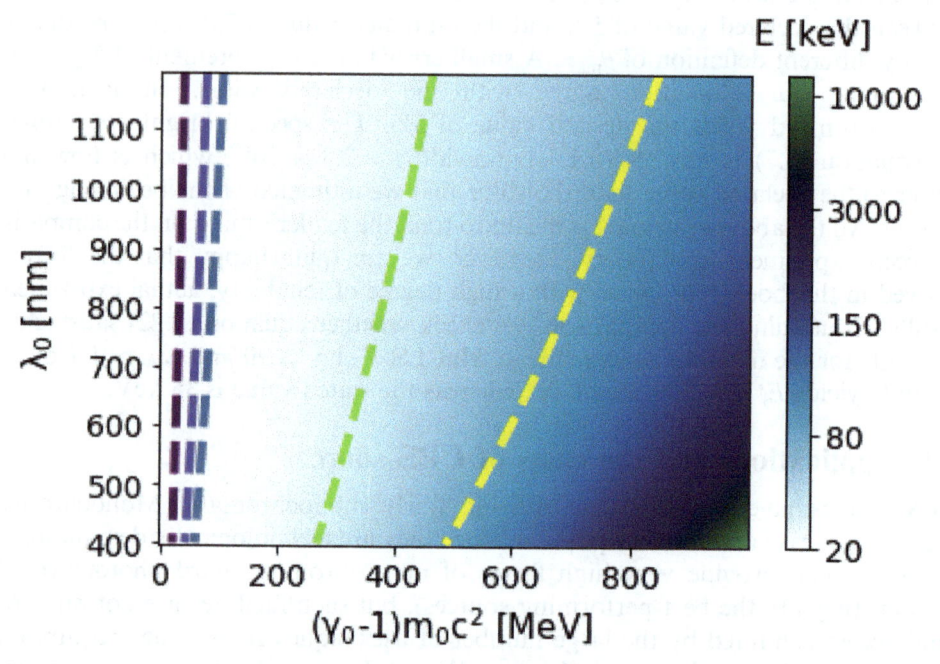

Figure 6.7. A plot of scattered photon energy $E = 4\gamma_0^2 \hbar \omega_0$ vs laser photon wavelength $\lambda_0 = 2\pi c/\omega_0$ and electron kinetic energy $(\gamma_0 - 1)m_0 c^2$. The dashed lines (from left to right) represent the photon energy values indicated in the colorbar (from the lowest to the highest).

physics of nuclei. In particular, we may think of studies of nuclear structure, studies of reactions for energy production, the generation of isotopes for nuclear medicine, and so on. We have seen that CBS sources can provide photon fluxes as high as 10^{10} photons s^{-1}, whereas today's synchrotron machines can exceed those values by over three orders of magnitude. However, there is a significant difference in cost between the two classes of machine, i.e. CBS and synchrotron sources. In fact, the cost of the first can be kept below 10 million Euros, while the second can be around several hundred million Euros. Furthermore, the accessibility of CBS sources may be greater than that of other sources because they are more compact (we estimate that an area of 100 m^2 is enough to include both laser and accelerator systems with sufficient performance) and can be produced in greater numbers and installed in more places (universities, research centers, industries, hospitals, etc).

6.7 Comments and exercises

6.7.1 X-ray tubes

In this book we have made reference to x-ray production using bremsstrahlung mechanisms in synchrotron radiation and FEL devices and CBS in laser electron sources.

Before the introduction of the aforementioned technologies, x-rays were produced by other means (albeit at significantly lower brightness) for medical, industrial, and scientific purposes. Figure 6.8 shows the evolution of x-ray brightness since the eve of the nineteenth century. The first sixty years were dominated by the technology of x-ray tubes, which culminated in the development of rotating-anode devices.

The paradigmatic steps characterizing the production of x-rays by Röntgen tubes are summarized below (see figure 6.9):

1. A filament of tungsten is warmed up by a current flowing through it;
2. Electrons are released through thermionic emission;
3. A superimposed difference of potential accelerates the electrons toward the anode so that they hit a tungsten target;
4. The electrons produce both heating of the target and the production of x-rays, which are released in a beam with a wide spectral range outside the window of the tube and then delivered to the experimental or imaging station.

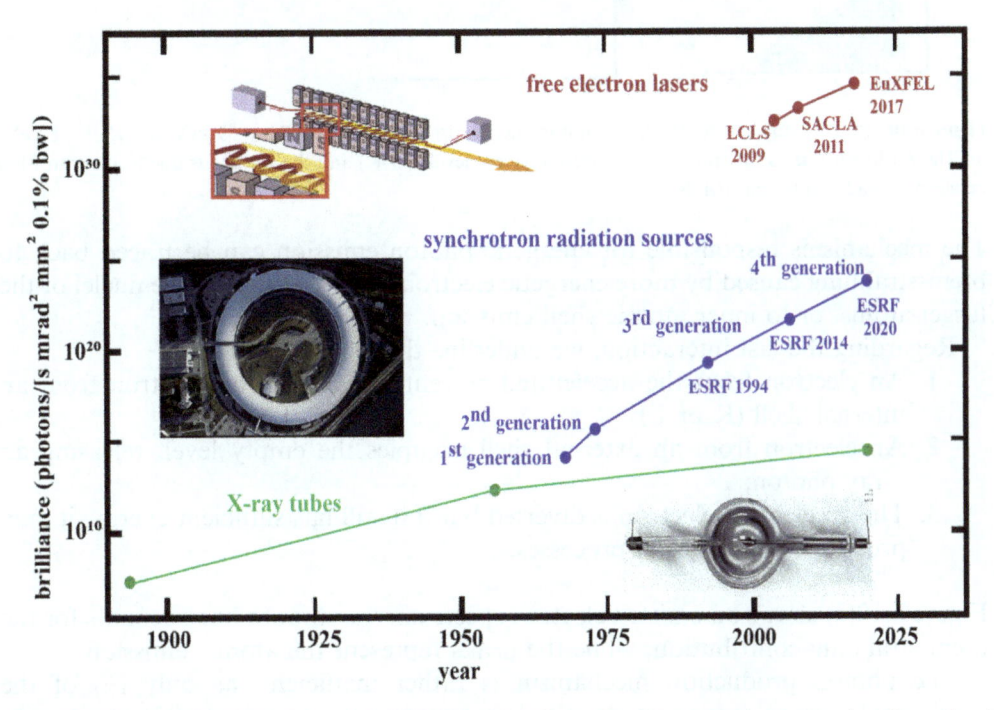

Figure 6.8. The development of x-ray brightness from the discovery of x-rays to the present day. Reproduced from Cerantola *et al* 2021 New frontiers in extreme conditions science at synchrotrons and free electron lasers *J. Phys. Condens. Matter* **33** 274003. Copyright The Authors. Published by IOP Publishing. CC BY 4.0.

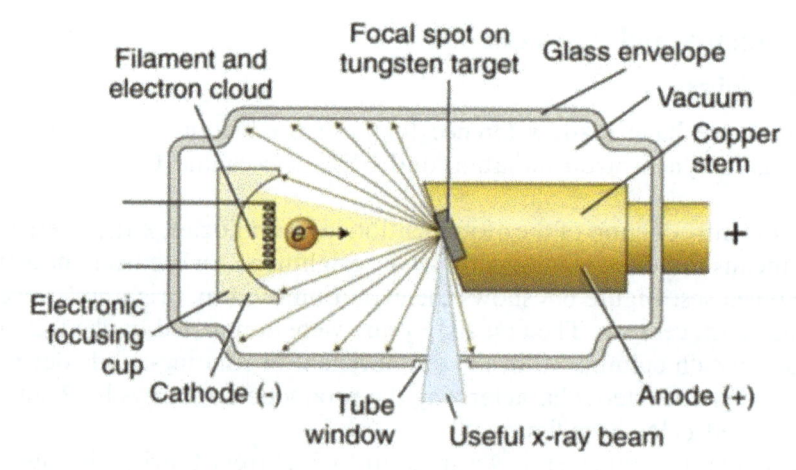

Figure 6.9. An x-ray tube and its components.

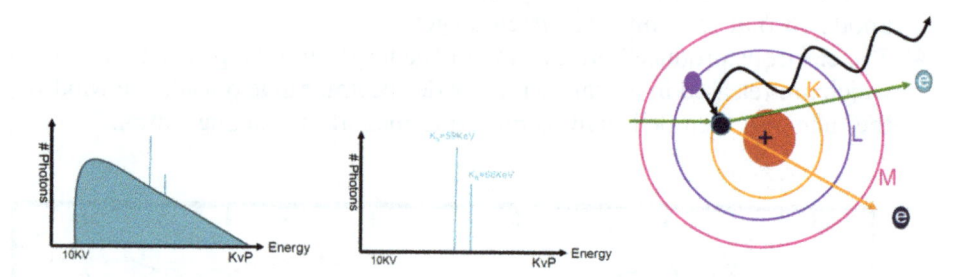

Figure 6.10. (a) An x-ray spectrum versus photon energy. (b) Transitions at $K_\alpha = 59$ keV and $K_\beta = 68$ keV. (c) The K_α line is due to the transition between adjacent levels (L–K) and the K_β line is due to the transition between non-adjacent shells (M–K).

The mechanisms responsible for energetic photon emission can be traced back to bremsstrahlung caused by more energetic electrons moving close to the nuclei of the target atoms, or to inner atomic shell emission.

Regarding the last interaction, we underline that
1. An electron from the accelerated current knocks out an electron from an internal shell (K or L);
2. An electron from an external shell occupies the empty level, releasing an x-ray photon;
3. The accelerated electron is diverted but if it still has sufficient energy, it may provide more emission processes.

Figure 6.10(a) shows an emitted electron spectrum: the smooth line accounts for the bremsstrahlung contribution, while the peaks represent the atomic emission.

The photon production mechanism is rather inefficient, as only 1% of the accelerated current energy is converted into x-rays; the remaining 99% turns into heat, which should be removed from the anode. A significant improvement that aimed to address this problem was provided by rotating-anode technology.

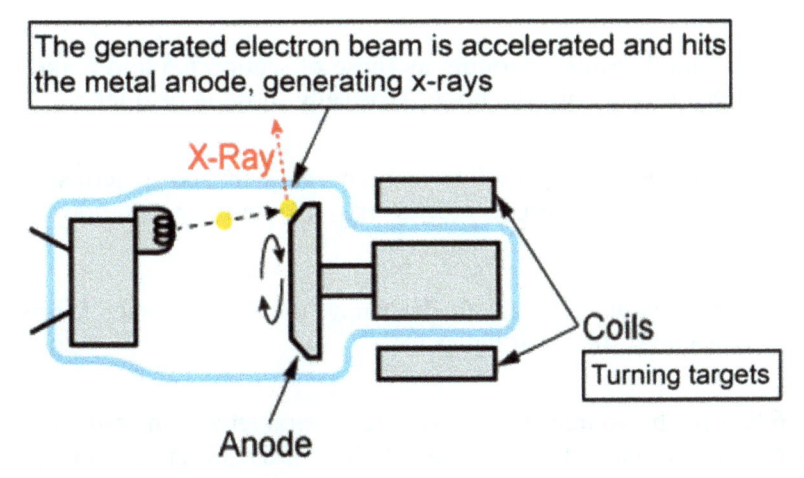

Figure 6.11. Sketch of a rotating-anode configuration.

Figure 6.11 is a sketch of a rotating-anode device, whose keynote is the use of a rotating target. The electrons hit a fixed geometric region and the deposited heat is dissipated more easily, since the deposition spot follows the disk rotation. Typical values of the tube current are hundreds of milliamps and the energy lies between 10 and 200 KV.

For further information, see the bibliography at the end of this chapter.

6.7.2 Exercises

Exercise 6.1. Estimate the scattered photon energies for the sources listed in tables 6.1 and 6.2.
Hint:
Information on the laser wavelengths of the above sources is reported in the text of section 6.3. The electron energy is directly reported in table 6.2. Use $\omega_s = 4\gamma_0^2 \omega_0$.

Exercise 6.2. Try to find information on as many as possible of the sources reported in figure 6.1 and calculate the corresponding scattered photon energies.
Hint:
Use $\omega_s = 4\gamma_0^2 \omega_0$.

Exercise 6.3. For the sources that have been characterized in terms of scattered photon energy via the previous exercises, discuss which applications are suitable for those sources.
Hint:
Consider table 6.3.

Exercise 6.4. For the sources listed in table 6.2, make a similar discussion to that of section 6.5.

Hint:
Follow the same lines of reasoning as those of section 6.5 and make use of the equations derived throughout the book, especially those in chapter 5.

Exercise 6.5. For the sources that have been characterized in terms of scattered photon energy via the previous exercises, try to obtain information on the laser intensity.
Hint:
Dig into the relevant literature to determine the a_0 parameter for as many as possible of the sources shown in figure 6.1.

Exercise 6.6. For the sources that have been characterized in terms of a_0 via the previous exercise, discuss the relevance of the relativistic effects and the Sauter–Schwinger effect on the scattered photon energy.
Hint:
Use equation (6.1).

Exercise 6.7. Try to design your own CBS source.
Hint: Use the material in chapters 5 and 6.

Exercise 6.8. Try to list the practical applications for your own CBS source.
Hint:
Consider table 6.3.

Exercise 6.9. Try to estimate the cost and the size of your CBS source.
Hint:
Dig into relevant sources of information for a relatively detailed budget plan. You need to find information on the cost of a laser system and of an accelerator system, based on the design of the source you carried out in the previous exercises. Please also consider the cost of the building (depending on the source size). Check whether the order of magnitude of the final cost agrees with the cost estimated for CBS sources in section 6.6.

Exercise 6.10. Correct the estimates of the previous exercise by taking into account the cost of 10 years' operation of your CBS source.
Hint:
First of all, consider your home country and the local cost of energy consumption. Make a reasonable guess at the energy/power need to run the facility. Moreover, try to estimate a reasonable number of people for the personnel needed at the CBS radiation facility, and calculate the cost of their salaries (considering their different roles).

6.7.3 CBS–FEL coupled devices

We have mentioned the possibility of operating CBS sources employing FEL photons. In figure 6.12 we show a layout exemplifying the operation of such an FEL–CBS device. Although we mentioned FEL–CBS storage rings, other devices employing linac–FEL combinations have been proposed (see the bibliography at the end of the chapter).

Exercise 6.11. Show that the frequency of an output Compton scattered photon in an FEL–CBS device scales according to γ_0^4, where γ_0 is the relativistic electron factor.

Hint:
Note that the FEL photon frequency is

$$\omega_F = \frac{2\gamma_0^2}{1 + \frac{K_u^2}{2}} \omega_u, \qquad (6.9)$$

$$\omega_u = \frac{2\pi c}{\lambda_u}.$$

Accordingly, the Compton backscattered photon frequency is

$$\omega_{F-C} = \frac{4\gamma_0^2}{1 + \frac{a_F^2}{2}} \omega_u = \frac{8\gamma_0^4}{\left(1 + \frac{a_F^2}{2}\right)\left(1 + \frac{K_u^2}{2}\right)} \omega_u \qquad (6.10)$$

where a_F is the strength parameter associated with the intracavity FEL power.

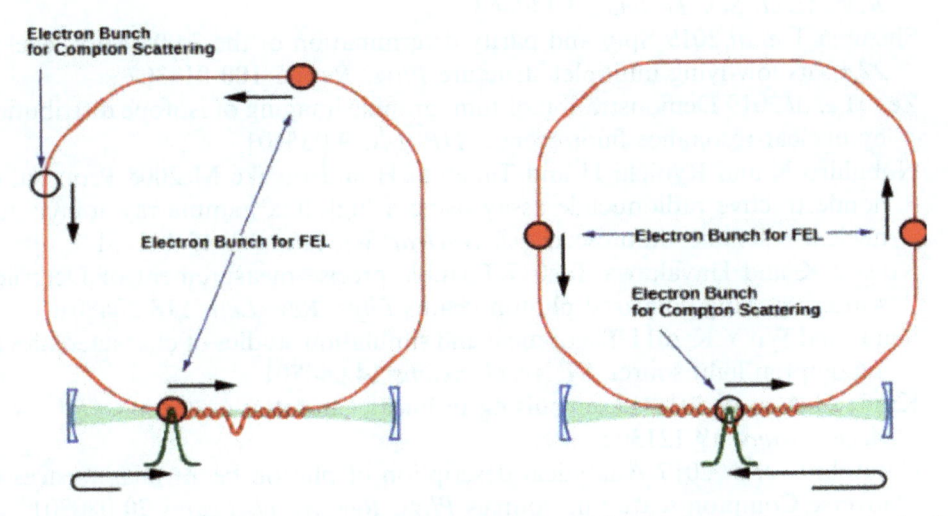

Figure 6.12. A layout of an FEL–CBS source. This device proposal foresees the use of a storage ring in which the circulating electron bunches are separately exploited to generate intracavity photons backscattered by a second bunch.

Exercise 6.12. Outline the design of a linac-based FEL–CBS source.

Hint:

The design of a FEL–CBS requires the specification of the FEL's characteristic features. The equilibrium intracavity power is the most representative figure of merit and is defined as

$$I_e = \sqrt{2} + 1 \left(\sqrt{\frac{1-\eta}{\eta}} G_M - 1 \right) I_S$$

$$I_S \left[\frac{MW}{cm^2} \right] = 6.9312 \times 10^2 \frac{1}{2} \left(\frac{\gamma_0}{N_u} \right)^4 \left(\lambda_u [cm] \frac{K_u}{\sqrt{2}} f_b \right)^{-2}$$

(6.11)

where η denotes the cavity losses, G_M is the FEL's maximum gain, and I_S represents the FEL's saturation intensity.

Equation (6.11) is a very useful tool for the design and the optimization of a linac-based FEL, since it contains all the quantities required to specify the engineering of the device.

We therefore have all the elements required to define the operating conditions of a linac-based FEL–CBS source.

For more details, the reader is referred to the bibliography and to the second volume.

References and further reading

For CBS devices, applications, and simulations, see the following:

Geoffrey A K and Priebe G 2010 Compton sources of electromagnetic radiation *Rev. Accel. Sci. Technol.* **3** 147–63

Shizuma T *et al* 2019 Spin and parity determination of the 3.004-MeV level in 27Al: its low-lying multiplet structure *Phys. Rev. C* **100** 014307

Zen H *et al* 2019 Demonstration of tomographic imaging of isotope distribution by nuclear resonance fluorescence *AIP Adv.* **9** 035101

Nobuhiro K and Ryoichi H and Takehito H and Eisuke M 2008 Proposal of nondestructive radionuclide assay using a high-flux gamma-ray source and nuclear resonance fluorescence *J. Nuclear Sci. Technol.* **45** 441–51

Koga J K and Hayakawa T 2017 Possible precise measurement of Delbrück scattering using polarized photon beams *Phys. Rev. Lett.* **118** 204801

Sun C and Wu Y K 2011 Theoretical and simulation studies of characteristics of a Compton light source *ST Accel. Beams* **14** 044701

Krafft G A *et al* 2016 Laser pulsing in linear Compton scattering *Phys. Rev. Accel. Beams* **19** 121302

Curatolo C *et al* 2017 Analytical description of photon beam phase spaces in inverse Compton scattering sources *Phys. Rev. Accel. Beams* **20** 080701

Ranjan N *et al* 2018 Simulation of inverse Compton scattering and its implications on the scattered linewidth *Phys. Rev. Accel. Beams* **21** 030701

Akagi T *et al* 2016 Narrow-band photon beam via laser Compton scattering in an energy recovery linac *Phys. Rev. Accel. Beams* **19** 114701

Gibson D J *et al* 2010 Design and operation of a tunable MeV-level Compton-scattering-based γ-ray source *Phys. Rev. ST Accel. Beams.* **13** 070703

Agostinelli S *et al* 2003 Geant4: a simulation toolkit *Nucl. Instrum. Methods Phys. Res.* **506** 250–303 https://www.sciencedirect.com/science/article/pii/S0168900203013688

Allison J *et al* 2006 Geant4 developments and applications *IEEE Trans. Nucl. Sci.* **53** 270–8

Hayakawa T *et al* 2010 Nondestructive assay of plutonium and minor actinide in spent fuel using nuclear resonance fluorescence with laser Compton scattering γ-rays *Nucl. Instrum. Methods Phys. Res.* **621** 695–700

Ohgaki H *et al* 2017 Nondestructive inspection system for special nuclear material using inertial electrostatic confinement fusion neutrons and laser Compton scattering gamma-rays *IEEE Trans. Nucl. Sci.* **64** 1635–40

Omer M and Hajima R 2017 Including Delbrück scattering in GEANT4 *Nucl. Instrum. Methods Phys. Res.* **405** 43–9 https://www.sciencedirect.com/science/article/pii/S0168583X17306092

Omer M and Hajima R 2019 Validating polarization effects in γ-rays elastic scattering by Monte Carlo simulation *New J. Phys.* **21** 113006

Chen P *et al* 1995 CAIN: Conglom érat d'ABEL et d'interactions non-linéaires *Nucl. Instrum. Methods Phys. Res.* **355** 107–10

Weller H R *et al* 2009 Research opportunities at the upgraded HIγS facility *Prog. Part. Nucl. Phys.* **62** 257–303

Utsunomiya H, Hashimoto S and Miyamoto S 2015 The γ-ray beam-line at NewSUBARU *Nucl. Phys. News* **25** 25–9

Horikawa K *et al* 2010 Measurements for the energy and flux of laser Compton scattering γ-ray photons generated in an electron storage ring: NewSUBARU *Nucl. Instrum. Methods Phys. Res.* **618** 209–15

Amano S *et al* 2009 Several-MeV γ-ray generation at NewSUBARU by laser Compton backscattering *Nucl. Instrum. Methods Phys. Res.* **602** 337–41 https://www.sciencedirect.com/science/article/pii/S0168900209000278

Utsunomiya H *et al* 2014 Energy calibration of the NewSUBARU storage ring for laser Compton-scattering gamma rays and applications *IEEE Trans. Nucl. Sci.* **61** 1252–8

Rinderknecht H G *et al* 2023 An electron-beam based Compton scattering x-ray source for probing high-energy-density physics arXiv:2207.01549[physics.plasm-ph]

Dattoli G, Gallardo J C and Ottaviani P L 1994 Free-electron laser intracavity light as a source of hard x-ray production by Compton backscattering *J. Appl. Phys.* **76** 1399–404

Sei N, Ogawa H and Jia Q 2020 Multiple-collision free-electron laser Compton backscattering for a high-yield gamma-ray source *Appl. Sci.* **10** 1418

Carlsten B E *et al* 2010 High repetition-rate inverse Compton scattering x-ray source driven by a free-electron laser *J. Phys. B* **47** 234012

Sei N *et al* 2002 Design of a new optical klystron for developing infrared free electron lasers in the storage ring NIJI-IV *Jpn. J. Appl. Phys.* **41** 1595

Sei N *et al* 2002 Study of expected performance of the hard x-ray beam for the FEL-X project *Nucl. Instrum. Methods Phys. Res.* **483** 429–33

Yamada K, Ogawa H and Sei N 2011 Characteritics of inverse Compton x-rays generated inside the NIJI-IV free electron laser oscillators *Proc. FEL 2010 (Malmö)* https://accelconf.web.cern.ch/FEL2010/html/auth0730.htm

Günther B *et al* 2020 The versatile x-ray beamline of the Munich compact light source: design, instrumentation and applications *J. Synchrotron Radiat.* **27** 1395–414

Eggl E *et al* 2016 The Munich compact light source: initial performance measures *J. Synchrotron Radiat.* **23** 1137–42

Gradl R *et al* 2017 Propagation-based phase-contrast x-ray imaging at a compact light source *Sci. Rep.* **7** 4908

Günther B *et al* 2018 The Munich compact light source: biomedical research at a laboratory-scale inverse-Compton synchrotron x-ray source *Microsc. Microanal.* **24** 984–5

Burger K *et al* 2020 Technical and dosimetric realization of *in vivo* x-ray microbeam irradiations at the Munich compact light source *Med. Phys.* **47** 5183–93

Dupraz K *et al* 2020 The ThomX ICS source *Phys. Open* **5** 100051

Petrillo V *et al* 2023 State of the art of high-flux Compton/Thomson x-rays sources *Appl. Sci.* **13** 752

Xu H *et al* 2011 Design of a compact storage ring for the TTX *Proc. IPAC 2011 (Joint Accelerator Conferences)* 3005 https://accelconf.web.cern.ch/IPAC2011/papers/thpc045.pdf

Tang C *et al* 2009 Tsinghua Thomson scattering x-ray source *Nucl. Instrum. Methods Phys. Res.* **608** S70–4

Ying-Chao D U *et al* 2008 Preliminary experiment of the Thomson scattering x-ray source at Tsinghua University *Chin. Phys. C* **32** 75

Chuan-Xiang T *et al* 2009 A simulation study of Tsinghua Thomson scattering x-ray source *Chin. Phys. C* **33** 146

Li-Xm Y *et al* 2009 TW Laser system for Thomson scattering x-ray light source at Tsinghua University *Chin. Phys. C* **33** 154

Du Y *et al* 2013 Generation of first hard x-ray pulse at Tsinghua Thomson scattering x-ray source *Rev. Sci. Instrum.* **84** 053301

Du Y *et al* 2011 Soft x-ray generation experiment at the Tsinghua Thomson scattering x-ray source *Nucl. Instrum. Methods Phys. Res.* **637** S168–71

Vaccarezza C *et al* 2016 The SPARC_LAB Thomson source *Nucl. Instrum. Methods Phys. Res.* **829** 237–42 https://www.sciencedirect.com/science/article/pii/S0168900216001303

Samsam S *et al* 2024 Progress in the energy upgrade of the Southern European Thomson back-scattering source (STAR) *Nucl. Instrum. Methods Phys. Res.* **1059** 168990

Faillace L *et al* 2019 Status of compact inverse Compton sources in Italy: BriXS and STAR *Proc. SPIE* **11110** 1111005

Bacci A *et al* 2016 Status of the STAR Project *IPAC 2016 Proc. 7th Int. Particle Accelerator Conf.* (Joint Accelerator Conferences) 1747–50 https://accelconf.web.cern.ch/ipac2016/papers/tupow004.pdf

Alesini D *et al* 2014 The STAR project *Proc. IPAC2014 (Dresden)* doi: https://dx.doi.org/10.18429/JACoW-IPAC2014-WEPRO115

Rinderknecht H G *et al* 2024 Electron-beam based Compton scattering x-ray source for probing high-energy-density physics *Phys. Rev. Accel. Beams* **27** 034701

Muşat V n.d. First ICS studies at CERN *CompactLight Complementary Use and Opportunities (8–9 November 2021)*

For FEL-based CBS devices, see the following:

Dattoli G, Gallardo J C and Ottaviani P L 1994 Free-electron laser intracavity light as a source of hard x-ray production by Compton backscattering *J. Appl. Phys.* **76** 1399–404

Glotin F *et al* 1996 Tunable x-ray generation in a free-electron laser by intracavity Compton backscattering *Phys. Rev. Lett.* **77** 3130–2

Litvinenko V N *et al* 1997 Gamma-ray production in a storage ring free-electron laser *Phys. Rev. Lett.* **78** 4569–72

Ciocci F *et al* 1997 Compton backscattering of free-electron laser photons and generation of high-energy monochromatic photon fluxes *IEEE J. Quant. Electron.* **33** 147–51

Torre A *et al* 2013 Double free-electron laser oscillator for photon-photon collisions *J. Opt. Soc. Am. B* **30** 2906–14

Placidi M *et al* 2017 Compact FEL-driven inverse compton scattering gamma-ray source *Nucl. Instrum. Methods Phys. Res.* **855** 55–60 https://www.sciencedirect.com/science/article/pii/S0168900217302723

Sei N, Ogawa H and Jia Q 2020 Multiple-collision free-electron laser Compton backscattering for a high-yield gamma-ray source *Appl. Sci.* **10** 1418

For x-ray production, see the following:

URL: https://www.radiologycafe.com/frcr-physics-notes/x-ray-imaging/production-of-x-rays/.

URL: https://ccah.vetmed.ucdavis.edu/sites/g/files/dgvnsk4586/files/local_resources/pdfs/rad-onc-matney-production-of-x-ray.pdf.

Khan F M 2010 *The Physics of Radiation Therapy* 4th edn (Philadelphia, PA: Lippincott Williams & Wilkins, A Wollers Kluwer Company)

Bushberg J T *et al* 1994 *The Essential Physics of Medical Imaging* 1st edn (Philadelphia, PA: Lippincott Williams & Wilkins, A Wollers Kluwer Company)

Milton Keynes UK
Ingram Content Group UK Ltd.
UKHW050048270824
447474UK00002B/6